Spaceplanes
The Future of Space Travel

Contents

Chapter 1

Spaceplane

A Space Shuttle rocketing into space, just after booster separation.

A **spaceplane** is an aerospace vehicle that operates as an aircraft in Earth's atmosphere, as well as a spacecraft when it is in space.*[1] It combines features of an aircraft and a spacecraft, which can be thought of as an aircraft that can endure and maneuver in the vacuum of space or likewise a spacecraft that can fly like an airplane. Typically, it takes the form of a spacecraft equipped with wings, although lifting bodies have been designed and tested as well. The propulsion to reach space may be purely rocket based or may use the assistance of airbreathing jet engines. The spaceflight is then followed by an unpowered glide return to landing.

Only five spaceplanes have successfully flown to date, having reentered Earth's atmosphere, returned to Earth, and safely landed —the North American X-15, Space Shuttle, Buran, SpaceShipOne, and Boeing X-37. All five are considered gliders. As of 2015, only these aircraft and rockets have succeeded in reaching space. Two of these five (X-15 and SpaceShipOne) are rocket-powered aircraft, having been carried up to an altitude of several tens of thousands of feet by an atmospheric mother ship before being released, and then flying beyond the boundaries of the earth's atmosphere under their own power. Three (Space Shuttle, Buran, and X-37) are vertical takeoff horizontal landing (VTHL) vehicles relying upon rocket lift for the ascent phase in reaching space and atmospheric lift for reentry, descent and landing. The three VTHL spaceplanes flew much further than the aircraft launched ones, not merely leaving the earth's atmosphere but also entering orbit around it, which requires at least 50 times more energy on the way up and heavy heat shielding for the trip back.*[2] Also, of the 5 vehicles, three have been piloted by astronauts, with the Buran and X-37 flying unmanned missions.

1.1 Description

Landing of NASA Space Shuttle Atlantis. *The American Space Shuttles were manned orbital spaceplanes.*

Significant features distinguish spaceplanes from spacecraft.

1.1.1 Aerodynamic lift

Main article: Lift (force)

All aircraft utilize aerodynamic surfaces in order to generate lift. For spaceplanes a variety of wing shapes can be used. Delta wings are common, but straight wings, lifting bodies and even rotorcraft have been proposed. Typically the force of lift generated by these surfaces is many times that of the drag that they induce.

1.1.2 Atmospheric reentry

Main article: Atmospheric entry

Because suborbital spaceplanes are designed for trajectories that do not reach orbital speed, they do not need the kinds of thermal protection orbital spacecraft required during the hypersonic phase of atmospheric reentry. The Space Shuttle thermal protection system, for example, protects the orbiter from surface temperatures that could otherwise reach as high as 1,650 °C (3,000 °F), well above the melting point of steel.[3]

1.1.3 Aircraft landing

Main article: Landing § Aircraft

A spaceplane operates as an aircraft in Earth's atmosphere. Aircraft may land on firm runways, helicopter landing pads, or even water (amphibious aircraft), snow or ice. To land, the airspeed and the rate of descent are reduced such that the aircraft descends at a slow enough rate to allow for a gentle touch down. Landing is accomplished by slowing down and descending. This speed reduction is accomplished by reducing thrust and/or inducing a greater amount of drag using flaps, landing gear or speed brakes. Splashdown is an easier technical feat to accomplish, requiring only the deployment of a parachute (or parachutes), rather than successfully aviating the atmosphere.[4] Project Gemini's original concept design was as a spaceplane, with paraglider and wheels (or skis) attached. However, this concept was abandoned in favor of parachute splashdowns, because of expensive technical failures during testing and development. Whereas Project Gemini's splashdown parachutes took only 5 months to develop in 1963, Gemini's spaceplane concept failed to materialize even after nearly 3 years of continued development.

1.1.4 Propulsion

Rocket engines

All spaceplanes to date have used rocket engines with chemical fuels. As the orbital insertion burn has to be done in space, orbital spaceplanes require rocket engines for at least that portion of the flight.

Airbreathing engines

Main article: Airbreathing jet engine

A difference between rocket based and air-breathing aerospace plane launch systems is that aerospace plane designs typically include minimal oxidizer storage for propulsion. Air-breathing aerospace plane designs include engine inlets so they can use atmospheric oxygen for combustion. Since the mass of the oxidizer is, at takeoff, the single largest mass of most rocket designs (the Space Shuttle's liquid oxygen tank weighs 629,340 kg, more than one of its solid rocket boosters[5]), this provides a huge potential weight savings benefit. However, air breathing engines are usually very much heavier than rocket engines and the empty weight of the oxidiser tank, and since, unlike oxidiser, this extra weight (which is not expended to add kinetic energy to the vessel, as is propellant mass) must be carried into space it may offset the overall system performance.

Types of air breathing engines proposed for spaceplanes include scramjet, liquid air cycle engines, precooled jet engines, pulse detonation engine and ramjets. Some engine designs combine several types of engines features into a combined cycle. For instance, the Rocket-based combined cycle (RBCC) engine uses a rocket engine inside a ramscoop so that at low speed, the rockets thrust is boosted by ejector augmented thrust. It then transitions to ramjet propulsion at near-supersonic speeds, then to supersonic combustion or scramjet propulsion, above Mach 6, then back to pure rocket propulsion above Mach 10.

1.1.5 Harsh flight environment

The flight trajectory required of air-breathing aerospace vehicles to reach orbit is to fly what is known as a 'depressed trajectory' which places the aerospace plane in the high-altitude hypersonic flight regime of the atmosphere. This environment induces high dynamic pressure, high temperature, and high heat flow loads particularly upon the leading edge surfaces of the aerospace plane. These loads typically require special advanced materials, active cooling, or both, for the structures to survive the environment.

However, even rocket-powered spaceplanes can face a significant thermal environment if they are burning for orbit, but this is nevertheless far less severe than air-breathing spaceplanes.

Suborbital space planes designed to briefly reach space do not require significant thermal protection, as they experience peak heating for only a short time during re-entry. Intercontinental suborbital trajectories require much higher speeds and thermal protection more similar to orbital spacecraft reentry.

1.1.6   Center of mass issues

A wingless launch vehicle has lower aerodynamic forces affecting the vehicle, and attitude control can be active perhaps with some fins to aid stability. For a winged vehicle the centre of lift moves during the atmospheric flight as well as the centre of mass; and the vehicle spends longer in the atmosphere as well. Historically, the X-33 and HOTOL spaceplanes were rear engined and had relatively heavy engines. This puts a heavy mass at the rear of the aircraft with wings that had to hold up the vehicle. As the wet mass reduces, the centre of mass tends to move rearward behind the centre of lift, which tends to be around the centre of the wings. This can cause severe instability that is usually solved by extra fins which add weight and decrease performance.

1.2   Flown spaceplanes

1.2.1   Orbital spaceplanes

All three of the *orbital* spaceplanes successfully flown to date utilize a VTHL (vertical takeoff, horizontal landing) design. They include the piloted United States Space Shuttle and two unmanned spaceplanes: the late-1980s Soviet Buran and the early-2010s Boeing X-37.

The early-1980s BOR-4 (subscale test vehicle for the Spiral spaceplane that was subsequently cancelled) was a spacecraft that did successfully reenter the atmosphere and fly like an aircraft. But it was not designed to sustain atmospheric flight. It was designed to stop flying, open a parachute and then splash in the ocean.

These vehicles have used wings to provide aerobraking to return from orbit and to provide lift, allowing them to land on a runway like conventional aircraft. These vehicles are still designed to ascend to orbit vertically under rocket power like conventional expendable launch vehicles. One drawback of spaceplanes is that they have a significantly smaller payload fraction than a ballistic design with the same takeoff weight. This is in part due to the weight of the wings —around 9-12% of the weight of the atmospheric flight weight of the vehicle. This significantly reduces the payload size, but the reusability is intended to offset this disadvantage.

While all spaceplanes have used atmospheric lift for the reentry phase, none to date have succeeded in a design that relies on aerodynamic lift for the ascent phase in reaching space (excluding a mother ship first stage). Efforts such as the Silbervogel and X-30/X-33 have all failed to materialize into a vehicle capable of successfully reaching space. The Pegasus winged booster has had many successful flights to deploy orbital payloads, but since its aerodynamic vehicle component operates only as a booster, and not operate in space as a spacecraft, it is not typically considered to be a spaceplane.

On the other hand, OREX[*][7] is a test vehicle of HOPE-X and launched into 450 km LEO using H-II in 1994. OREX succeeded to reenter, but it was only hemispherical head of HOPE-X, that is, not plane-shaped.

1.2.2   Suborbital spaceplanes

Main article: Suborbital spaceplane

 Other spaceplane designs are suborbital, requiring far less energy for propulsion, and can use the vehicle's wings to provide lift for the *ascent to* space in addition to the rocket. As of 2010, the only such craft to have successfully flown to and from space, back to earth, have been the North American X-15 and SpaceShipOne. Neither of these craft was capable of entering orbit. The X-15 and SpaceShipOne both began their independent flight only after being lifted to high altitude by a carrier aircraft.

Scaled Composites and Virgin Galactic unveiled on December 7, 2009, the SpaceShipTwo space plane, the VSS Enterprise, and its WhiteKnightTwo mothership, "Eve". SpaceShipTwo is designed to carry two pilots and six passengers on suborbital flights. On 29 April 2013, after three years of unpowered testing, the spacecraft successfully performed its first powered test flight.[8]

XCOR Aerospace signed a $30 million contract with Yecheon Astro Space Center to build and lease its Lynx Mark II spaceplane, which would be designed to take off from a runway under its own rocket power, and to reach the same altitude and speed range as SpaceShipOne and SpaceShipTwo, due to the fact that Lynx is propelled by higher specific impulse fuels. Lynx is designed to only carry a pilot and one passenger, although tickets are expected to be around half those quoted for Virgin Galactic services.[9]

Hyflex[10][11] was a miniaturized suborbital demonstrator of HOPE-X launched in 1996. Hyflex flew to 110 km altitude and succeeded in atmospheric reentry, subsequently achieving hypersonic flight. Though Hyflex achieved a controlled aircraft descent, it was not designed for a planned aircraft landing, the engineers opting instead for a splashdown without a parachute. The Hyflex that flew failed to recover and sank in the Pacific Ocean.

See also: Rocket-powered aircraft

1.3 Other projects

Various types of spaceplanes have been suggested since the early twentieth century. Notable early designs include Friedrich Zander's spaceplane equipped with wings made of combustible alloys that it would burn during its ascent, and Eugen Sänger's Silbervogel bomber design. Also in Nazi Germany and then in the USA, winged versions of the V-2 rocket were considered during and after World War II, and when public interest in space exploration was high in the 1950s and '60s, winged rocket designs by Wernher von Braun and Willy Ley served to inspire science fiction artists and filmmakers.

1.3.1 United States

The U.S. Air Force invested some effort in a paper study of a variety of spaceplane projects under their Aerospaceplane efforts of the late 1950s, but later ended these when they decided to use a modified version of Sänger's design. The result, Boeing X-20 Dyna-Soar, was to have been the first orbital spaceplane, but was canceled in the early 1960s in lieu of NASA's Project Gemini and the U.S. Air Force's Manned Orbiting Laboratory program.

In 1961, NASA originally planned to have the Gemini spacecraft land on a firm, solid ground runway[12] with a Rogallo wing airfoil,[13] rather than as a splashdown with parachute.[13] The test vehicle became known as the Paraglider Research Vehicle. Development work on both Gemini's splashdown parachute and spaceplane paraglider began in 1963.[14] By December 1963, the parachute was already to undergo full-scale deployment testing.[14] On the other hand, by December 1963 the paraglider spaceplane concept was running into technical difficulties[12] and subsequently became replaced by the parachute splashdown concept.[14] Though attempts to revive Gemini's paraglider spaceplane concept persisted within NASA and North American Aviation as late as 1964,[15] NASA Headquarters Gemini Chief William Schneider discontinued development as technical hurdles became too expensive.[15]

The Rockwell X-30 National Aero-Space Plane (NASP), begun in the 1980s, was an attempt to build a scramjet vehicle capable of operating like an aircraft and achieving orbit like the shuttle. It was canceled due to increasing technical challenges, growing budgets, and the loss of public interest. In 1994 Mitchell Burnside Clapp proposed a single stage to orbit peroxide/kerosene spaceplane called "Black Horse".[16] It was to take off almost empty and undergo mid-air refueling before launching to orbit.[17]

The Lockheed Martin X-33 was a prototype made as part of an attempt by NASA to build a SSTO hydrogen-fuelled spaceplane VentureStar that failed when the hydrogen tank design proved to be unconstructable in the planned way. The March 5, 2006 edition of Aviation Week & Space Technology published a story purporting to be "outing" a highly classified U.S. military two-stage-to-orbit spaceplane system with the code name Blackstar, SR-3/XOV among other nicknames.

In 1999 NASA started the Boeing X-37 project, an unmanned, remote controlled spaceplane. The project was transferred to the U.S. Department of Defense in 2004.

Boeing has proposed that a larger variant of the X-37B, the X-37C could be built to carry up to six passengers up to LEO. The spaceplane would also be usable for carrying cargo, with both upmass and downmass (return to Earth) cargo capacity. The ideal size for the proposed derivative "is approximately 165 to 180 percent of the current X-37B." *[18]

In December 2010, Orbital Sciences made a commercial proposal to NASA to develop the Prometheus, a lifting-body spaceplane vehicle about one-quarter the size of the Space Shuttle, in response to NASA's Commercial Crew Development (CCDev) phase 2 solicitation. The vehicle would be launched on a human-rated (upgraded) Atlas V rocket but would land on a runway.*[19] For the same solicitation, Sierra Nevada Corporation proposed phase 2 extensions of its Dream Chaser spaceplane technology, partially developed under the first phase of NASA's CCDev program.*[20] Both the Orbital Sciences proposal and the Dream Chaser are lifting body designs.*[21] Sierra Nevada will utilize Virgin Galactic to market Dream Chaser commercial services and may use "Virgin's WhiteKnightTwo carrier aircraft as a platform for drop trials of the Dream Chaser atmospheric test vehicle" *[20]*[22] NASA expects to make approximately $200 million of phase 2 awards by March 2011, for technology development projects that could last up to 14 months.*[23]

National Aerospace Plane

Main article: Rockwell X-30
President Ronald Reagan described NASP in his 1986 State of the Union address as "...a new Orient Express that could, by the end of the next decade, take off from Dulles Airport and accelerate up to twenty-five times the speed of sound, attaining low earth orbit or flying to Tokyo within two hours..." *[24]

There were six identifiable technologies which were considered critical to the success of the NASP project. Three of these "enabling" technologies were related to the propulsion system, which would consist of a hydrogen-fueled scramjet.*[24] The NASP program became the Hypersonic Systems Technology Program (HySTP) in late 1994.

HySTP was designed to transfer the accomplishments made in hypersonic technologies by the National Aero-Space Plane (NASP) program into a technology development program. On January 27, 1995 the Air Force terminated participation in (HySTP).*[24]

1.3.2   Soviet Union and Russia

The Soviet Union firstly considered a preliminary design of rocket-launch small spaceplane Lapotok in early 1960s. Then the Spiral airspace system with small orbital spaceplane and rocket as second stage was widely developed in the 1960s-1980s. **Mikoyan-Gurevich MiG-105** was a manned test vehicle to explore low-speed handling and landing.*[25]

Cosmoplane

In recent times, an orbital spaceplane, called *cosmoplane* (Russian: космоплан) capable of transporting passengers has been proposed by Russia's Institute of Applied Mechanics. According to researchers, it could take about 20 minutes to fly from Moscow to Paris, using hydrogen and oxygen-fueled engines.*[26]*[27]

1.3.3   United Kingdom

The Multi-Unit Space Transport And Recovery Device (MUSTARD) was a concept explored by the British Aircraft Corporation (BAC) around 1964-1965 for launching payloads weighing as much as 5,000 lb into orbit. It was never constructed.*[28] The British Government also began development of a SSTO-spaceplane, called HOTOL, but the project was canceled due to technical and financial issues.*[29]

The lead engineer from the HOTOL project has since set up a private company dedicated to creating a similar plane called Skylon with a different combined cycle rocket/turbine precooled jet engine called SABRE. This vehicle is intended to be capable of a single stage to orbit launch carrying a 15,000 kg payload into Low Earth Orbit. If successful it would be far in advance of anything currently in operation.*[30]

The British company Bristol Spaceplanes has undertaken design and prototyping of three potential spaceplanes since its founding by David Ashford in 1991. The European Space Agency has endorsed these designs on several occasions.*[31]

1.3.4 France and the European Space Agency

France worked on the Hermes manned spaceplane launched by Ariane rocket in the late 20th century, and proposed in January 1985 to go through with Hermes development under the auspices of the ESA.[*][32] Hopper was one of several proposals for a European reusable launch vehicle (RLV) planned to cheaply ferry satellites into orbit by 2015.[*][33] One of those was 'Phoenix', a German project which is a one-seventh scale model of the Hopper concept vehicle.[*][34] The suborbital Hopper was a FESTIP (Future European Space Transportation Investigations Programme) system study design[*][35] A test project, the Intermediate eXperimental Vehicle (IXV), has demonstrated lifting reentry technologies and will be extended under the PRIDE programme.[*][36]

1.3.5 Japan

HOPE was a Japanese experimental spaceplane project designed by a partnership between NASDA and NAL (both now part of JAXA), started in the 1980s. It was positioned for most of its lifetime as one of the main Japanese contributions to the International Space Station, the other being the Japanese Experiment Module. The project was eventually cancelled in 2003, by which point test flights of a sub-scale testbed had flown successfully.

1.3.6 Germany

After the German Sänger-Bredt RaBo and Silbervogel of the 1930s and 1940s, Eugen Sänger worked for time on various space plane projects, coming up with several designs for Messerschmitt-Bölkow-Blohm such as the MBB Raumtransporter-8.[*][37] In the 1980s, West Germany funded design work on the MBB Sänger II with the Hypersonic Technology Program. Development continued on MBB/Deutsche Aerospace Sänger II/HORUS until the late 1980s when it was canceled. Germany went on to participate in the Ariane rocket, Columbus space station and Hermes spaceplane of ESA, Spacelab of ESA-NASA and *Deutschland* missions (non-U.S. funded Space Shuttle flights with Spacelab). The Sänger II had predicted cost savings of up to 30 percent over expendable rockets.[*][38][*][39] The Daimler-Chrysler Aerospace RLV was a much later small reusable spaceplane prototype for ESA FLPP/FLTP program.

1.3.7 India

AVATAR (Sanskrit: अवतार) (from "**A**erobic **V**ehicle for Hypersonic **A**erospace **Tr**AnspoRtation") is an early-2000s concept of a manned single-stage reusable spaceplane capable of horizontal takeoff and landing, by India's Defense Research and Development Organization, the Indian Space Research Organization (ISRO) and other research institutions, intended for both military and civilian satellite launches. The ISRO plans to test the concept with a scaled-down suborbital **Reusable Launch Vehicle-Technology Demonstrator** (RLV-TD) spaceplane in 2015,[*][40] and aims to fly the full prototype by 2025.[*][41]

1.3.8 China

Main article: Shenlong (spacecraft)

Shenlong (Chinese: 神龙; pinyin: *shén lóng*; literally: "divine dragon") is a proposed Chinese robotic space plane that is similar to the American Boeing X-37.[*][42] Only a few images have been released since late 2007.[*][43][*][44][*][45]

1.4 See also

- Ansari X Prize

- List of manned spacecraft

- List of private spaceflight companies#Crew and cargo transport vehicles

- Spaceflight

1.4.1 Spaceplane vehicles and projects

- Third Reich:
  - Sänger-Bredt RaBo
  - Silbervogel

- United States:
  - X-plane variations & test vessels:
    - **North American X-15**
    - Boeing X-20 Dyna-Soar
    - Martin X-23 PRIME
    - Martin Marietta X-24A
    - Martin Marietta X-24B
    - Lockheed X-24C
    - Rockwell X-30 (NASP)
    - Lockheed Martin X-33
    - Orbital Sciences X-34
    - **Boeing X-37**
    - NASA X-38 (Spacewedge)
    - Boeing X-40
    - X-41 Common Aero Vehicle
    - NASA X-43
    - **Boeing X-51**
  - Gemini Spaceplane
  - **ASSET**
  - North American DC-3
  - **Space Shuttle**
  - HL-20 Personnel Launch System
  - VentureStar
  - **DARPA Falcon Project**
  - Prometheus
  - **SpaceShipOne, SpaceShipTwo** & SpaceShipThree
  - Chrysler SERV
  - Martin Marietta Spacemaster
  - XCOR Aerospace Lynx
  - XCOR Aerospace Xerus
  - Rocketplane XP
  - Silver Dart
  - Prometheus
  - Black Horse
  - Dream Chaser
  - Blackstar
  - TR-3A Black Manta
  - TAW-50
  - Aurora aircraft
  - Military flying saucers

- Soviet Union/ Russia:

  - Keldysh bomber
  - Tsybin's Lapotok
  - Tupolev's Zvezda (Tu-136/139)
  - Chelomey's LKS (Kosmolyot)
  - Chelomey's Uragan
  - **Mikoyan-Gurevich MiG-105** (part of a program known as *Spiral*) & **BOR-4**
  - **Buran Shuttle & BOR-5**
  - MAKS (Molniya)
  - Kliper
  - Tupolev's RAKS (Tu-444/2000)
  - Cosmopolis XXI (Explorer)

- esa European Space Agency:

  - Hermes
  - Hopper
  - Airbus Space and Defence SpacePlane
  - Intermediate eXperimental Vehicle (IXV) & PRIDE

- United Kingdom:

  - MUSTARD
  - Bristol's Ascender, Spacebus & Spacecab
  - HOTOL & HOTOL-2
  - Skylon

- France:

  - Astrobus
  - ARES

- Germany:

  - MBB Raumtransporter-8
  - MBB/Deutsche Aerospace HORUS
  - Sanger II
  - Falke
  - SHEFEX
  - Daimler-Chrysler Aerospace RLV

- Ukraine:

  - Svityaz
  - Oril
  - Sura

- Switzerland: S3 SOAR

- Romania: Orizont

- ▌✦▌ Canada: Wild Fire

- ▨ People's Republic of China:

  - Shenlong Space Plane
  - Project 921-3

- ● Japan:

  - Yamato
  - **Hyflex, OREX & HIMES**
  - HOPE-X

- ▨ India:

  - AVATAR
  - Hyperplane & Indian Shuttle

1.5   References

[1] Chang, Kenneth (20 October 2014). "25 Years Ago, NASA Envisioned Its Own 'Orient Express' ". *New York Times*. Retrieved 21 October 2014.

[2] Hoffman, Carl (22 May 2007). "Betting on a Mission Beyond Earth's Orbit". *WIRED MAGAZINE* (Conde Nast) (15.06). Retrieved 12 June 2015.

[3] "ORBITER THERMAL PROTECTION SYSTEM". NASA KSC. 1989.

[4] Hacker, Barton C., and Grimwood, James M., pp. xvi-xvii, 145-148, 171-173

[5] Space Shuttle external tank#Technical data

[6] David, Leonard (October 7, 2011). "Secretive US X-37B Space Plane Could Evolve to Carry Astronauts". *Space.com*. Retrieved August 5, 2015.

[7] "OREX". Space Transportation System Research and Development Center, JAXA. Retrieved 2011-05-15.

[8] "Sir Richard Branson's Virgin Galactic spaceship ignites engine in flight". BBC. 29 April 2013. Retrieved 29 April 2013.

[9] Andy Pasztor (December 17, 2009). "XCOR Aerospace Gets First Lease Customer for Its Space Plane". *The Wall Street Journal*.

[10] "Hyflex". astronautix.com. Retrieved 2011-05-15.

[11] "HYFLEX". Space Transportation System Research and Development Center, JAXA. Retrieved 2011-05-15.

[12] Hacker, Barton C., and Grimwood, James M. *On the Shoulders of Titans: A History of Project Gemini* (1975) "Preface," pp. xvi, xvii, 1975. Published as *NASA Special Publication-4203*, 1977.

[13] Please refer to Project Gemini#Spacecraft.

[14] Hacker, Barton C., and Grimwood, James M., pp. 145-148.

[15] Hacker, Barton C., and Grimwood, James M., pp. 171-173.

[16] Black Horse. astronautix.com

[17] Robert M. Zubrin; Mitchell Burnside Clapp (June 1995). "Black Horse: One Stop to Orbit". *Analog Magazine*. Archived from the original on 2001-11-19. Retrieved 2009-04-22. working link

[18] David, Leonard (2011-10-07). "Secretive US X-37B Space Plane Could Evolve to Carry Astronauts". *space.com*. Retrieved 2011-10-13.

[19] "Orbital Proposes Spaceplan for Astronauts". *Wall Street Journal*, December 14, 2010. Accessed: December 15, 2010.

[20] Orbital Aims For Station With Lifting Body, *Aviation Week*, 2010-12-17, accessed 2010-12-20. "will use Virgin to market its services. But Sierra is also in discussions about using Virgin's WhiteKnightTwo carrier aircraft as a platform for drop trials of the Dream Chaser atmospheric test vehicle"

[21] Companies submit plans for new NASA spacecraft, *Daily Record*, 2010-12-17, accessed 2010-12-20.

[22] Virgin joins forces with two companies on CCDev, *NewSpace Journal*, 2010-12-16, accessed 2010-12-18.

[23] "NASA Seeks More Proposals On Commercial Crew Development". *press release 10-277*. NASA. October 25, 2010.

[24] "X-30 National Aerospace Plane (NASP)". *Federation of American Scientists*. Retrieved 2010-04-30.

[25] Soviet X-planes; Yefim Gordon, Bill Gunston

[26] "Russia Develops New Aircraft – Cosmoplane :: Russia-InfoCentre". *russia-ic.com*. Retrieved 13 June 2015.

[27] RusUsa.com Космоплан – самолет будущего

[28] David Darling (2010). "MUSTARD INFO". Retrieved 29 September 2010.

[29] "HOTOL History". Reaction Engines Limited. 2010. Retrieved 29 September 2010.

[30] "Skylon FAQ". Reaction Engines Limited. 2010. Retrieved 29 September 2010.

[31] "Bristol Spaceplanes Company Information". Bristol Spaceplanes. 2014. Retrieved 26 September 2014.

[32] Martin Bayer, *Hermes: Learning from our mistakes*, Space Policy, Volume 11, Number 3, August 1995, pp. 171-180(10)

[33] Europe's space shuttle passes early test | 10 May 2004

[34] Launching the next generation of rockets - BBC News, 2004.

[35] Possible Future European Launchers, A Process of Convergence | ESA Bulletin Number 97 | March 1999

[36] Jeremy Hsu, 15 October 2008, Europe Aims For Re-entry Spacecraft

[37] "Saenger I". *astronautix.com*. Retrieved 13 June 2015.

[38] "Saenger Article in Astronautix Encyclopedia". *Astronautix*. Retrieved 26 September 2014.

[39] "FAS Germany Page". *www.fas.org*. Retrieved 26 September 2014.

[40] "ISRO's design of reusable launch vehicle approved". *DNA India* (Bangalore, India). 5 January 2012.

[41] "Wednesday, August 03, 2011 India's Space Shuttle [Reusable Launch Vehicle (RLV)] program". *AA Me, IN*. 2011. Retrieved 2014-10-22.

[42] "Shenlong Space Plane: China's Answer To U.S. X-37B Drone?". *The Huffington Post*. Retrieved 13 June 2015.

[43] "And Races Into Space". StrategyCenter.net.

[44] "Shenlong Space Plane Advances China's Military Space Potential". International Assessment and Strategy Center.

[45] "Invoking China to keep the shuttle alive". Space Politics.

1.6 Bibliography

- Hacker, Barton C.; Grimwood, James M. (1977). *On the Shoulders of Titans: A History of Project Gemini*. Washington, D.C.: NASA. OCLC 3821896. NASA SP-4203.

- Kuczera, Heribert; Sacher, Peter W. (2011). *Reusable Space Transportation Systems*. Berlin: Springer. ISBN 978-3-540-89180-2.

1.7 External links

- Encyclopedia Astronautica article on Uragan / Zenit

- Russianspacweb: Russian Reusable Spacecraft

- Popular Science article: Space Shuttle proposals written by Wernher von Braun - July 1970

- Popular Science article: VentureStar, X-34, MAKS, Burlak and other - October 1996

- Popular Science article: Space Access' Space Plane - January 1998

- Popular Science article: Space planes - May 1999

- Popular Science article: Space plane replacement of Space Shuttle and info on past designs including NASP and Clipper - May 2003

- MSNBC - Classic design inspires futuristic space glider

Buran orbiter rear showing rocket engine nozzles, for maneuvering in low Earth orbit and thin air

First Spaceplanes

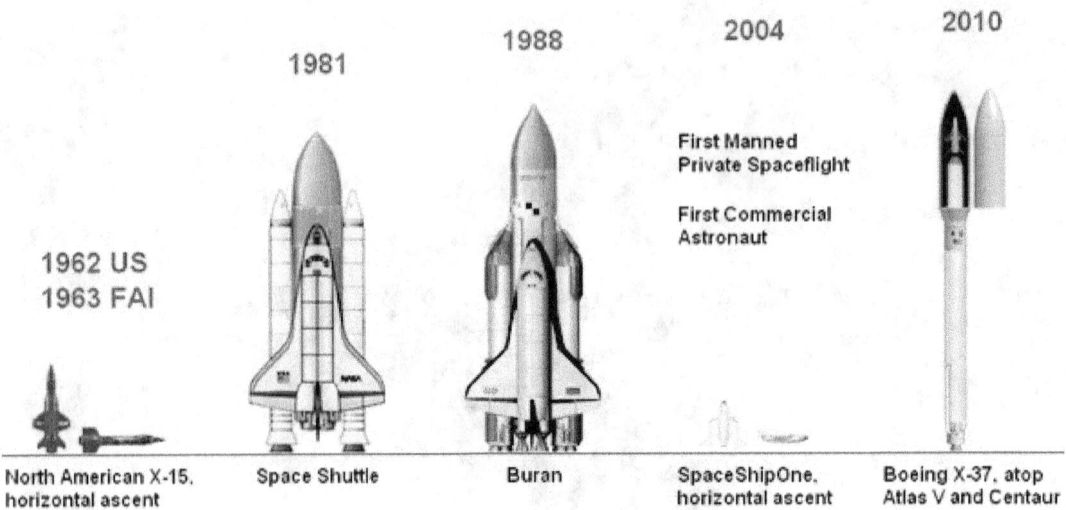

World's first spaceplanes: North American X-15, Space Shuttle, Buran, SpaceShipOne, Boeing X-37. The X-15 reached space in 1962/1963 (USAF/FAI Kármán line classifications). SpaceShipOne was piloted by the first commercial astronaut. Both X-15 and SpaceShipOne ascend horizontally from a mother ship. Both Buran and X-37 spaceflights were unmanned. The X-37 launches atop an Atlas V 501 launch vehicle.[6]

The X-15's rocket engine used ammonia and liquid oxygen.

SpaceShipOne Space plane

United States Gemini spaceplane concept testing, August 1964.

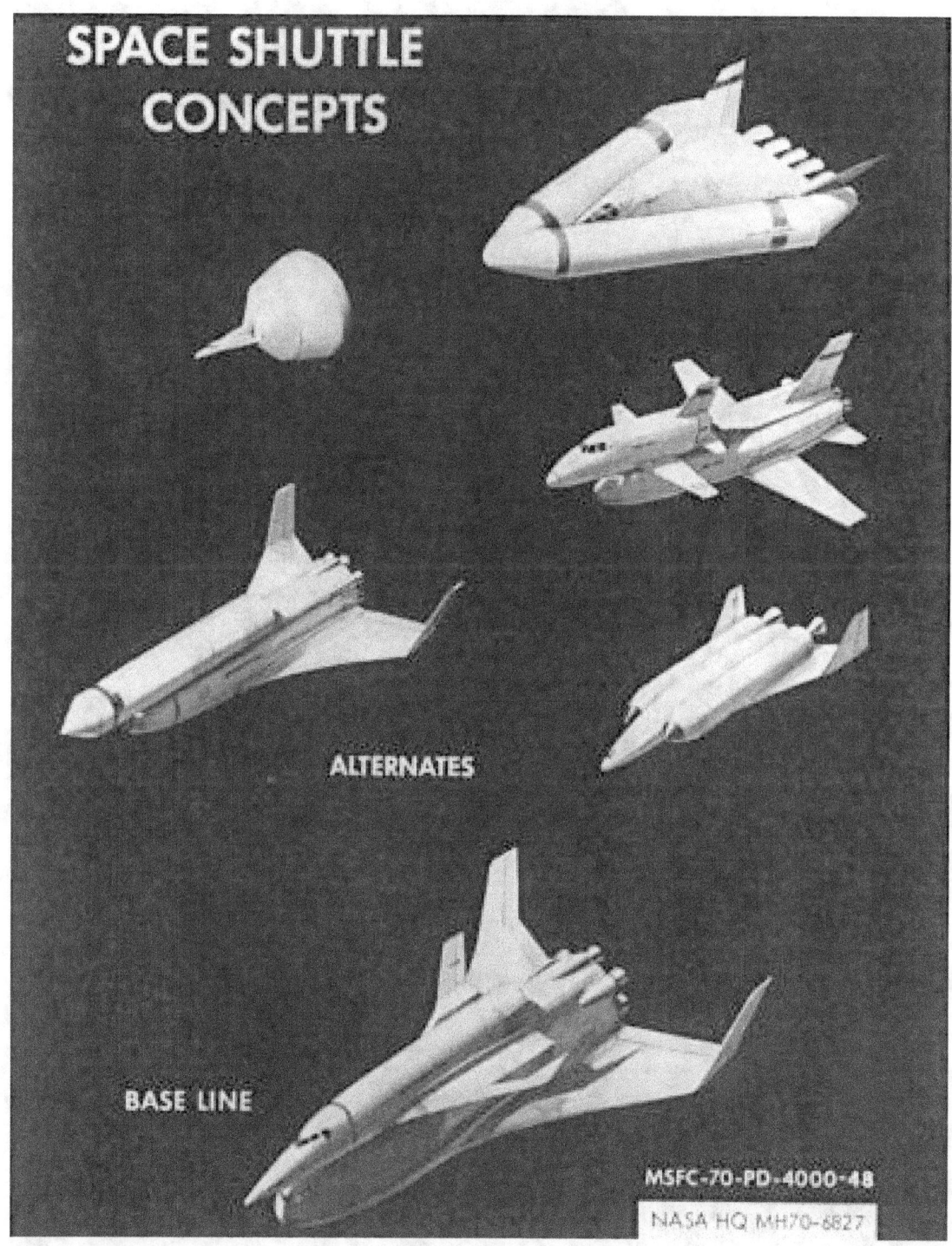

United States STS Space shuttle concepts circa 1970s

Boeing X-37B being prepared for launch in 2010 on an expendable orbital rocket

NASP taking off

Buran orbiter being transported via An-225

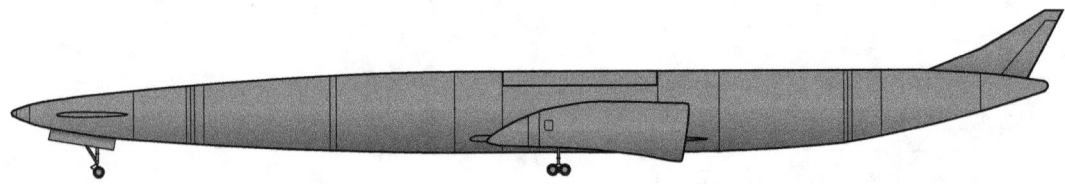

The Skylon spaceplane is designed as a two-engine, "tailless" aircraft, which is fitted with a steerable canard.

Chapter 2

Aerospaceplane

Aerospaceplane 1 *(artist conception). Lab research showed hydrogen fueled airbreathers could be used for space launch.*

The US Air Force's **aerospaceplane** project encompassed a variety of projects from 1958 until 1963 to study a fully reusable spaceplane. A variety of designs were studied during the lifetime of the project, including most of the early efforts on liquid air cycle engines (LACE) and even a nuclear-powered ramjet.

The effort was started largely due to the work of Weldon Worth at the Wright-Patterson AFB, who published a short work outlining a manned spaceplane. AF officials were interested enough to start SR-89774 (study requirement-) for a reusable spaceplane in 1957. By 1959 this work had resulted in the **Recoverable Orbital Launch System**, or **ROLS**, based around a LACE engine, known at the time as a *Liquid Air Collection System*, or *LACES*.

Further work showed that more performance could be gained by extracting only the oxygen from the liquid air, a system they referred to as *Air Collection and Enrichment System*, or *ACES*. A contract to develop an ACES testbed was placed with Marquardt and General Dynamics, with Garrett AiResearch building the heat exchanger for cooling the air. The original ACES design was fairly complex; the air was first liquified in the heat exchanger cooled by liquid hydrogen fuel, then pumped into a low pressure tank for short term storage. From there it was then pumped into a high pressure tank where the oxygen was separated and the rest (mostly nitrogen) was dumped overboard. In late 1960 and early 1961 a 125 N demonstrator engine was being operated for up to five minutes at a time.

In early 1960 Air Force offered a development contract to build a spaceplane with a crew of three that could take off from any runway and fly directly into orbit and return. They wanted the design to be in operation in 1970 for a total development cost of only $5 billion. Boeing, Douglas, Convair, Lockheed, Goodyear, North American, and Republic all responded. Most of these designs ignored the ACES system and instead used a scramjet for power. The scramjet had first been outlined at about the same time as the original LACES design in a NASA paper of 1958, and many companies were highly interested in seeing it develop, perhaps none more than Marquardt, whose ramjet business was dwindling with the introduction of newer jet engines and who had already started work on the scramjet. Both Alexander Kartveli and Antonio Ferri were proponents of the scramjet approach. Ferri successfully demonstrated a scramjet producing net thrust in November 1964, eventually producing 517 lbf, about 80% of his goal.

Later that year a review suggested that the basic concepts of the aerospaceplane were far too new for development of an operational system to begin. They pointed out that far too much was being spent on development of the aircraft, and not nearly enough on basic research. Moreover, the designs were all extremely sensitive to weight, and any increase (and there always is one) could result in all of the designs not working. In 1963 the Air Force changed their priorities in SR-651, and focused entirely on development of a variety of high-speed engines. Included were LACES and ACES engines, as well scramjets, turboramjets and a "normal" (subsonic combustion) ramjet with an intake suitable for use up to Mach 8. In October a further review concluded that the technology was simply too new for anyone to predict when any such aerospaceplane could ever be built, and funding was wound down in 1964.

2.1   References

- *Aerospaceplane - 1961*. Aerospace Projects Review, Volume 2, No 5.

- *Aspects of the Aerospace Plane*. Flight International, 2 January 1964, pages 36-37.

Chapter 3

Airbus Defence and Space Spaceplane

Mock-up of the vehicle at Paris Air Show 2007

The **Airbus Defence and Space Spaceplane**, also called **EADS Astrium TBN** according to some sources,[*][1] is a suborbital spaceplane concept for carrying space tourists, proposed by EADS Astrium (currently Airbus Defence and Space), the space subsidiary of the European consortium EADS (currently Airbus Group). A full-size mockup was officially unveiled in Paris, France, on June 13, 2007,[*][2] and is now on display in the Concorde hall of the Musée de l'Air et de l'Espace. The project is the first space tourism entry by a major aerospace contractor.

It is a rocket plane with a large wingspan, straight rearwards wing and a pair of canards.[*][3] Propulsion is ensured by classical turbofan jet engines for the atmospheric phase[*][4] and a methane-oxygen rocket engine for the space tourism phase. It can carry a pilot and four passengers. The dimensions and looks are somewhat similar to those of a business jet.

As of 2007, EADS Astrium hoped to start development of this rocket plane by 2008, with the objective of a first flight in 2011. There was also a possibility that the Tunisian area of Tozeur might be used for the initial flights.[*][5]

Demonstrator test flight regarding conditions encountered in the end-of-flight phase of a return from space occurred on June 5, 2014.[*][6]

EADS Astrium plans to raise public and private money for its project.

3.1 Origin of the project

The origin of the project is a proposal by a group of young French, German, British and Spanish engineers from EADS Astrium. It was studied in great secrecy for two years and finally approved by the chairman of EADS Astrium, François Auque. The design is similar in concept to the Rocketplane XP. They looked at the main concepts under development and their studies showed that Rocketplane's jet and rocket combination made the most sense.[*][7]

In the following months, a core team came up with a detailed concept and assembled the required expertise from different areas of Astrium and other EADS subsidiaries, such as Socata, as well as several external industry partners. Australian designer Marc Newson,[*][8] who earned his reputation in the field of aviation as Creative Director of Qantas, was also invited to join the project.[*][9]

3.2 Flight profile

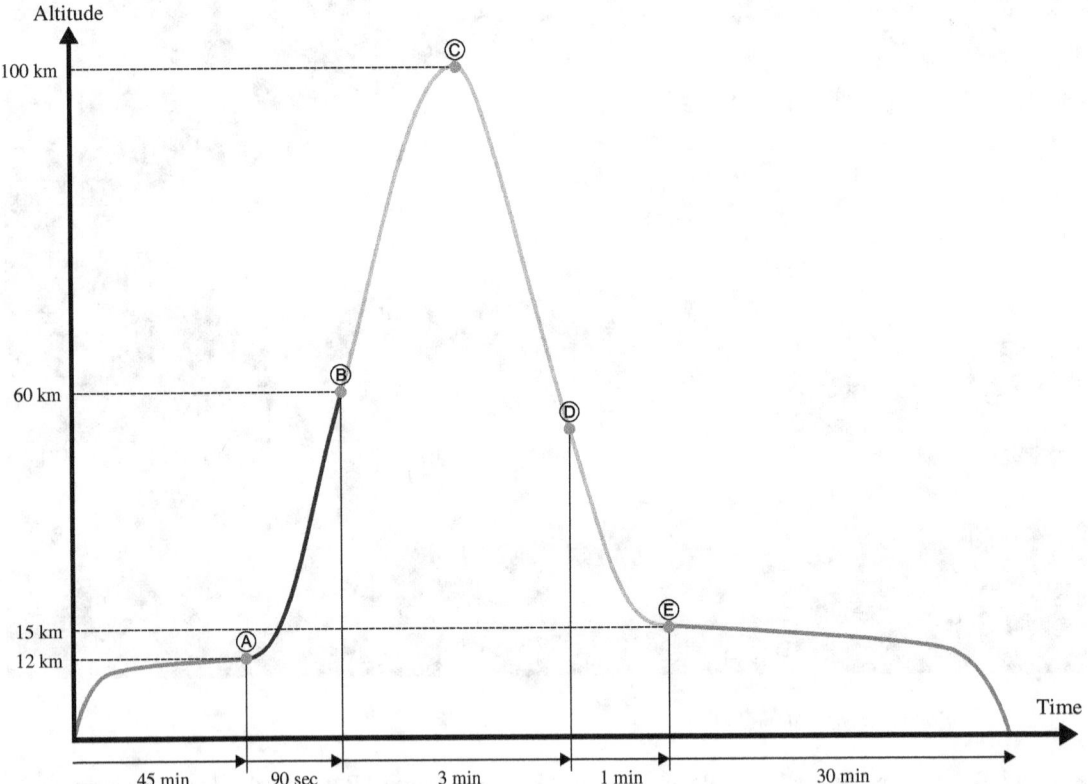

Turbofan propulsion
Rocket propulsion
No propulsion
A: *Ignition of rocket engine followed by turbofan shutdown*
B: *Shutdown of rocket engine. Acceleration 3 g. Start of weightlessness phase.*
C: *Culmination*
D: *Beginning of atmospheric deceleration. Max acceleration 4.5 g.*
E: *Turbofan ignition. Transition to aeronautical mode.*

After takeoff the plane reaches an altitude of 12 km. This classical aeronautical phase can last for 45 minutes. The pilot shuts down the jets and starts the methane oxygen rocket engine at the rear of the vehicle. The plane then raises along a vertical trajectory. For 90 seconds of flight with a top speed of Mach 3, the plane is rocketed upwards. The maximum acceleration is 3 g (30 m/s^2). At an altitude of 60 km, the rocket engine is shut down and the plane continues to climb up to a maximum altitude of 100 km. This is the weightlessness phase.

Then the plane gets down to 15 km at a high angle of attack, being progressively decelerated by the atmosphere; at this altitude, after transition to aeronautical mode, the jets are reignited to bring the plane back to a classical landing strip.[*][10]

3.3 Characteristics

Interior layout of the vehicle at Paris Air Show 2007

The total mass of the vehicle is 18 metric tons (39,700 lb) at takeoff. The plane has two jet engines, and an oxygen-methane engine with a thrust of 30 tons.[*][11] The rocket engine uses the technology of the Vulcain (the main engine of Ariane 5), but is reusable thirty times and burns methane instead of hydrogen (hydrogen would require too much tank volume, as the density of methane is 667.2 kg/m^3 and the density of hydrogen is 89.9 kg/m^3).

The cabin has a diameter of 2.3 m (7 ft 6 in), and provides 3 m^3 (106 ft^3) of cabin space to each passenger. The seats are attached to a pendular system which allows the acceleration to be perpendicular to the back of the passengers. They pivot around the attachment points so that the passengers are aligned rearside to the spacecraft x-axis (body aligned on Gx-axis) during launch acceleration and they are rearside on the negative z-axis during weightlessness and reentry.[*][12]

The plane is designed for ten years of service at a flight rate of once a week.[*][11]

3.4 Industrial organisation

The development will be led by EADS Astrium. Its technical responsibility currently resides with the CTO Robert Lainé.

In 2007, development cost of $1 billion was projected by some sources. EADS Astrium plans to raise mostly private money for its project. One of the possible public investors mentioned by François Auque is the southern German state of Bavaria, where the engines are to be produced.[13] Astrium could produce up to 5 planes a year and have a fleet of 20 planes, which would require a production of 20 rocket engines a year. They do not exclude selling models to other entrepreneurs such as Sir Richard Branson from Virgin Galactic.

The final assembly would be in France, while the other industrial facilities of Astrium would provide the rocket engines (Ottobrunn, Germany[4]) or the carbon fiber structures (Spain). Other European industrial partners are associated with the project.

The target of Astrium is to secure 30% of the market of space tourism by 2020, 5000 passengers a year. [14]

The ticket price will be 200,000 euros, including a round trip to the spaceport, training, and luxury accommodation in a theme park/resort.

3.5 Competition

The closest concept is the Rocketplane XP of Rocketplane Limited, Inc., which shares the same overall rocket plane principle. Other competitors include the SpaceShipTwo, New Shepard, Lynx and Dream Chaser.

3.6 Criticisms

Burt Rutan, founder of Scaled Composites, a competitor in space tourism to EADS, expressed scepticism towards the EADS Project.[15]

3.7 See also

- Rocketplane XP
- SpaceShipTwo
- New Shepard
- Lynx
- Dream Chaser
- Zero Emission Hyper Sonic Transport

3.8 References

[1] Radio interview of French ESA astronaut Jean-François Clervoy, during the show of Jacques Pradel on Europe 1, June 21, 2007

[2] "Planned Jet to Take Tourists Into Space" . Forbes. 2007-06-13. Retrieved 2007-06-14.

[3] Astrium dévoile son projet d'avion-fusée *Le Figaro, June 14, 2007, Page 18*

[4] Space jet' s turbofans can cope with vacuum says EADS Rob Coppinger Flightglobal.com June 21, 2007

[5] "Tunis espère que le site de Tozeur sera retenu pour le lancement du futur avion spatial" . Canada.com. 2007-06-13. Retrieved 2007-06-14.

[6] "Airbus tests SpacePlane demonstrator" .

[7] Europe's Tourist Rocket, Popular Science Magazine, October 2007, pg 41

[8] The technical challenge of making space travel easy *International Herald Tribune*, June 17, 2007, By Alice Rawsthorn

[9] Planet AeroSpace Issue 4/2007: EADS Astrium plans space tourism

[10] Description of the project - Presentation at IAC 2007 Hyderabad by Christophe Chavagnac - Flight profile provided on page 7

[11] EADS Astrium se lance dans le tourisme spatial *Air & Cosmos*, June 15, 2007, n° 2082 - Pages 150-151

[12] EADS reinvents Rocketplane

[13] "European company unveils space plane". CNN. 2007-06-14. Retrieved 2007-06-14.

[14] L'entreprise veut associer capitaux publics et privés pour le financement *Le Figaro, June 14, 2007, Page 18*

[15] "Burt Rutan: Reaction to European Spaceliner". Blog of Leonard David on livescience.com. 2007-06-13. Retrieved 2007-06-15.

3.9 External links

3.9.1 Text

- Astrium web site
- Brochure
- Marc Newson Ltd.
- Presentation of the project at the IAC 2007 in Hyderabad
- Analysis of the project by *The Space Review*
- Popsci article (EADS Spaceplane) - October 2007

3.9.2 Video

- Video animation - Astrium's Spaceplane
- Video - Presentation of The Space Tourism project at IAC 2007. 28 min

Chapter 4

ASSET (spacecraft)

ASSET, or **Aerothermodynamic Elastic Structural Systems Environmental Tests** was an experimental US space project involving the testing of an unmanned sub-scale reentry vehicle.

4.1 Development and Testing

Begun in 1960, ASSET was originally designed to verify the superalloy heat shield of the X-20 Dyna-Soar prior to full-scale manned flights. The vehicle's biconic shape and low delta wing were intended to represent Dyna-Soar's forward nose section, where the aerodynamic heating would be the most intense; in excess of an estimated 2200 °C (4,000 °F) at the nose cap. Following the X-20 program's cancellation in December 1963, completed ASSET vehicles were used in reentry heating and structural investigations with hopes that data gathered would be useful for the development of future space vehicles, such as the Space Shuttle.

Built by McDonnell, each vehicle was launched on a suborbital trajectory from Cape Canaveral's Pad 17B at speeds of up to 6,000 m/s before making a water landing in the South Atlantic near Ascension Island. Originally, a Scout launch vehicle had been planned for the tests, but this was changed after a large surplus of Thor and Thor-Delta missiles (returned from deployment in the United Kingdom) became available.

Of the six vehicles built, only one was successfully recovered and is currently on display at the National Museum of the United States Air Force in Dayton, Ohio.[*][1]

4.2 Flights

4.3 Specifications

General characteristics

- **Crew:** None
- **Length:** 5.74 ft (1.75 m)
- **Wingspan:** 4.57 ft (1.39 m)
- **Height:** 2.73 ft (0.83 m)
- **Loaded weight:** 1,190 lb (540 kg)
- **Powerplant:** × Hydrogen peroxide reaction control thrusters

Performance

- **Maximum speed:** Mach 25

ASSET pre-launch checkout.

- **Range:** 2,700 miles (4300 km)

- **Service ceiling:** 50 miles (80 km)

- **Hypersonic L/D Ratio:** 1:1

4.4 Related content

ASSET-Thor combination on Pad 17B.

4.4.1   Comparable Aircraft

- Molniya BOR-4

- Martin X-23 PRIME

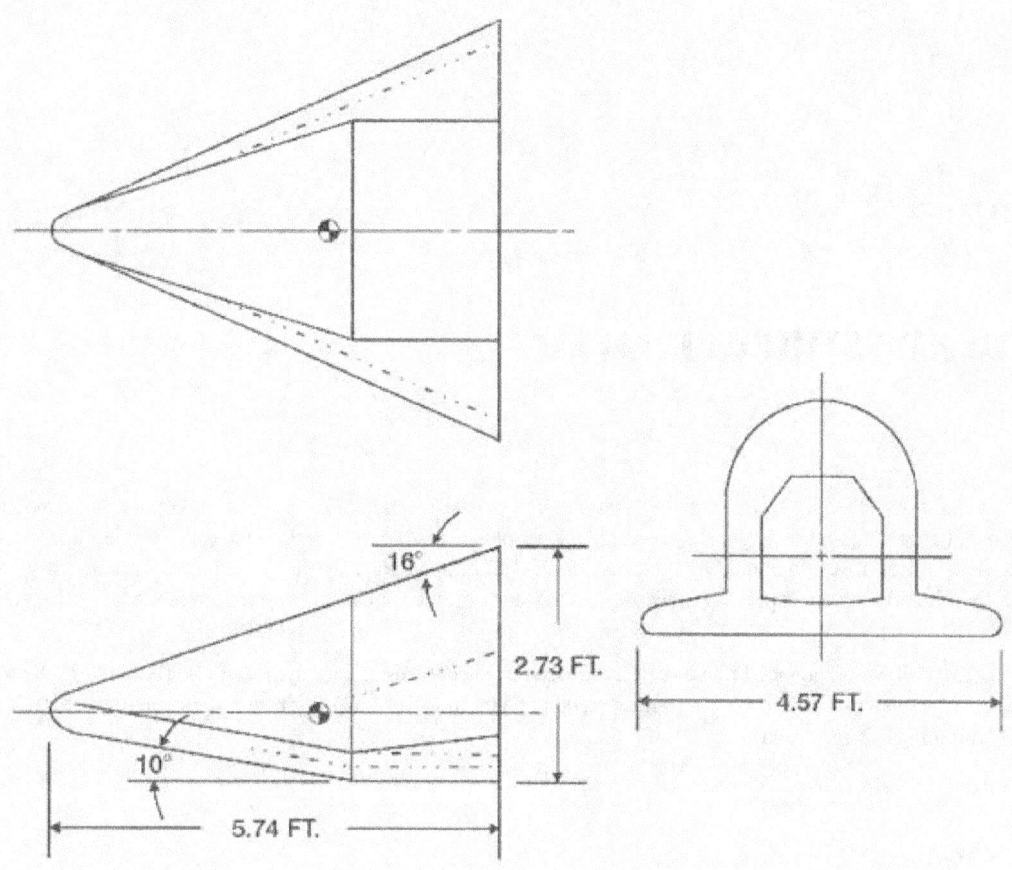

ASSET 3-view

4.4.2 Winged Gemini

In the mid-1960s, McDonnell proposed a variant of the Gemini capsule which retained the original spacecraft's internal subsystems and crew compartment, but dispensed with the tail-first ballistic reentry, parachute recovery and water landing.

Instead, the vehicle would be heavily modified externally into an ASSET-like lifting-reentry configuration. Post-reentry, a pair of stowed swing-wings would be deployed, giving the spacecraft sufficient L/D to make a piloted glide landing on a concrete runway using a skid-type landing gear (reinstated from the planned, but cancelled paraglider landing system), much like the Space Shuttle.

According to Mark Wade's Encyclopedia Astronautica, the intent seems to have been to field a manned military spaceplane at a minimal cost following the cancellation of the Dyna-Soar program.[*][2]

4.5 References

[1] ASSET at Encyclopedia Astronautica

[2] Winged Gemini at Encyclopedia Astronautica

Chapter 5

Avatar (spacecraft)

AVATAR (Sanskrit: अवतार) (from "**A**erobic **V**ehicle for Transatmospheric Hypersonic **A**erospace **Tr**Anspo**R**tation")
) is a concept for a unmanned single-stage reusable spaceplane capable of horizontal takeoff and landing, by India's Defence Research and Development Organization along with Indian Space Research Organization and other research institutions. The mission concept is for low cost military and commercial satellite space launches, as well as for space tourism.[3][4][5]

In January 2012, it was announced that a scaled prototype, called 'Reusable Launch Vehicle-Technology Demonstrator' (RLV-TD), was approved to be built and tested.[1] The first RLV-TD flight test is planned for 2016,[6] and the first orbital flight is proposed for 2025.[2]

5.1 Concept

In Sanskrit, an Avatar (अवतार *avatāra*) is a deliberate descent of a deity to Earth, or a descent of the Supreme Being.

The idea is to develop a spaceplane vehicle that can takeoff from conventional airfields. Its liquid air cycle engine would collect air in the atmosphere on the way up, liquefy it, separate oxygen and store it on board for subsequent flight beyond the atmosphere. The AVATAR, a reusable launch vehicle (RLV) was first announced in May 1998 at the Aero India 98 exhibition held at Bangalore.[7]

AVATAR is projected to weigh 25 tons, of which 60% of that mass would be liquid hydrogen fuel.[4] The oxygen required by the vehicle for combustion in outer space would be collected from the atmosphere during takeoff, thus reducing the need to carry oxygen during launch.[4] The notional specification is for a payload weighing up to 1,000 kg to low Earth orbit and to withstand up to 100 launches and reentries.[3][4]

If built, AVATAR would takeoff horizontally like a conventional airplane from a conventional airstrip using turbo-ramjet engines that burn hydrogen and atmospheric oxygen.[4] Once at a cruising altitude, the vehicle would use scramjet propulsion to accelerate from Mach 4 to Mach 8.[2] During this cruising phase, an on-board system would collect air from the atmosphere, from which liquid oxygen would be separated and stored.[4] The liquid oxygen collected would then be used to burn the stored hydrogen in the final flight phase to attain orbit. The vehicle would be designed to permit at least one hundred launches and atmospheric reentries.[4]

5.2 Development

AVATAR is being developed by India's Defence Research and Development Organization.[3] Air Commodore Raghavan Gopalaswami, who is heading the project, made a presentation on the spaceplane at the global conference on propulsion at Salt Lake City, USA on July 10, 2001.[3][4] Gopalaswami said the idea for AVATAR originated from the work published by the RAND Corporation of the United States in 1987.[4]

In January 2012, ISRO announced that a scaled prototype, called **Reusable Launch Vehicle-Technology Demonstrator (RLV-TD)**, was approved to be built and tested.[1]

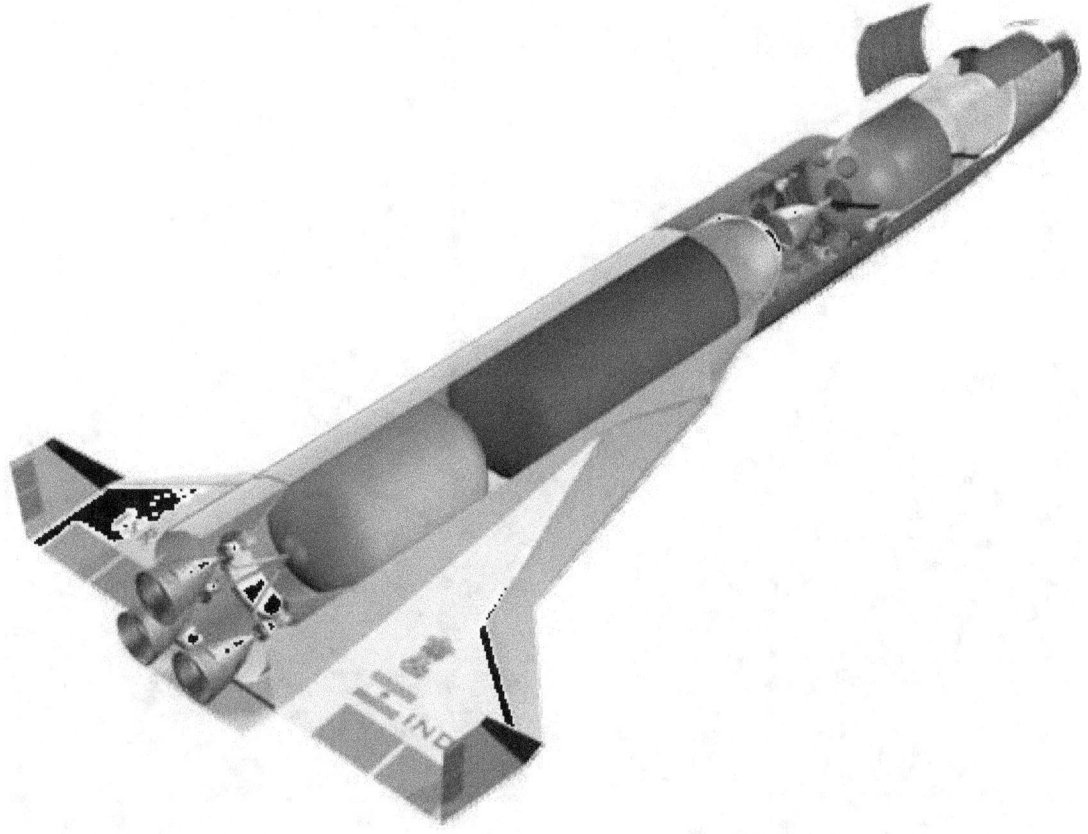

AVATAR RLV-TSTO

5.2.1 Scaled-down tests

The aerodynamics characterization of the RLV-TD prototype was done by National Aerospace Laboratories in India. The unmanned scaled-down prototype has a diameter of 0.56 m and a length of 10 m.[2] The RLV-TD is in the last stages of construction by a Hyderabad-based private company called CIM Technologies.

By May 2015, engineers at the Vikram Sarabhai Space Centre (VSSC) in Thumba Equatorial Rocket Launching Station were installing thermal tiles on the outer surface of the 'RLV-TD', so it can withstand the intense heat during atmospheric reentry.[1] This prototype weighs around 1.5 tonnes and would fly up to an altitude of 70 km.[1] ISRO has tentatively slated the prototype's test flight from the first launchpad of Satish Dhawan Space Centre for 2016. The RLV-TD will be mounted on top of a rocket and launched beyond the atmosphere, after which the RLV-TD will separate and reenter the atmosphere while traveling through the hypersonic regime.[8] The rocket is expendable while the RLV would glide back to Earth and fall in Bay of Bengal as there are no airstrips that are 5 km long in India that could be used to land such aircraft. ISRO has made detailed reports to construct an airstrip greater than 4 km long in the Sriharikota island and it will be built in near future. [9]

5.3 Manned version

There is also a proposal to develop a larger advanced manned version of the spacecraft after the successful deployment of the unmanned SSTO.[10]

5.4 See also

Aircraft of comparable role, configuration and era

- Boeing X-37

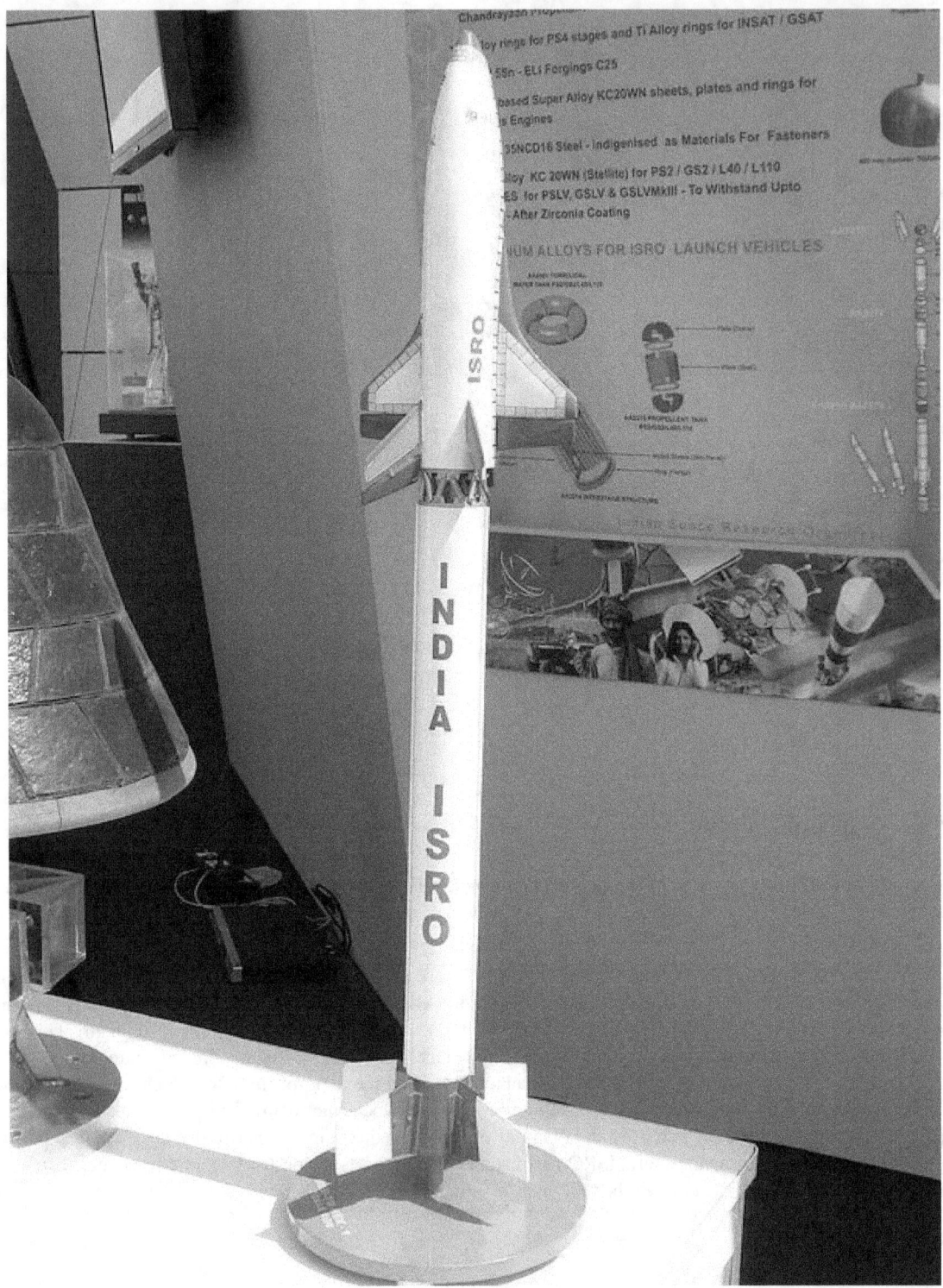

A model of the RLV-TD on the top of a launcher

- Buran (spacecraft)

- Dream Chaser

- Skylon

- Space Shuttle

5.5 References

[1] "India's Futuristic Unmanned Space Shuttle Getting Final Touches". *EXPRESS NEWS SERVICE* (Indian Defence Research Wing). 20 May 2015. Retrieved 2015-05-20.

[2] "Wednesday, August 03, 2011India's Space Shuttle [Reusable Launch Vehicle (RLV)] program". *AA Me, IN*. 2011. Retrieved 2014-10-22.

[3] "Indian Scientists unveils space plane Avatar in US". *Gujarat Science City*. 10 July 2001. Retrieved 2014-10-22.

[4] "India Eyes New Spaceplane Concept". *Space Daily* (New Delhi). August 8, 2001. Retrieved 2014-10-22.

[5] "AVATAR- Hyper Plane to be built by INDIA". *India's Military and Civilian Technological Advancements*. December 19, 2011.

[6] http://www.newindianexpress.com/states/kerala/RLV-TD-Mission-to-be-Delayed-At-least-Till-the-Year-End/2015/10/02/article3058357.ece

[7] "AVATAR- Hyper Plane to be built by INDIA". *Abin Puthiyath*. Indian Defence Research. Retrieved 19 December 2011.

[8] "Reusable Launch Vehicles". *Brahmand.com*. November 25, 2010. Retrieved 2014-10-22.

[9] "ISRO Gears up for 6 Major Missions This Year". *Express News Service*. 30 May 2015.

[10] Jan 6, 2015. "Another leap in space: India to test reusable vehicle in March". *Arun Ram*. Times Of India. Retrieved Jan 6, 2015.

5.6 External links

- Concept of AVATAR

Chapter 6

Blackstar (spacecraft)

Blackstar is the reported codename of a secret United States orbital spaceplane system. The possible existence of the Blackstar program was reported in March 2006 by *Aviation Week & Space Technology* (*Aviation Week*, *AWST*) magazine; the magazine reported that the program had been underway since at least the early 1990s, and that the impetus for Blackstar was to allow the United States government to retain orbital reconnaissance capabilities jeopardized following the 1986 *Challenger* disaster. The article also said that the United States Air Force's Space Command was unaware of Blackstar, suggesting it was operated by an intelligence agency such as the National Reconnaissance Office.[1][2]

Aviation Week speculated that such a spacecraft could also have offensive military capabilities, a concept colloquially known as "The Space Bomber".[3][4] The magazine also stated that it was likely that Blackstar would be mothballed, although it is unclear whether this is due to cost or failure of the program.

The *Aviation Week* report was dismissed a few days later as "almost certainly bogus" and the project termed a "technical absurdity" by Jeffrey F. Bell in an article in Spacedaily.[5]

6.1   The Blackstar system

Aviation Week describes Blackstar as a two stage to orbit system, comprising a high-speed jet "mothership" aircraft (which *Aviation Week* referred to as the **SR-3**). Its description of SR-3 is similar to the North American B-70 Valkyrie Mach 3 strategic bomber, and to patents filed in the 1980s by Boeing. The SR-3 would carry a second, smaller airframe, codenamed the **XOV** (eXperimental Orbital Vehicle) underneath, between its two laterally separated engine-banks, containing each 2 or 3 engines. This rocket-powered spaceplane, with similarities to the X-20 Dyna-Soar project, would be released by its mothership at an altitude of around 100,000 feet. The XOV would then light its rocket motor (aerospike engines, similar to those used by the Lockheed Martin X-33), and could achieve both suborbital and orbital flight; one source quoted by *Aviation Week* estimates the XOV could reach an orbit of 300 miles (480 km) above the Earth, depending on payload and mission profile. The XOV would then reenter the atmosphere and glide back to any landing site where it would land horizontally on a conventional runway. This combination of jet-powered mothership and a smaller rocket-powered spaceplane resembles the civilian Tier One spaceplane system as well as NASA's X-15, but capable of much higher velocities and of thus attaining orbit. Readers are cautioned to examine the challenges involved in supersonic separation of vehicles as opposed to the more common subsonic separation of ordnance from aircraft, but this separation from the belly might be easier than from the top, which proved to be problematic on the Lockheed D-21/M-21.

6.2   The program

The primary use of a military spaceplane such as Blackstar would be to conduct high-altitude or orbital reconnaissance, allowing surprise overflights of foreign locations with very low risk of the spyplane being successfully engaged by existing air-defense systems. This is similar to the goals of the earlier U-2 and SR-71 Blackbird reconnaissance aircraft; in some circumstances such an overflight yields more information than a pass by a reconnaissance satellite, as the satellite's path is predictable, allowing sensitive material to be hidden.

The March 6, 2006 cover of Aviation Week & Space Technology *depicting the rumored "Blackstar" project vehicles*

Military analysts have suggested that a military spaceplane could also be used to place small satellites in orbit, to retrieve them, to provide a means of launching nuclear weapons from orbit, or to serve as a platform for exotic orbit-to-ground hypervelocity weapons. The small spaceplane described by *Aviation Week* appears to have only a very modest cargo capacity, limiting its use in such missions.

Aviation Week suggests that the huge costs of the Blackstar program were borne both by the Department of Defense's own black budget and by hiding the costs of Blackstar inside the procurement costs attached to acknowledged military purchases. To assist in this, and to allow politicians to deny the USAF operates such a vehicle, the Blackstar assets may nominally be owned and operated by the civilian defense contractors who built it. The magazine suggests that a consortium of Boeing and Lockheed are responsible for Blackstar.

It is unclear if the Blackstar program became fully operational, although it may have been so since the mid-1990s. *Aviation Week's* article speculated that the success of Blackstar explains the Government's willingness to cancel the SR-71 Blackbird and Air Force satellite-launch programs.

6.3   Discussions of similar aircraft

During the 1970s, when studies were underway which led to the specification of the Space Shuttle, most leading US aerospace contractors explored orbital spaceplane designs, some based on a two-stage design. The most serious of these was the Lockheed HGV under the X-24C program, which was a manned hypersonic vehicle dropped from underwing a B-52, even to the point of rumors that it had actually been flight tested, according to Encyclopedia Astronautica. With the adoption of the Space Shuttle design, these avenues appear to have been abandoned. The use of a spaceplane as part of the launching system to replace the Space Shuttle has been suggested in programs such as VentureStar.

Some of the details of the SR-3 resemble the rumored Brilliant Buzzard or "Mothership" aircraft, but these were supposed to carry their second stage aircraft on top, rather than on the bottom as with the SR-3. This second stage was rumored to be Aurora, (a high-speed, high-altitude delta-winged aircraft), and the lengthening of runways at facilities such as Area 51 (taken by some as evidence of Aurora) could instead be necessary either to support SR-3's takeoff or XOV's landing. Most descriptions of Aurora, however, describe it as a hypersonic plane with exotic engine technology; the SR-3 described by *Aviation Magazine* is similar to existing rocket-powered aircraft. Pulse Detonation Engine (PDE) technology, visually apparent by donuts-on-a-rope contrail - and audibly by its deep bass pulsing boom noise, has been associated with these programs from eyewitness accounts during the 1990s.

In the late 1960s the North American Aircraft Corporation studied conceptual designs using the B-70 bomber for small space launch of an X-15 type rocket plane. These were abandoned as unpromising.

What is known, and a matter of public record, is that, through the 1980s and 1990s, the USAF did undertake a series of projects to study, research, develop and test demonstrator vehicles capable of SSTO (single-stage-to-orbit) and TSTO (two-stage-to-orbit) missions. These programs were code-named, in order, SCIENCE DAWN, SCIENCE REALM, COPPER CANYON, and COPPER COAST, and involved the development of three different competitive demonstrator vehicles. It was at the conclusion of COPPER CANYON's design phase that President Reagan proposed the X-30 NASP, which is claimed by the Blackstar story to have been used to pay for development of this spaceplane.

According to one declassified RAND Corp. report, two of the three vehicles failed to achieve their full flight envelope (i.e. couldn't make orbit), while the third, an "assisted SSTO", did achieve orbital capability. Furthermore, three code-named programs to design the stealthing of these three vehicles fell under the programs known as HAVE BLINDERS I, HAVE BLINDERS II, and HAVE BLINDERS III. All of these programs can be found in US military budget documents, with associated budget account numbers for years in the 1980s up into the late 1990s in the case of COPPER COAST, though the code name was dropped from the account number in the mid-1990s, even though many millions were budgeted up until recent years.

Whether any of these vehicles were individually code named "BLACKSTAR" is unknown at this time.

6.3.1   Blackswift

Main article: DARPA Falcon Project § Blackswift

Details emerged in 2008 of an unmanned hypersonic platform called Blackswift, otherwise known as HTV-3X or X-43A, part of the DARPA Falcon Project.*[6]

Computer-generated concept videos of the tests of this vehicle were made available by NASA / Lockheed Martin in June 2008.

6.4 May 2006 UK Defence report on "Black"aircraft sightings

In May 2006, the British Ministry of Defence (MoD) released an extensive report on Unexplained Aerial Phenomena (UAPs) in the UK air defence area . It was written by the Defence Intelligence Staff in 2000 and was originally classified "SECRET UK eyes only". One of the Working Papers is entitled ""BLACK" AND OTHER AIRCRAFT AS UAP EVENTS" . It says "it is acknowledged that some UAP sightings can be attributed to covert aircraft programmes". The report lists three "Western" programmes which might result in this – all of which appear to be American. The first – not surprisingly – is the SR-71. Programme 2 and Programme 3 are redacted from the report – even their names are withheld. Two photos are also redacted. This was reported on June 14, 2006 by BBC Newsnight.*[7]*[8]

6.5 See also

- VentureStar

- X-20 Dyna-Soar

- SR-71 Blackbird

- XB-70 Valkyrie

- X-30 National Aerospace Plane

- ISINGLASS a cancelled in 1960s USAF atmospheric suborbital skip spyplane

6.6 References

[1] "Two-Stage-to-Orbit 'Blackstar' System Shelved at Groom Lake?." Scott, W., *Aviation Week & Space Technology*. March 5, 2006.

[2] "Did Pentagon create orbital space plane?." Oberg, J., MSNBC. March 6, 2006.

[3] "Bush plans 'space bomber'." Vulliamy, E., *The Observer*. July 29, 2001.

[4] "Pentagon planning for space bomber." Windrem, R., MSNBC. August 14, 2001.

[5] Jeffrey F. Bell (2006-03-10). "Blackstar A False Messiah From Groom Lake" . Space Daily. Retrieved 2007-09-21.

[6] http://www.darpa.mil/tto/solicit/PS08-02.pdf

[7] http://webarchive.nationalarchives.gov.uk/20121026065214/http://www.mod.uk/NR/rdonlyres/6A30B96E-35AD-4F73-93B1-863F59A3A0E0/uap_vol2_pgs76to90.pdf

[8] http://webarchive.nationalarchives.gov.uk/20121026065214/http://www.mod.uk/NR/rdonlyres/EDAB29D1-BBE2-4811-A62B-45D6C9608B0/uap_vol2_pgs91to105.pdf

Miller, Jay. The X-Planes: X-1 to X-45. Hinckley, UK: Midland, 2001.

Rose, Bill, 2008. Secret Projects: Military Space Technology. Hinckley, England: Midland Publishing.

6.7 External links

- BBC Newsnight 14 June 2006 Links to UK MoD "Black" aircraft working paper

- The Space Review: Six blind men in a zoo: Aviation Week's mythical Blackstar Dwayne A. Day, *The Space Review*, Monday, March 13, 2006

- robotpig.net - TSTO spaceplanes presentation of a Boeing TSTO patent, the blackstar tsto and the respective technologies

- Blackstar: the US space conspiracy that never was?

Chapter 7

Boeing X-20 Dyna-Soar

The **Boeing X-20 Dyna-Soar** ("Dynamic Soarer") was a United States Air Force (USAF) program to develop a spaceplane that could be used for a variety of military missions, including reconnaissance, bombing, space rescue, satellite maintenance, and as a space interceptor to sabotage enemy satellites.[*][1] The program ran from October 24, 1957 to December 10, 1963, cost US$660 million ($5.08 billion today[*][2]), and was cancelled just after spacecraft construction had begun.

Other spacecraft under development at the time, such as Mercury or Vostok, were based on space capsules that returned on ballistic re-entry profiles. Dyna-Soar was more like the much later Space Shuttle. It could not only travel to distant targets at the speed of an intercontinental ballistic missile, it was designed to glide to earth like an aircraft under control of a pilot. It could land at an airfield, rather than simply falling to earth and landing with a parachute. Dyna-Soar could also reach earth orbit, like Mercury or Gemini.[*][3]

These characteristics made Dyna-Soar a far more advanced concept than other human spaceflight missions of the period. Research into a spaceplane was realized much later, in other reusable spacecraft such as the Space Shuttle,[*][4][*][5] which had its first orbital flight in 1981, and, more recently, the Boeing X-40 and X-37B spacecraft.

7.1 Background

Following World War II, many German scientists were taken to the United States by the Office of Strategic Services's "Operation Paperclip". Among them was Dr. Walter Dornberger, the former head of Germany's wartime rocket program, who had detailed knowledge of Eugen Sänger's Silbervogel project.[*][6] Working for Bell, he attempted to create interest in a boost-glide system in the USAF, and elsewhere. This resulted in the USAF requesting a number of feasibility and design studies – carried out by Bell, Boeing, Convair, Douglas, Martin, North American, Republic, and Lockheed – for boost-glide vehicles during the early 1950s:

- Bomi (bomber missile);[*][7][*][8]

- Hywards (HYpersonic Weapons Research and Development Supporting system);[*][9]

- The Brass Bell reconnaissance vehicle;[*][10][*][11] and

- Rocket Bomber "Robo".[*][12][*][13]

The development of Dyna Soar can be traced back to the Silbervogel bomber project of World War II.[*][14] The concept was a rocket-powered bomber that could travel vast distances by gliding to its target after being boosted to high speed (>5.5 km/s) and high altitude (50–150 km) by A-4 or A-9 rockets.[*][14]

7.1.1 Lifting re-entry method

Essentially, these rockets would place the vehicle onto an exoatmospheric intercontinental ballistic missile-like trajectory and then fall away. When the vehicle reentered the atmosphere, instead of fully reentering, bleeding off its speed and landing, the vehicle would use the lift from its wings to redirect its glide angle upward while bleeding off

Artist's impression of the X-20 after test flight

speed in the process. In this way, the vehicle would be "bounced" back into space again. This skip-glide[15] method would repeat until the speed was low enough that the pilot of the vehicle would need to pick a landing spot and glide the vehicle to a landing. This use of hypersonic atmospheric lift meant that the vehicle could greatly extend its range over a ballistic trajectory using the same engines.[14]

Such boost-glide systems could potentially strike at targets anywhere in the world (so called "antipodal bombers") at hypersonic speeds, be very difficult to intercept, and the aircraft itself could be small and lightly armed, compared to a typical heavy bomber. In addition, a boost-glide aircraft may be recoverable, acting as a manned bomber, or as an unmanned non-recoverable missile.

7.2 Development

Boeing mock-up of X-20 Dyna-Soar

On October 10, 1957 ARDC (USAF Air Research and Development Command) headquarters consolidated Hywards, Brass Bell, and Robo studies into a three-step abbreviated development plan for System 464L, Dyna-Soar[16](alternative date October 24, 1957[17]). The proposal drew together the existing boost-glide proposals, as the USAF believed a single vehicle could be designed to carry out all the bombing and reconnaissance tasks intended for the separate studies, and act as successor to the X-15 research program. The Dyna-Soar program was to be conducted in three stages: a research vehicle (**Dyna-Soar I**), a reconnaissance vehicle (**Dyna-Soar II**, previously Brass Bell), and a vehicle that added strategic bombing capability (**Dyna-Soar III**, previously Robo). The first glide tests for Dyna-Soar I were expected to be carried out in 1963, followed by powered flights, reaching Mach 18, the following year. A robotic glide missile was to be deployed in 1968, with the fully operational weapons system (Dyna-Soar III) expected by 1974.[18]

In March 1958, nine U.S. aerospace companies tendered for the Dyna-Soar contract. Of these, the field narrowed to proposals from Bell and Boeing. Even though Bell had the advantage of six years' worth of design studies, the contract for the spaceplane was awarded to Boeing in June 1959 (by which time their original design had changed markedly and now closely resembled what Bell had submitted). In late 1961, the Titan III was chosen as the launch vehicle.[19] The Dyna-Soar was to be launched from Cape Canaveral Air Force Station, Florida.

7.3   Design

The overall design of the X-20 Dyna-Soar was outlined in March 1960. It had a low-wing delta shape, with winglets for control rather than a more conventional tail. The framework of the craft was to be made from the René 41 *super alloy*, as were the upper surface panels. The bottom surface was to be made from molybdenum sheets placed over insulated René 41, while the nose-cone was to be made from graphite with zirconia rods.[20]

Due to the changing requirements, various forms of the Dyna-Soar were designed. All variants shared the same basic shape and layout. A single pilot sat at the front, with an equipment bay situated behind. This bay contained data-collection equipment, weapons, reconnaissance equipment, or (in the X-20X "shuttle space vehicle") a four-man mid-deck.

A transition-stage rocket engine, located behind the equipment bay, would maneuver the craft in orbit or fire during launch as part of an abort sequence. This trans-stage would be jettisoned before descent into the atmosphere. While falling through the atmosphere an opaque heat shield made from a refractory metal would protect the window at the front of the craft. This heat shield would then be jettisoned after aerobraking so the pilot could see, and safely land.[21]

A drawing in *Space/Aeronautics* magazine from before the project's cancellation depicts the craft dipping down into the atmosphere, skimming the surface, to change its orbital inclination. It would then fire its rocket to resume orbit. This would be a unique ability for a spacecraft, for the laws of celestial mechanics mean it requires an enormous expenditure of energy for a rocket to change its orbital inclination once it has reached orbit. Hence the Dyna-Soar could have had a military capacity of being launched into one orbit and rendezvousing with a satellite, even if the target were to expend all its propellant in changing its orbit. Acceleration forces on the pilot would be severe in such a maneuver.

Unlike the later Space Shuttle, Dyna-Soar did not have wheels on its tricycle undercarriage as the rubber tyres required cooled compartments or they would burn during re-entry. Instead Goodyear developed retractable wire-brush skids made of the same René 41 alloy as the airframe.[22]

7.4   Operational history

In April 1960, seven astronauts were secretly chosen for the Dyna-Soar program:[23]

- Neil Armstrong (1930–2012; NASA) 1960–62

- Bill Dana (1930–2014; NASA) 1960–62

- Henry C. Gordon (1925–96; Air Force) 1960–63

- Pete Knight (1929–2004; Air Force) 1960–63

- Russell L. Rogers (1928–67; Air Force) 1960–63

An artist's impression of Dyna-Soar being launched using a Titan booster, with large fins added to the Titan's first stage

- Milt Thompson (1926–93; NASA) 1960–63

- James W. Wood (1924–90; Air Force) 1960–63

Neil Armstrong and Bill Dana left the program in mid-1962.

On September 19, 1962, Albert Crews was added to the Dyna-Soar program and the names of the six remaining Dyna-Soar astronauts were announced to the public.

Artist's impression of the X-20 on landing approach at Edwards Air Force Base

By the end of 1962, Dyna-Soar had been designated "X-20", the booster (to be used in the Dyna Soar I drop-tests) successfully fired, and the USAF had held an "unveiling" ceremony for the X-20 in Las Vegas.*[24]*[25]

Minnesota Honeywell Corporation completed flight tests on an inertia guidance sub-system for the X-20 project at Eglin Air Force Base, Florida, utilizing an NF-101B Voodoo by August 1963.*[26]

Boeing B-52C-40-BO Stratofortress, *53-0399*,*[27] was assigned to the programme for air-dropping the X-20, similar to the X-15 launch profile. When the X-20 was cancelled, it was used for other air-drop tests including that of the B-1A escape capsule.*[28]

7.4.1 Problems

Besides the funding issues that often accompany research efforts, the Dyna-Soar program suffered from two major problems: uncertainty over the booster to be used to send the craft into orbit, and a lack of a clear goal for the project.

Many different boosters were proposed to launch Dyna-Soar into orbit. The original USAF proposal suggested LOX/JP-4, fluorine-ammonia, fluorine-hydrazine, or RMI (X-15) engines. Boeing, the principal contractor, favored an Atlas-Centaur combination. Eventually the Air Force stipulated a Titan, as suggested by failed competitor Martin, but the Titan I was not powerful enough to launch the five-ton X-20 into orbit.

The Titan II and Titan III boosters could launch Dyna-Soar into Earth orbit, as could the Saturn C-1 (later renamed the Saturn I), and all were proposed with various upper-stage and booster combinations. While the Titan IIIC was eventually chosen to send Dyna-Soar into space, the vacillations over the launch system delayed the project as it complicated planning.

The original intention for Dyna-Soar, outlined in the Weapons System 464L proposal, called for a project combining aeronautical research with weapons system development. Many questioned whether the USAF should have a manned space program, when that was the primary domain of NASA. It was frequently emphasized by the U.S. Air Force that, unlike the NASA programs, Dyna-Soar allowed for controlled re-entry, and this was where the main effort in the X-20 program was placed. On January 19, 1963, the Secretary of Defense, Robert McNamara, directed the U.S. Air Force to undertake a study to determine whether Gemini or Dyna-Soar was the more feasible approach to a space-based weapon system. In the middle of March 1963, after receiving the study, Secretary McNamara "stated that the Air Force had been placing too much emphasis on controlled re-entry when it did not have any real objectives for orbital flight" .*[29] This was seen as a reversal of the Secretary's earlier position on the Dyna-Soar program. Dyna-Soar was also an expensive program that would not launch a manned mission until the mid-1960s at the earliest. This high cost and questionable utility made it difficult for the U.S. Air Force to justify the program. Eventually, the X-20 Dyna-Soar program was canceled on December 10, 1963.*[4]

On the day that X-20 was canceled, the U.S. Air Force announced another program, the Manned Orbiting Laboratory, a spin-off of Gemini. This program was also eventually canceled. Another black program ISINGLASS, which was to be airlaunched from a B-52 bomber, was evaluated and some engine work done. This eventually was cancelled also.*[30]

7.4.2 Legacy

Despite cancellation of the X-20, the affiliated research on spaceplanes influenced the much larger Space Shuttle. The final design also used delta wings for controlled landings. The later, and much smaller Soviet BOR-4 was closer in design philosophy to the Dyna-Soar,*[31] while NASA's Martin X-23 PRIME and Martin Marietta X-24A/HL-10 research aircraft also explored aspects of sub-orbital and space flight.*[32] The ESA proposed Hermes manned space craft took the design and expanded its scale.

7.5 Specifications (as designed)

General characteristics

- **Crew:** one pilot

- **Length:** 35 ft 4 in (10.77 m)

- **Wingspan:** 20 ft 10 in (6.34 m)

- **Height:** 8 ft 6 in (2.59 m)

- **Wing area:** 345 ft^2 (32 m^2)

- **Empty weight:** 10,395 lb (4,715 kg)

- **Max. takeoff weight:** 11,387 lb (5,165 kg)

- **Powerplant:** 1 × Martin Trans-stage rocket engine, 72,000 lbf (323 kN)

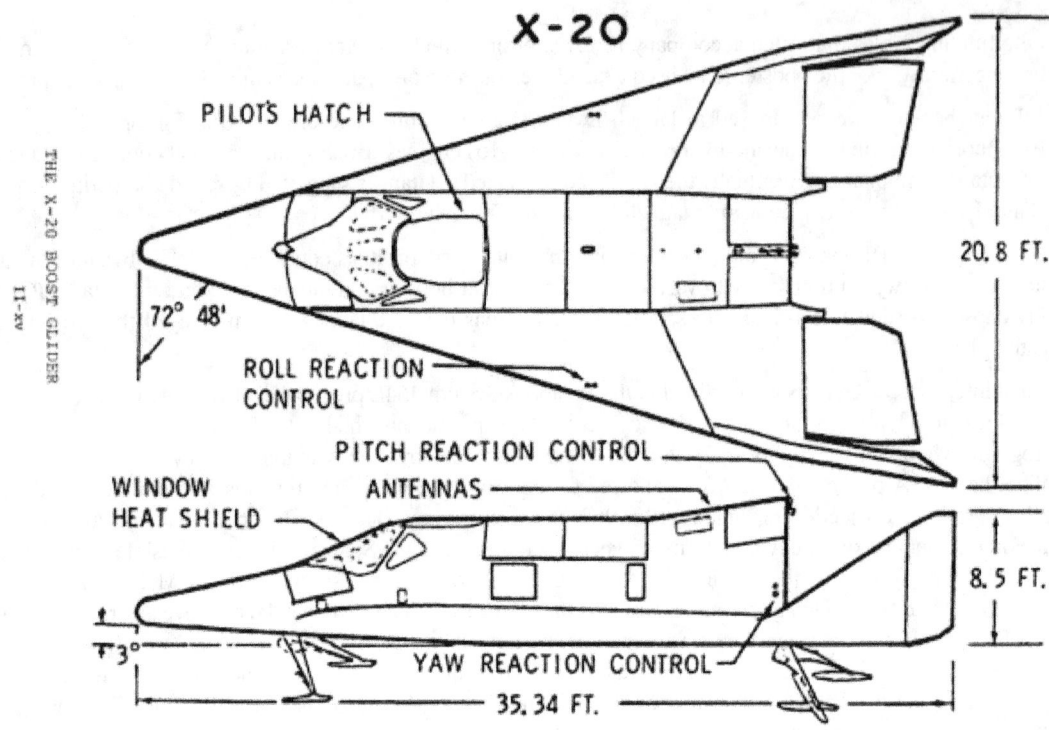

Orthographically projected diagram of the X-20.

Performance

- **Maximum speed:** 17,500 mph (28,165 km/h)

- **Range:** earth orbit 22,000 nautical miles, (40,700 km)

- **Service ceiling:** 530,000 ft (160 km)

- **Rate of climb:** 100,000 ft/min (510 m/s)

- **Wing loading:** 33 lb/ft^2 (161 kg/m^2)

7.6   Media

- The 1959 *Twilight Zone* season 1 episode titled "And When the Sky Was Opened" made reference to a spacecraft called the X20 which had a similar profile but could carry a crew of three.

- John Berryman's 1963 short story "The Trouble with Telstar" featured a Dyna-Soar being used to intercept communications satellites for repair.[*][33]

- The 1969 Hollywood film drama *Marooned* featured a rescue craft modeled somewhat after the Dyna-Soar (called the X-RV for eXperimental Rescue Vehicle) being hurriedly deployed to rescue astronauts aboard a crippled Apollo command capsule.

7.7   See also

- Boeing X-37

- Saturn-Shuttle

- Hermes

- Hypersoar

Related development

- Manned Orbital Laboratory

Aircraft of comparable role, configuration and era

- ASSET – a subscale reentry test vehicle designed to verify the superalloy heatshield of the DynaSoar.

- BOR-4

- BOR-5

- Mikoyan-Gurevich MiG-105

- North American X-15

- Silbervogel

7.8 References

7.8.1 Notes

[1] Goebel, Greg. "The X-15, Dyna-Soar, & The Lifting Bodies – [1.2] The USAF and DYNA-SOAR". *VectorSite.net.* Vectors by Greg Goebel. Retrieved January 16, 2015.

[2] Consumer Price Index (estimate) 1800–2014. Federal Reserve Bank of Minneapolis. Retrieved February 27, 2014.

[3] "History: X-20 Dyna-Soar Space Vehicle." *Boeing.* Retrieved: September 24, 2010.

[4] Yenne 1985, p. 136.

[5] Bilstein, Roger E. (2003). *Testing aircraft, exploring space : an illustrated history of NACA and NASA.* Baltimore: Johns Hopkins Univ. Press. p. 90. ISBN 0801871581. Retrieved January 19, 2015.

[6] Dornberger 1956, pp. 19–37.

[7] http://www.astronautix.com/craft/bomi.htm

[8] MX-2276 http://www.dtic.mil/dtic/tr/fulltext/u2/073754.pdf

[9] http://www.astronautix.com/craft/hywards.htm

[10] http://www.astronautix.com/craft/brasbell.htm

[11] http://www.dtic.mil/dtic/tr/fulltext/u2/136057.pdf

[12] Neufeld 1995, pp. 19, 33, 55.

[13] http://www.astronautix.com/craft/robo.htm

[14] Duffy 2004, p. 124.

[15] Launius, Roger D.; Jenkins, Dennis R. (2012). *Coming home : reentry and recovery from space.* Washington, DC: National Aeronautics and Space Administration. p. 170. ISBN 0160910641. Retrieved 20 June 2015.

[16] History of the X-20A Dyna-Soar, Clarence J. Geiger, Sep 1963 www.dtic.mil/cgi-bin/GetTRDoc?AD=ADA951933

[17] Godwin 2003, p. 38.

[18] Godwin 2003, p. 65.

[19] Godwin 2003, p. 286.

[20] Godwin 2003, p. 186.

[21] Launius, Roger D.; Jenkins, Dennis R. (2012). *Coming home : reentry and recovery from space*. Washington, DC: National Aeronautics and Space Administration. p. 178. ISBN 9780160910647. Retrieved January 16, 2015.

[22] Heppenheimer, T.A. (September 2007). *Facing the Heat Barrier: A History of Hypersonics* (PDF). Washington, DC: National Aeronautics and Space Administration – History Division. p. 150. ISBN 978-1493692569. Retrieved January 16, 2015.

[23] Pelt, Michel van (2012). *Rocketing into the future : the history and technology of rocket planes*. New York: Springer. p. 269. ISBN 978-1461431992. Retrieved January 16, 2015.

[24] Peebles, Curtis (1997). *High frontier : the U.S. Air Force and the Military Space Program* (Air Force 50th anniversary commemorative ed.). Washington, DC: Air Force History and Museums Program. p. 19. ISBN 0160489458. Retrieved January 16, 2015.

[25] Jenkins, compiled by Dennis R. (2004). *X-planes photo scrapbook*. North Branch, MN: Specialty Press. p. 95. ISBN 978-1580070768. Retrieved January 16, 2015.

[26] "Fiery Crash of Drone Plane Kills Two, Injures One – Four Firemen Overcome in Wake of Blaze." *Playground Daily News* (Fort Walton Beach, Florida), Volume 16, Number 271, August 20, 1963, p. 1.

[27] http://www.joebaugher.com/usaf_serials/1953.html

[28] Spahr, Greg, "Might have beens" ,*B-52 Stratofortress: Celebrating 60 Remarkable Years*, Key Publishing Ltd., Stamford, Lincs., UK, 2014, page 38.

[29] Geiger 1963, pp. 349–405.

[30] "The U-2's intended successor: Project OXCART, 1956–1968." *Central Intelligence Agency,* December 31, 1968, p. 49. Retrieved: August 10, 2010.

[31] Marks, Paul. "Cosmonaut: Soviet space shuttle was safer than NASA's." *New Scientist,* July 7, 2007. Retrieved: August 28, 2011.

[32] Jenkins, Dennis R., Tony Landis and Jay Miller. *American X-Vehicles: An Inventory—X-1 to X-50*. Washington, DC: Monographs in Aerospace History No. 31, SP-2003-4531, June 2003.

[33] Berryman, John (June 1963). "The Trouble with Telstar". *Analog Science Fact & Fiction*. Retrieved 14 May 2015.

7.8.2 Bibliography

- Caidin, Martin. *Wings into Space: The History and Future of Winged Space Flight*. New York: Holt, Rinehart and Winston Inc., 1964.

- Dornberger, Walter R. "The Rocket-Propelled Commercial Airliner". *Dyna-Soar: Hypersonic Strategic Weapons System, Research Report No 135.*. Minneapolis, Minnesota: University of Minnesota, Institute of Technology, 1956.

- Duffy, James P. *Target: America, Hitler's Plan to Attack the United States*. Santa Barbara, California: Praeger, 2004. ISBN 0-275-96684-4.

- *Dyna-Soar: Hypersonic Strategic Weapons System: Structure Description Report*. Andrews Air Force Base, Maryland: Air Force Systems Command, 1961, pp. 145–189.

- Geiger, Clarence J. *History of the X-20A Dyna-Soar. Vol. 1: AFSC Historical Publications Series 63-50-I, Document ID ASD-TR-63-50-I*. Wright Patterson AFB, Ohio: Aeronautical Systems Division Information Office, 1963.

- Godwin, Robert, ed. *Dyna-Soar: Hypersonic Strategic Weapons System*. Burlington, Ontario, Canada: Apogee Books, 2003. ISBN 1-896522-95-5.

- Houchin, Roy. *U.S. Hypersonic Research and Development: The Rise and Fall of Dyna-Soar, 1944–1963*. London: Routledge, 2006. ISBN 0-415-36281-4.

- Neufeld, Michael J. *The Rocket and the Reich: Peenemünde and the Coming of the Ballistic Missile Era.* New York: The Free Press, 1995. ISBN 978-0-674-77650-0.

- Strathy, Charlton G. *Dyna-Soar: Hypersonic Strategic Weapons System: Weapon System 464L Abbreviated Development Plan,* 1957, pp. 38–75.

- Yenne, Bill. *The Encyclopedia of US Spacecraft.* London: Bison Books, 1985. ISBN 978-5-551-26650-1.

7.9 External links

- Dyna Soar at Encyclopedia Astronautica

- Official United States Air Force film from the 1960 describing the spacecraft.

- Dynasoar

- Tsien Space Plane 1949

- Tsien Space Plane 1978

- Transonic aerodynamic characteristics of the Dyna-Soar glider and Titan 3 launch vehicle configuration with various fin arrangements (PDF format) NASA report – April 1963

- *American X-Vehicles: An Inventory X-1 to X-50*, SP-2000-4531 – June 2003; NASA online PDF Monograph

- *Deepcold: Secrets of the Cold War in Space, 1959–1969*

Chapter 8

Boeing X-37

The **Boeing X-37**, also known as the **Orbital Test Vehicle** (**OTV**), is a reusable unmanned spacecraft. It is boosted into space by a launch vehicle, then re-enters Earth's atmosphere and lands as a spaceplane. The X-37 is operated by the United States Air Force for orbital spaceflight missions intended to demonstrate reusable space technologies.[5] It is a 120%-scaled derivative of the earlier Boeing X-40.

The X-37 began as a NASA project in 1999, before being transferred to the U.S. Department of Defense in 2004. It conducted its first flight as a drop test on 7 April 2006, at Edwards Air Force Base, California. The spaceplane's first orbital mission, USA-212, was launched on 22 April 2010 using an Atlas V rocket. Its successful return to Earth on 3 December 2010 was the first test of the vehicle's heat shield and hypersonic aerodynamic handling. A second X-37 was launched on 5 March 2011, with the mission designation USA-226; it returned to Earth on 16 June 2012. A third X-37 mission, USA-240, launched on 11 December 2012 and landed at Vandenberg AFB on 17 October 2014. The fourth X-37 mission, USA-261, launched on 20 May 2015 and is in progress.

8.1 Development

8.1.1 Origins

In 1999, NASA selected Boeing Integrated Defense Systems to design and develop an orbital vehicle, built by the California branch of Boeing's Phantom Works. Over a four-year period, a total of $192 million was contributed to the project, with NASA contributing $109 million, the U.S. Air Force $16 million, and Boeing $67 million. In late 2002, a new $301-million contract was awarded to Boeing as part of NASA's Space Launch Initiative framework.[6]

The X-37's aerodynamic design was derived from the larger Space Shuttle orbiter, hence the X-37 has a similar lift-to-drag ratio, and a lower cross range at higher altitudes and Mach numbers compared to DARPA's Hypersonic Technology Vehicle.[7] An early requirement for the spacecraft called for a delta-v of 7,000 mph (3.1 km/s) to change its orbit.[8] An early goal for the program was for the X-37 to rendezvous with satellites and perform repairs.[9] The X-37 was originally designed to be carried into orbit in the Space Shuttle's cargo bay, but underwent redesign for launch on a Delta IV or comparable rocket after it was determined that a shuttle flight would be uneconomical.[10]

The X-37 was transferred from NASA to the Defense Advanced Research Projects Agency (DARPA) on 13 September 2004.[11] Thereafter, the program became a classified project. DARPA promoted the X-37 as part of the independent space policy that the United States Department of Defense has pursued since the 1986 *Challenger* disaster.

8.1.2 Glide testing

The vehicle that was used as an atmospheric drop test glider had no propulsion system. Instead of an operational vehicle's payload bay doors, it had an enclosed and reinforced upper fuselage structure to allow it to be mated with a mothership. In September 2004, DARPA announced that for its initial atmospheric drop tests the X-37 would be launched from the Scaled Composites White Knight, a high-altitude research aircraft.[12]

On 21 June 2005, the X-37A completed a captive-carry flight underneath the White Knight from Mojave Spaceport in

1999 artist's rendering of the X-37 spacecraft

The Scaled Composites White Knight was used to launch the X-37A on glide tests.

Mojave, California.[13][14] Through the second half of 2005, the X-37A underwent structural upgrades, including the reinforcement of its nose wheel supports. Further captive-carry flight tests and the first drop test were initially expected to occur in mid-February 2006. The X-37's public debut was scheduled for its first free flight on 10 March 2006, but was canceled due to an Arctic storm.[15] The next flight attempt, on 15 March 2006, was canceled due to high winds.[15]

On 24 March 2006, the X-37 flew again, but a datalink failure prevented a free flight, and the vehicle returned to the ground still attached to its White Knight carrier aircraft. On 7 April 2006, the X-37 made its first free glide flight. During landing, the vehicle overran the runway and sustained minor damage.[16] Following the vehicle's extended downtime for repairs, the program moved from Mojave to Air Force Plant 42 (KPMD) in Palmdale, California for the remainder of the flight test program. White Knight continued to be based at Mojave, though it was ferried to Plant 42 when test flights were scheduled. Five additional flights were performed,[N 1] two of which resulted in X-37 releases with successful landings. These two free flights occurred on 18 August 2006 and 26 September 2006.[17]

8.1.3   X-37B Orbital Test Vehicle

On 17 November 2006, the U.S. Air Force announced that it would develop its own variant from NASA's X-37A. The Air Force version was designated the X-37B Orbital Test Vehicle (OTV). The OTV program was built on earlier industry and government efforts by DARPA, NASA and the Air Force, and was led by the U.S. Air Force Rapid Capabilities Office, in partnership with NASA and the Air Force Research Laboratory. Boeing was the prime contractor for the OTV program.[8][18][19] The X-37B was designed to remain in orbit for up to 270 days at a time.[20] The Secretary of the Air Force stated that the OTV program would focus on "risk reduction, experimentation, and operational concept development for reusable space vehicle technologies, in support of long-term developmental space objectives." [18]

The X-37B was originally scheduled for launch in the payload bay of the Space Shuttle, but following the 2003 *Columbia* disaster, it was transferred to a Delta II 7920. The X-37B was subsequently transferred to a shrouded configuration on the Atlas V rocket, following concerns over the unshrouded spacecraft's aerodynamic properties during launch.[21] Following their missions, X-37B spacecraft land on a runway at Vandenberg Air Force Base, California, with Edwards Air Force Base as an alternate site.[22] In 2010, manufacturing work began on the second X-37B, OTV-2,[23] which conducted its maiden launch in March 2011.[24]

On 8 October 2014, NASA confirmed that X-37B vehicles would be housed at Kennedy Space Center in Orbiter Processing Facilities (OPF) 1 and 2, hangars previously occupied by the Space Shuttle. Boeing had said the space planes would use OPF-1 in January 2014, and the Air Force had previously said it was considering consolidating X-37B operations, housed at Vandenberg Air Force Base in California, nearer to their launch site at Cape Canaveral. NASA also stated that the program had completed tests to determine whether the X-37B, one-fourth the size of the Space Shuttle, could land on the former Shuttle runways.[25] NASA furthermore stated that renovations of the two hangars would be completed by the end of 2014; the main doors of OPF-1 were marked with the message "Home of the X-37B" by this point.[25]

Most of the activities of the X-37B project are secret. The official U.S. Air Force statement is that the project is "an experimental test program to demonstrate technologies for a reliable, reusable, unmanned space test platform for the U.S. Air Force." [5] The primary objectives of the X-37B are twofold: reusable spacecraft technology, and operating experiments which can be returned to Earth.[5] The Air Force states that this includes testing avionics, flight systems, guidance and navigation, thermal protection, insulation, propulsion, and re-entry systems.[26]

8.1.4   Speculation regarding purpose

In 2010, Tom Burghardt wrote for *Space Daily* that the X-37B could be used as a spy satellite or to deliver weapons from space.[27] The Pentagon subsequently denied claims that the X-37B's test missions supported the development of space-based weapons.[27] In January 2012, allegations were made that the X-37B was being used to spy on China's Tiangong-1 space station module.[28]

Former U.S. Air Force orbital analyst Brian Weeden later refuted this claim, emphasizing that the different orbits of the two spacecraft precluded any practical surveillance fly-bys.[29] In October 2014, *The Guardian* reported the claims of security experts that the X-37B was being used "to test reconnaissance and spy sensors, particularly how they hold up against radiation and other hazards of orbit." [30]

8.2   Design

The X-37 Orbital Test Vehicle is a reusable robotic spaceplane. It is a 120%-scale derivative of the Boeing X-40,[6][22] measuring over 29 feet (8.8 m) in length, and features two angled tail fins.[5][31] The X-37 launches

First Spaceplanes

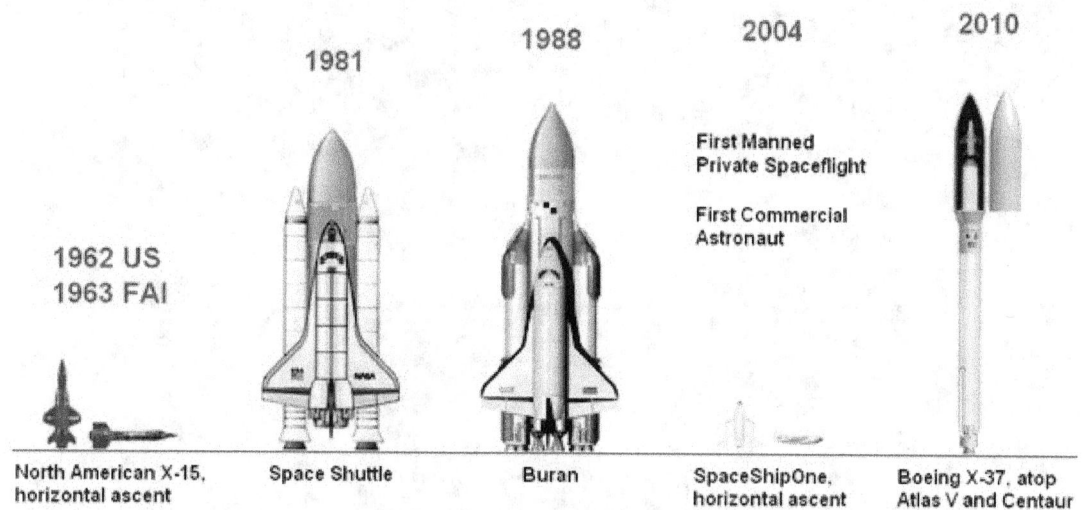

At the time of its maiden launch, the X-37 (far right) was the smallest and lightest orbital spaceplane yet flown. Both the North American X-15 and SpaceShipOne were suborbital. Of the spaceplanes shown, only the X-37 and Buran conducted unmanned spaceflights.

atop an Atlas V version 501 rocket with a Centaur second stage.[*][5][*][19] The X-37 is designed to operate in a speed range of up to Mach 25 on its reentry.[*][32][*][33]

The technologies demonstrated in the X-37 include an improved thermal protection system, enhanced avionics, an autonomous guidance system and an advanced airframe.[*][10] The spaceplane's thermal protection system is built upon previous generations of atmospheric reentry spacecraft,[*][34] incorporating silica ceramic tiles.[*][35] The X-37's avionics suite was used by Boeing to develop its CST-100 manned spacecraft.[*][36] According to NASA, the development of the X-37 will "aid in the design and development of NASA's Orbital Space Plane, designed to provide a crew rescue and crew transport capability to and from the International Space Station".[*][37]

The X-37 is independently powered by one Aerojet AR2-3 engine using storable propellants, providing thrust of 6,600 pounds-force (29.341 kN).[*][38] The human-rated AR2-3 engine had been used on the dual-power NF-104A astronaut training vehicle, and was given a new flight certification for use on the X-37 with hydrogen peroxide/JP-8 propellants.[*][39]

The X-37 lands automatically upon returning from orbit, and is the second reusable spacecraft to have such a capability, after the Soviet Buran shuttle.[*][40] The X-37 is the smallest and lightest orbital spaceplane flown to date; with a launch mass of around 11,000 pounds (5,000 kg), it is approximately a quarter the size of the Space Shuttle orbiter.[*][41] In 2013, Guinness World Records recognised the X-37 as the world's smallest orbital spaceplane.[*][42]

The Space Foundation awarded the X-37 team on 13 April 2015 with the 2015 Space Achievement Award "for significantly advancing the state of the art for reusable spacecraft and on-orbit operations, with the design, development, test and orbital operation of the X-37B space flight vehicle over three missions totaling 1,367 days in space." [*][43]

8.3 Operational history

As of 17 October 2014, the two operational X-37Bs have conducted three orbital missions, spending a combined total 1,367 days in space.[*][44]

8.3.1 OTV-1

Main article: USA-212

OTV-1, the first X-37B, launched on its first mission – USA-212 – on an Atlas V rocket from Cape Canaveral Air

OTV-1 sits on the runway after landing at Vandenberg AFB at the close of its USA-212 mission on 3 December 2010.

Force Station, Florida, on 22 April 2010 at 23:58 UTC. The spacecraft was placed into low Earth orbit for testing.[*][19] While the U.S. Air Force revealed few orbital details of the mission, amateur astronomers claimed to have identified the spacecraft in orbit and shared their findings. A worldwide network of amateur astronomers reported that, on 22 May 2010, the spacecraft was in an inclination of 39.99 degrees, circling the Earth once every 90 minutes on an orbit 249 by 262 miles (401 by 422 km).[*][45][*][46] OTV-1 reputedly passed over the same given spot on Earth every four days, and operated at an altitude of 255 miles (410 km), which is typical for military surveillance satellites.[*][47] Such an orbit is also common among civilian LEO satellites, and the spaceplane's altitude was the same as that of the ISS and most other manned spacecraft.

The U.S. Air Force announced on 30 November 2010 that OTV-1 would return for a landing during the 3–6 December timeframe.[*][48][*][49] As scheduled, OTV-1 de-orbited, reentered Earth's atmosphere, and landed successfully at Vandenberg AFB on 3 December 2010, at 09:16 UTC,[*][50][*][51][*][52] conducting America's first autonomous orbital landing onto a runway; the first spacecraft to perform such a feat was the Soviet Buran shuttle in 1988. In all, OTV-1 spent 224 days in space.[*][53] OTV-1 suffered a tire blowout during landing and sustained minor damage to its underside.[*][23]

8.3.2 OTV-2

Main article: USA-226

OTV-2, the second X-37B, launched on its inaugural mission, designated USA-226,[*][54] aboard an Atlas V rocket from Cape Canaveral on 5 March 2011.[*][55] The mission was classified and described by the U.S. military as an effort to test new space technologies.[*][56] On 29 November 2011, the U.S. Air Force announced that it would extend the mission of USA-226 beyond the 270-day baseline design duration.[*][53] In April 2012, General William L. Shelton of the Air Force Space Command declared the ongoing mission a "spectacular success".[*][57]

On 30 May 2012, the Air Force stated that OTV-2 would complete its mission and land at Vandenberg AFB in June 2012.[*][58][*][59] The spacecraft landed autonomously on 16 June 2012, having spent 469 days in space.[*][1][*][60]

8.3.3 OTV-3

Main article: USA-240

OTV-3, the second mission for the first X-37B and the third X-37B mission overall, was originally scheduled to launch on 25 October 2012,[*][61] but was postponed because of an engine issue with the Atlas V launch vehicle.[*][62] The X-37B was successfully launched from Cape Canaveral on 11 December 2012.[*][41][*][63][*][64] The launch was designated USA-240.[*][65][*][66] The OTV-3 mission ended with a landing at Vandenberg AFB on 17 October 2014 at 16:24 UTC, after a total time in orbit of just under 675 days.[*][3][*][67]

8.3.4 OTV-4

The Air Force launched a fourth X-37B mission, designated OTV-4 and codenamed AFSPC-5, aboard an Atlas V rocket from Cape Canaveral Air Force Station on 20 May 2015.[*][44][*][68] The launch was designated USA-261 and is the second flight of the second X-37B vehicle.[*][21] The mission will test a Hall effect thruster in support of the Advanced Extremely High Frequency communications satellite program,[*][68] and conduct a NASA investigation for testing various materials in space.[*][4][*][21][*][43] The mission is expected to last at least 200 days.[*][21]

8.4 Variants

8.4.1 X-37A

The X-37A was the initial NASA version of the spacecraft; the X-37A Approach and Landing Test Vehicle (ALTV) was used in drop glide tests in 2005 and 2006.[*][14][*][69]

8.4.2 X-37B

The X-37B is a modified version of the NASA X-37A, intended for the U.S. Air Force.[*][5] It conducted orbital test missions in 2010, 2011 and 2012.[*][63]

8.4.3 X-37C

In 2011, Boeing announced plans for a scaled-up variant of the X-37B, referring to it as the X-37C. The X-37C spacecraft would be between 165% and 180% of the size of the X-37B, allowing it to transport up to six astronauts inside a pressurized compartment housed in the cargo bay. Its proposed launch vehicle is the Atlas V Evolved Expendable Launch Vehicle.[*][70] In this role, the X-37C could potentially compete with Boeing's CST-100 commercial space capsule.[*][71]

8.5   Specifications

Boeing X 37B OTV (Orbital Test Vehicle) 2

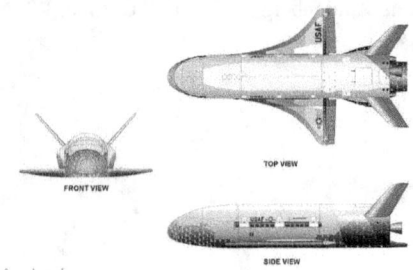

Boeing X 37B OTV (Orbital Test Vehicle) 2

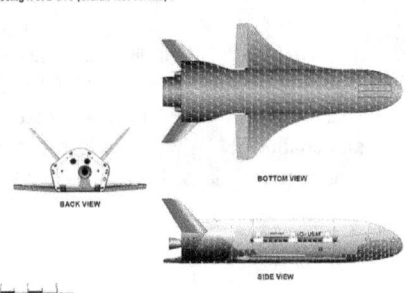

Three-views of the X-37B

8.5.1   X-37B

Data from USAF,[*][5][*][34] Boeing,[*][72] *Air & Space Magazine*,[*][69] and *PhysOrg*.[*][73]

General characteristics

- **Crew:** none

- **Length:** 29 ft 3 in (8.92 m)

- **Wingspan:** 14 ft 11 in (4.55 m)

- **Height:** 9 ft 6 in (2.90 m)

- **Max takeoff weight:** 11,000 lb (4,990 kg)

- **Electrical power:** Gallium arsenide solar cells with lithium-ion batteries[*][5]

- **Payload bay:** 7 × 4 ft (2.1 × 1.2 m)[*][72]

- **Powerplant:** 1 × Aerojet AR2-3 rocket engine, 6,596 lbf (29.341 kN) thrust using hydrogen peroxide and JP-8 fuel[*][38]

Performance

- **Orbital speed:** 28,044 km/h (17,426 mph)[*][74]

- **Orbit:** Low Earth orbit

- **Orbital time:** 270 days (design)[*][75][*][N 2]

8.6 See also

- Boeing X-20 Dyna-Soar, the U.S. Air Force's previous spaceplane, which was canceled in the 1960s

- DARPA Falcon Project, a hypersonic missile-delivery and satellite-launch project

- Intermediate eXperimental Vehicle (IXV), an ESA design for an experimental reentry vehicle

- Programme for Reusable In-orbit Demonstrator in Europe (PRIDE), ESA's unmanned spaceplane follow-up to IXV

- Orbital Sciences X-34, a proposed unmanned suborbital reusable-rocket technology testbed

- Saturn-Shuttle, a proposed Space Shuttle launch configuration

Related development

- Boeing X-40, a subsonic test glider, direct predecessor to the X-37B

Aircraft of comparable role, configuration and era

- Avatar, an Indian design intended for horizontal takeoff

- Dream Chaser, a crewed spaceplane being developed by Sierra Nevada Corporation

- Shenlong spaceplane, a Chinese design dating from around 2007

- Skylon, a British reusable unmanned spaceplane in development

8.7 Notes

[1] Source of flights: mission markings posted on side of White Knight aircraft.

[2] This figure is based on pre-launch design estimates; it does not reflect the spacecraft's actual performance capacity. During its 2012–2014 test mission, the OTV-3 X-37B spent over 670 days in space.

8.8 References

[1] "X-37B lands this morning at Vandenberg AFB". *Santa Maria Times*. 16 June 2012. Archived from the original on 26 June 2014.

[2] Thornhill, Ted (8 March 2012). "Revealed: How America's secret space plane has been in orbit for over a year – and no one knows what it's doing". *Daily Mail*. Retrieved 29 April 2012.

[3] "Secret space plane lands at US air force base after unknown two-year mission". *The Guardian*. Associated Press. 17 October 2014. Retrieved 17 October 2014.

[4] Wall, Mike (20 May 2015). "US Air Force Launches X-37B Space Plane on 4th Mystery Mission". *Space.com*. Retrieved 21 May 2015.

[5] "Fact Sheet: X-37 Orbital Test Vehicle". U.S. Air Force. 21 May 2010. Archived from the original on 26 June 2014.

[6] "X-37 Technology Demonstrator: Blazing the trail for the next generation of space transportation systems" (PDF). NASA Facts. NASA. September 2003. FS-2003-09-121-MSFC. Retrieved 23 April 2010.

[7] "Air Force Bloggers Roundtable: Air Force set to launch first X-37B Orbital Test Vehicle". United States Department of Defense. 20 April 2010. Retrieved 23 April 2010.

[8] Clark, Stephen (2 April 2010). "Air Force spaceplane is an odd bird with a twisted past". *Spaceflight Now*. Retrieved 3 April 2010.

[9]   "NASA, Boeing Enter Cooperative Agreement to Develop and Fly X-37 Technology Demonstrator" (Press release). NASA. 14 July 1999. Press Release 99-139. Retrieved 20 October 2014.

[10] Yenne 2005, p. 277.

[11] Berger, Brian (15 September 2004). "NASA Transfers X-37 Project to DARPA". *Space.com.*

[12] Berger, Brian (16 September 2004). "DARPA takes on space plane project". *NBC News.* Retrieved 19 October 2014.

[13] David, Leonard (23 June 2005). "White Knight carries X-37 aloft". *CNN.* Retrieved 15 October 2012.

[14] Parsch, Andreas (November 2009). "Boeing X-37 / X-40". *Designation-Systems.net.* Retrieved 2 August 2012.

[15] "Mojave web log entries". *MojaveWeblog.com.* Archived from the original on 4 June 2006. Retrieved 15 October 2012.

[16] David, Leonard (7 April 2006). "X-37 Flies At Mojave But Encounters Landing Problems". *Space.com.*

[17] "X-37 Test Flight B-Roll (No Audio)". *YouTube.com.* U.S. Air Force. 22 April 2010.

[18] David, Leonard (17 November 2006). "U.S. Air Force Pushes For Orbital Test Vehicle". *Space.com.* Retrieved 17 November 2006.

[19] Clark, Stephen (22 April 2010). "Atlas rocket delivers Air Force spaceplane to orbit". *Spaceflight Now.*

[20] Clark, Stephen (25 February 2010). "Air Force X-37B spaceplane arrives in Florida for launch". *Spaceflight Now.* Retrieved 3 March 2010.

[21] Krebs, Gunter. "X-37B". *Gunter's Space Page.* Retrieved 5 August 2008.

[22] Covault, Craig (3 August 2008). "USAF To Launch First Spaceplane Demonstrator". *Aviation Week.* Archived from the original on 12 August 2011.

[23] Norris, Guy (7 December 2010). "Second X-37B Prepared For Launch". *Aviation Week.*

[24] "Second secret space shuttle blasts into orbit ... but what does it actually do up there?". *Daily Mail.* 7 March 2011. Retrieved 15 January 2012.

[25] Dean, James (8 October 2014). "NASA: Secret X-37B space plane will use shuttle hangars". *Florida Today.* Retrieved 9 October 2014.

[26] Smith-Strickland, Kiona (27 May 2015). "Now We Know at Least Two Payloads on the X-37B". *Air & Space/Smithsonian.* Retrieved 27 May 2015.

[27] Burghardt, Tom (11 May 2010). "The Militarization of Outer Space: The Pentagon's Space Warriors". *Space Daily.* Retrieved 15 October 2012.

[28] Parnell, Brid-Aine (6 January 2012). "US 'space warplane' may be spying on Chinese spacelab". *The Register.* Retrieved 13 January 2012.

[29] Corocoto, Genalyn (9 January 2012). "Expert: U. S. Secret Space Plane Not Likely 'Spying' on China Module". *International Business Times.* Retrieved 13 January 2012.

[30] Yuhas, Alan (27 October 2014). "X-37B secret space plane's mission remains mystery outside US military". *The Guardian.* Retrieved 28 October 2014.

[31] Miller 2001, p. 377.

[32] David, Leonard (27 May 2005). "Mothership adopts a new space plane". *NBC News.* Retrieved 2 February 2013.

[33] "X-37 Demonstrator to Test Future Launch Technologies in Orbit and Reentry Environments". NASA. May 2003. FS-2003-05-65-MSFC. Retrieved 2 February 2013.

[34] Jameson, Austin D. (April 2001). "X-37 Space Vehicle: Starting a New Age in Space Control?" (PDF). Defense Technical Information Center. ADA407255.

[35] "Mr. Gary Payton, Under Secretary of the Air Force for Space Programs, Media Teleconference (Pentagon), X-37B Launch" (PDF). *Defense.gov.* 20 April 2010. Retrieved 3 December 2010.

[36] David, Leonard (5 April 2012). "Boeing's New Crew-Carrying Spaceship Taking Shape". *Space.com.* Retrieved 17 April 2012.

[37] "Pilotless US space plane lands after 469 days in orbit". *Space Daily.* AFP. 17 June 2012. Retrieved 18 June 2012.

[38] "Boeing X-37 Technology Demonstrator, USA". *Airforce-Technology.com*. Retrieved 12 December 2012.

[39] Herbert, Bartt (24 July 2009). "Peroxide (H2O2) Test Programs, AR2-3 Flight Certification". NASA Engineering and Test Directorate.

[40] Chertok, Boris E. (2005). Siddiqi, Asif A., ed. *Rockets and People* Ракеты и люди (PDF). NASA History Series **1**. NASA. p. 179. SP-2005-4110.

[41] Halvorson, Todd (9 December 2012). "AF's X-37B mini-shuttle shrouded in secrecy". *Florida Today* via *Military Times*. Retrieved 10 December 2012.

[42] Glenday, Craig, ed. (2012). *Guinness World Records 2013*. Guinness. p. 29. ISBN 9781904994879.

[43] David, Leonard (8 May 2015). "Inside the US Air Force's Next X-37B Space Plane Mystery Mission". *Space.com*. Retrieved 11 May 2015.

[44] Parsons, Dan (17 October 2014). "US Air Force's shadowy space plane lands after 674 days in orbit". *Flight Global*. Retrieved 18 October 2014.

[45] David, Leonard (22 May 2010). "Secret X-37B Space Plane Spotted by Amateur Skywatchers". *Space.com*.

[46] "Amateur astronomers unravel X37-B orbit, say likely use for deploying spy satellites". *News.com.au*. 24 May 2010.

[47] Broad, William J. (21 May 2010). "Surveillance is Suspected as Spacecraft's Main Role". *The New York Times*. Retrieved 22 May 2010.

[48] "Preparations underway for first landing of X-37B". U.S. Air Force. 30th Space Wing Public Affairs. 30 November 2010. Retrieved 21 May 2012.

[49] "USAF X-37B Landing Slated for Dec. 3–6". *Aviation Week*. 30 November 2010. Retrieved 1 December 2010.

[50] Clark, Stephen (3 December 2010). "Home again: U.S. military space plane returns to Earth". *Spaceflight Now*. Retrieved 3 December 2010.

[51] Warwicj, Graham (3 December 2010). "USAF's X-37B Spaceplane Returns to Earth". *Aviation Week*. Retrieved 2 June 2012.

[52] Rincon, Paul (3 December 2010). "X-37B US military spaceplane returns to Earth". *BBC News*. Retrieved 3 December 2010.

[53] Clark, Stephen (29 November 2011). "Military space shuttle receives mission extension". *Spaceflight Now*. Retrieved 1 December 2011.

[54] McCants, Mike. "OTV 2 (USA 226) Satellite details 2011-010A NORAD 37375". *N2YO.com*. Retrieved 8 April 2012.

[55] Hennigan, W. J. (5 March 2011). "Air Force hopes to launch X-37B space plane after weather delay". *Los Angeles Times*. Retrieved 2 August 2011.

[56] Wall, Mike (5 March 2011). "Secretive X-37B Space Plane Launches on New Mystery Mission". *Space.com*. Retrieved 8 April 2012.

[57] David, Leonard (9 May 2012). "Air Force's secret X-37B a 'spectacular success'". *NBC News*. Retrieved 10 May 2012.

[58] "Preparations underway for X-37B landing". U.S. Air Force. 30th Space Wing Public Affairs. 30 May 2012. Retrieved 1 June 2012.

[59] David, Leonard (30 May 2012). "Air Force's Secretive X-37B Space Plane Will Land Soon". *Space.com*. Retrieved 2 June 2012.

[60] Wall, Mike (16 June 2012). "Air Force's secret X-37B space plane lands in Calif. after mystery mission". *Space.com*.

[61] Ray, Justin (18 September 2012). "Another Atlas 5 readied to launch mini space shuttle". *Spaceflight Now*. Retrieved 4 October 2012.

[62] McCullough, Amy (25 October 2012). "X-37B Launch Delayed". *Air Force Magazine*. Archived from the original on 20 February 2014. Retrieved 11 December 2012.

[63] Atkinson, Nancy (11 December 2012). "Air Force's Secret X-37B Space Plane Launches on Third Mission". *Universe Today*. Retrieved 11 December 2012.

[64] David, Leonard (21 November 2012). "New Delay for Mysterious Military Space Plane Launch". *Space.com*. Retrieved 26 November 2012.

[65] "NSSDC ID: 2012-071A". *National Space Science Data Center*. NASA. Retrieved 24 December 2012.

[66] Badger, Eric (11 December 2012). "Air Force launches 3rd X-37B Orbital Test Vehicle". U.S. Air Force. Archived from the original on 26 June 2014.

[67] Wall, Mike (17 October 2014). "X-37B Military Space Plane Lands After Record-Shattering Secret Mission". *Space.com*. Retrieved 17 October 2014.

[68] Ray, Justin (27 April 2015). "X-37B launch date firms up as new details emerge about experiment". *Spaceflight Now*. Retrieved 27 April 2015.

[69] Klesius, Michael (January 2010). "Space Shuttle Jr.". *Air & Space Magazine*.

[70] David, Leonard (7 October 2011). "Secretive X-37B US Space Plane Could Evolve to Carry Astronauts". *Space.com*. Retrieved 10 October 2011.

[71] "Boeing vs. Boeing". Citizens in Space. 25 March 2012. Retrieved 21 December 2012.

[72] "X-37B Orbital Test Vehicle". *Boeing.com*. Archived from the original on 21 March 2015. Retrieved 6 December 2010.

[73] Antczak, John (3 April 2010). "Air Force to launch robotic winged space plane". *PhysOrg*. Associated Press. Retrieved 16 February 2013.

[74] Molczan, Ted (22 May 2012). "Re: X-37B OTV 2-1 search elements". *Satobs.org*. Retrieved 2 June 2012.

[75] Evans, Michael (24 April 2010). "Launch of secret US space ship masks even more secret launch of new weapon". *The Times*. Archived from the original on 29 May 2010. Retrieved 25 April 2010.

8.8.1 Bibliography

- Bentley, Matthew A. (2008). *Spaceplanes: From Airport to Spaceport*. New York: Springer. ISBN 978-0-387-76509-9.

- Gump, David P. (1989). *Space Enterprise: Beyond NASA*. Westport, CT: Praeger. ISBN 978-0-275-93314-2.

- Miller, Jay (2001). *The X-Planes: X-1 to X-45*. Hinckley, UK: Midland. ISBN 1-85780-109-1.

- Yenne, Bill (2005). *The Story of the Boeing Company*. Minneapolis, MN: Zenith. ISBN 978-0-7603-2333-5.

8.9 External links

- X-37 Orbital Test Vehicle fact sheet from the U.S. Air Force

- X-37B Orbital Test Vehicle page at Boeing.com

- X-37 fact sheet at NASA.gov

- X-37, X-37A and X-37B pages at GlobalSecurity.org

Chapter 9

Boeing X-40

The **Boeing X-40A Space Maneuver Vehicle** was a test platform for the X-37 Future-X Reusable Launch Vehicle.

9.1 History

The unpiloted X-40 was built to 85 percent scale to test aerodynamics and navigation of the X-37 Future-X Reusable Launch Vehicle project.

After the first drop test in August 1998 the vehicle was transferred to NASA, which modified it. Between April 4 and May 19, 2001 the vehicle successfully conducted seven free flights.*[3] In 2001 it successfully demonstrated the glide capabilities of the X-37's fat-bodied, short-winged design and validated the proposed guidance system.

9.2 Testing

The first X-40 drop test occurred at Holloman AFB, New Mexico on August 11, 1998 at 06:59. This was a joint Air Force/Boeing project known as Space Maneuver Vehicle. It was released from an altitude of approximately 9,200 feet (2,800 m) and 2.5 miles (4.0 km) away from the end of Runway 04 by a UH-60 Black Hawk helicopter*[1]*[2]*[3] (later tests used the CH-47 Chinook helicopter).*[2]*[3] The vehicle dove to the runway in an approach similar to the space shuttle's, flared, and landed left of the runway centerline. Its drag chutes successfully deployed, and the vehicle tracked to within seven feet of the centerline and stopped at a distance of slightly more than 7,000 feet (2,100 m).

9.3 Specifications (X-40)

General characteristics

- **Crew:** None
- **Length:** 6.5 m (21 ft)
- **Wingspan:** 3.5 m (11 ft)
- **Height:** 2.3 m (7.5 ft)
- **Empty weight:** 2,500 lb (1,100 kg)
- **Useful load:** 1,200 lb (540 kg)

Performance

- **Maximum speed:** 480 km/h (300 mph)

Avionics
Honeywell 12-channel Space Integrated GPS/INS (SIGI) system

9.4   See also

- Boeing X-37

Related lists

- List of experimental aircraft

9.5   References

[1] X-40 Space Maneuver Vehicle Integrated Tech Testbed at FAS.org

[2] X-40 Space Maneuver Vehicle (SMV) at GlobalSecurity.org

[3] Boeing X-37 / X-40 page at Designation-Systems.Net

9.6   External links

- NASA Dryden X-40A Image Gallery

- X-40A Test Flight, Boeing press release

- X-40 Space Maneuver Vehicle Integrated Tech Testbed at FAS.org

- X-40 Space Maneuver Vehicle (SMV) at GlobalSecurity.org

- Boeing X-37 / X-40 page at Designation-Systems.Net

Chapter 10

Boost-glide

Silbervogel *introduced the boost-glide concept as part of an* "*antipodal bomber*" *able to attack New York.*

Boost-glide trajectories are a class of spacecraft guidance and reentry trajectories that extend the range of suborbital spaceplanes by employing aerodynamic lift in the high upper atmosphere. In most examples, boost-glide roughly doubles the range over the purely ballistic trajectory. In others, a series of *skips* allows range to be further extended, and leads to the alternate term **skip-glide**.

The concept was introduced as a way to extend the range of ballistic missiles, but has not been used operationally in this form. The concepts have been used to produce maneuverable reentry vehicles, or MARV, to increase the accuracy of some missiles. More recently the traditional form with an extended gliding phase has been considered as a way to reach targets while flying below their radar coverage.

10.1   Early concepts

The conceptual basis for the boost-glide concept was first noticed by German artillery officers, who found that their *Peenemünder Pfeilgeschosse* arrow shells travelled much further when fired from higher altitudes. This was not entirely unexpected due to geometry and thinner air, but when these factors were accounted for they still could not explain the much greater ranges being seen. Investigations at Peenemünde led them to discover that the longer trajectories resulted in the shell having an angle of attack that produced aerodynamic lift. At the time this was considered highly undesirable because it made the trajectory very difficult to calculate, but its possible application for extending range was not lost on the observers.[*][1]

In June 1939, Kurt Patt of Klaus Riedel's design office at Peenemünde proposed wings for converting rocket speed and altitude into aerodynamic lift and range.[*][2] He calculated that this would roughly double range of the A-4 rockets from 275 kilometres (171 mi) to about 550 kilometres (340 mi). Early development was considered under the A-9 name, although little work other than wind tunnel studies at the Zeppelin-Staaken company would be carried out for the next few years. Low-level research continued until 1942 when it was cancelled.[*][3]

The earliest known use of the boost-glide concept for truly long-range use dates to the 1941 *Silbervogel* proposal by Eugen Sänger for a rocket powered bomber able to attack New York City from bases in Germany and then fly on for landing somewhere in the Pacific Ocean held by the Empire of Japan. The idea would be to use the vehicle's wings to generate lift and pull up into a new ballistic trajectory, exiting the atmosphere again and giving the vehicle time to cool off between the skips.[*][4] It was later demonstrated that the heating load during the skips was much higher than initially calculated, and would have melted the spacecraft.[*][5]

In 1943 the A-9 work was dusted off again, this time under the name A-4b. It has been suggested this was either because it was now based on an otherwise unmodified A-4,[*][3] or because the A-4 program had "national priority" by this time, and placing the development under the A-4 name guaranteed funding.[*][6] A-4b used swept wings in order to extend the range of the V2 enough to allow attacks on UK cities in The Midlands or to reach London from areas deeper within Germany.[*][1] The A-9 was originally similar, but later featured long ogive delta shaped wings instead of the more conventional swept ones. This design was adapted as a manned upper stage for the A-9/A-10 intercontinental missile, which would glide from a point over the Atlantic with just enough range to bomb New York before the pilot bailed out.[*][6][*][lower-alpha 1]

10.2   Post-war development

In the immediate post-war era, Soviet engineers heard of the *Silbervogel* when rocket engineer Alexey Isayev found a copy of an updated August 1944 report on the concept. The paper was translated to Russian, where it eventually came to the attention of Joseph Stalin who was intensely interested in the concept of an antipodal bomber. In 1946, he sent his son Vasily Stalin and scientist Grigori Tokaty, who had also worked on winged rockets before the war, to visit Sänger and Irene Bredt in Paris and attempt to convince them to join a new effort in the Soviet Union.[*][8]

Sänger and Bredt turned down the invitation, so in November 1946 the Soviets formed the NII-1 design bureau under Mstislav Keldysh to develop their own version.[*][9] Their early work convinced them to convert from a rocket powered hypersonic skip-glide concept to a ramjet powered supersonic cruise missile, not unlike the Navaho being developed in the United States during the same period. Development continued for a time as the Keldysh bomber, but improvements in conventional ballistic missiles ultimately rendered the project unneeded.[*][8][*][lower-alpha 2]

In the United States the concept was advocated by the many German scientists who moved there, primarily Walter Dornberger and Krafft Ehricke at Bell Aircraft. In 1952 Bell proposed a bomber concept that was essentially a vertical launch version of *Silbervogel* known as Bomi. This led to a number of follow-on concepts during the 1950s, including Robo, Hywards, Brass Bell, and ultimately the Boeing X-20 Dyna-Soar.[*][10] Earlier designs were generally bombers, while later models were aimed at reconnaissance or other roles. The two also collaborated on a 1955 *Popular Science* article pitching the idea for airliner use.[*][11][*][12]

The introduction of successful intercontinental ballistic missiles (ICBMs) in the offensive role ended any interest in the bomber concepts, as did the reconnaissance satellite for the spyplane roles. The X-20 space fighter saw continued interest through the 1960s, but was ultimately the victim of budget cuts.

To date, the X-20 Dyna Soar is the project that has come closest to actually building a manned boost-glide vehicle. This illustration shows the Dyna Soar during reentry.

10.3   Production use

The boost-glide concept relies on the lift generated during the hypersonic flight of reentry. Through the 1960s this concept saw interest not as a way to extend range, which was not longer a concern with modern missiles, but as possible maneuverable reentry vehicles for ICBMs. The first known example were the Alpha Draco tests of 1959, followed by the Boost Glide Reentry Vehicle (BGRV) test series, ASSET*[13] and PRIME.*[14]

This research was eventually put to use in the Pershing II's MARV reentry vehicle. In this case there is no extended gliding phase; the warhead uses lift only for short periods to adjust its trajectory. This is used late in the reentry process, combining data from a Singer Kearfott inertial navigation system with a Goodyear Aerospace active radar.*[15] Similar concepts have been developed for most nuclear armed nation's theatre ballistic missiles.

In contrast to these maneuvering warhead concepts, there has been growing interest in the traditional boost-glide concept not to extend range *per-se*, but to allow it to reach a given range while flying at a much lower altitude. The goal in this case is to keep the reentry vehicle below radar coverage until it enters the terminal phase. Such a system is assumed to be used on the Chinese DF-21D anti-ship ballistic missile, which is also believed to maneuver during the terminal phase to make interception more difficult.*[16]

In the early 21st century, boost-glide became the topic of some interest as a possible solution to the Prompt Global Strike (PGS), which seeks a weapon that can hit a target anywhere on the Earth within one hour of launch from the United States. PGS does not define the mode of operation, and current studies include Advanced Hypersonic Weapon boost-glide warhead, Falcon HTV-2 hypersonic aircraft, and submarine launched missiles.*[17] According to reports from The Pentagon, the Chinese have started development of a similar weapon known as WU-14 that now is under test flights, and Russia having Kholod and Igla hypersonic test projects earlier but now carring the test flights of similar Yu-71 hypersonic warhead.*[18]

10.4   Notes

[1] Yengst's chronology of the A-series weapons differs considerably from most accounts. For instance, he suggests the A-9 and A-10 were two completely separate developments, as opposed to the upper and lower stages of a single ICBM design. He also states that the A-4b was the SLBM development, as opposed to the winged A-4.[*][7]

[2] Navaho met the same fate in 1958, when it was cancelled in favor of the Atlas missile.

10.5   References

10.5.1   Citations

[1] Yengst 2010, p. 29.

[2] Neufeld 1995, p. 92.

[3] Neufeld 1995, p. 93.

[4] Duffy, James (2004). *Target: America —Hitler's Plan to Attack the United States.* Praeger. p. 124. ISBN 0-275-96684-4.

[5] Reuter, Claus (2000). *The V2 and the German, Russian and American Rocket Program.* German - Canadian Museum of Applied History. p. 99. ISBN 9781894643054.

[6] Yengst 2010, pp. 30-31.

[7] Yengst 2010, p. 31.

[8] Westman, Juhani (2006). "Global Bounce". *PP.HTV.fi.* Retrieved 2008-01-17.

[9] Wade, Mark. "Keldysh". *Encyclopedia Astronautica.*

[10] Godwin, Robert (2003). *Dyna-Soar: Hypersonic Strategic Weapons System.* Apogee Books. p. 42. ISBN 1-896522-95-5.

[11] "Rocket Liner Would Skirt Space to Speed Air Travel". *Popular Science*: 160–161. February 1955.

[12] Dornberger, Walter (1956). *The Rocket-Propelled Commercial Airliner* (Technical report). University of Minnesota Institute of Technology.

[13] Wade, Mark. "ASSET". *Encyclopedia Astronautica.*

[14] Jenkins, Dennis; Landis, Tony; Miller, Jay (June 2003). *AMERICAN X-VEHICLES An Inventory—X-1 to X-50* (PDF). NASA. p. 30.

[15] Wade, Mark. "Pershing". *Encyclopedia Astronautica.*

[16] "Chinese Develop "Kill Weapon" to Destroy US Aircraft Carriers". *US Naval Institute.* 21 March 2009.

[17] Woolf, Amy (6 February 2015). *Conventional Prompt Global Strike and Long-Range Ballistic Missiles: Background and Issues* (PDF) (Technical report). Congressional Research Service.

[18] Gertz, Bill (13 January 2014). "Hypersonic arms race: China tests high-speed missile to beat U.S. defenses". *The Washington Free Beacon.*

10.5.2   Bibliography

• Neufeld, Michael (1995). *The Rocket and the Reich: Peenemünde and the Coming of the Ballistic Missile Era.* Simon and Schuster. ISBN 9780029228951.

• Yengst, William (April 2010). *Lightning Bolts: First Maneuvering Reentry Vehicles.* Tate Publishing. ISBN 9781615665471.

Chapter 11

Buran programme

This article is about the Soviet/Russian reusable space programme. For the orbiter launched in 1988 see *Buran* (spacecraft). For other uses, see Buran (disambiguation).

The **Buran** (Russian: Бура́н, IPA: [bʊˈran], *Snowstorm* or *Blizzard*) programme, also known as the **VKK Space Orbiter** (Russian: Воздушно Космический Корабль, *Air Space Ship*) programme,*[3] was a Soviet and later Russian reusable spacecraft project that began in 1974 at the Central Aerohydrodynamic Institute and was formally suspended in 1993.*[4] In addition to being the designation for the whole Soviet/Russian reusable spacecraft project, *Buran* was also the name given to Orbiter K1, which completed one unmanned spaceflight in 1988 and remains the only Soviet reusable spacecraft to be launched into space. The Buran-class space shuttle orbiters used the expendable *Energia rocket* as a launch vehicle. They are generally treated as a Soviet equivalent of the United States' Space Shuttle but in the Buran project, only the airplane-shaped orbiter itself was theoretically reusable, and while Orbiter K1 was recovered successfully after its first orbital flight in 1988, it was never reused.

The Buran programme was started by the Soviet Union as a response to the United States Space Shuttle programme.*[5] The project was the largest and the most expensive in the history of Soviet space exploration.*[4] Development work included sending BOR-5 test vehicles on multiple sub-orbital test flights, and atmospheric flights of the OK-GLI aerodynamic prototype. Buran completed one unmanned orbital spaceflight in 1988 before its cancellation in 1993.*[4] Orbiter K1, which flew the test flight in 1988 was crushed in a hangar collapse on 12 May 2002 in Kazakhstan. The OK-GLI resides in Technikmuseum Speyer. Although Soviet/Russian Buran spacecraft was similar in appearance to NASA's Space Shuttle, and could similarly operate as a re-entry spaceplane, its internal and functional design was distinct. For example, the main engines during launch were on the Energia rocket and were not taken into orbit by the spacecraft. Smaller rocket engines on the craft's body provided propulsion in orbit and de-orbital burns. Thus the Buran programme matched an expendable rocket to a reusable spaceplane.

11.1 Introduction

The Buran orbital vehicle programme was developed in response to the U.S. Space Shuttle programme, which in the 1980s raised considerable concerns among the Soviet military and especially Defense Minister Dmitriy Ustinov. An authoritative chronicler of the Soviet and later Russian space programmes, the academic Boris Chertok, recounts how the programme came into being.*[6] According to Chertok, after the U.S. developed its Space Shuttle programme, the Soviet military became suspicious that it could be used for military purposes, due to its enormous payload, several times that of previous U.S. launch vehicles. The Soviet government asked the TsNIIMash (ЦНИИМАШ, Central Institute of Machine-building, a major player in defense analysis) for an expert opinion. Institute director, Yuri Mozzhorin, recalls that for a long time the institute could not envisage a civilian payload large enough to require a vehicle of that capacity.

Officially, the Buran orbital vehicle was designed for the delivery to orbit and return to Earth of spacecraft, cosmonauts, and supplies. Both Chertok and Gleb Lozino-Lozinskiy (Chief Designer of RKK Energia) suggest that from the beginning, the programme was military in nature; however, the exact military capabilities, or intended capabilities, of the *Buran* programme remain classified. Commenting on the discontinuation of the programme in his interview to *New Scientist*, Russian cosmonaut Oleg Kotov confirms their accounts:

> We had no civilian tasks for *Buran* and the military ones were no longer needed. It was originally designed as a military system for weapon delivery, maybe even nuclear weapons. The American shuttle also has military uses.*[7]

Like its American counterpart, the Buran orbital vehicle, when in transit from its landing sites back to the launch complex, was transported on the back of a large jet aeroplane —the Antonov An-225 Mriya transport aircraft, which was designed in part for this task and remains the largest aircraft in the world to fly multiple times.*[8]. Before the *Mriya* was ready (after the *Buran* had flown), the Myasishchev VM-T *Atlant*, a variant on the Soviet Myasishchev M-4 *Molot* (Hammer) bomber (NATO code: Bison), fulfilled the same role.

11.2   History of the Buran programme

11.2.1   Background

The Soviet reusable space-craft programme has its roots in the very beginning of the space age, the late 1950s. The idea of Soviet reusable space flight is very old, though it was neither continuous, nor consistently organized. Before Buran, no project of the programme reached production.

The idea saw its first iteration in the Burya high-altitude jet aircraft, which reached the prototype stage. Several test flights are known, before it was cancelled by order of the Central Committee. The Burya had the goal of delivering a nuclear payload, presumably to the United States, and then returning to base. The cancellation was based on a final decision to develop ICBMs. The next iteration of the idea was Zvezda from the early 1960s, which also reached a prototype stage. Decades later, another project with the same name was used as a service module for the International Space Station. After Zvezda, there was a hiatus in reusable projects until Buran.

11.2.2   Programme development

The development of the Buran began in the early 1970s as a response to the U.S. Space Shuttle program. Soviet officials were concerned about a perceived military threat posed by the U.S. Space Shuttle. In their opinion, the Shuttle's 30-ton payload-to-orbit capacity and, more significantly, its 15-ton payload return capacity, were a clear indication that one of its main objectives would be to place massive experimental laser weapons into orbit that could destroy enemy missiles from a distance of several thousands of kilometers. Their reasoning was that such weapons could only be effectively tested in actual space conditions and that to cut their development time and save costs it would be necessary to regularly bring them back to Earth for modifications and fine-tuning.*[9] Soviet officials were also concerned that the U.S. Space Shuttle could make a sudden dive into the atmosphere to drop bombs on Moscow.*[10]

Soviet engineers were initially reluctant to design a spacecraft that looked superficially identical to the Shuttle, but subsequent wind tunnel testing showed that NASA's design was already ideal.*[11] Even though the Molniya Scientific Production Association proposed its Spiral programme design (halted 13 years earlier), it was rejected as being altogether dissimilar from the American shuttle design. While NPO Molniya conducted development under the lead of Gleb Lozino-Lozinskiy, the Soviet Union's Military-Industrial Commission, or VPK, was tasked with collecting all data it could on the U.S. Space Shuttle. Under the auspices of the KGB, the VPK was able to amass documentation on the American shuttle's airframe designs, design analysis software, materials, flight computer systems and propulsion systems. The KGB targeted many university research project documents and databases, including Caltech, MIT, Princeton, Stanford and others. The thoroughness of the acquisition of data was made much easier as the U.S. shuttle development was unclassified.*[12]

The construction of the shuttles began in 1980, and by 1984 the first full-scale Buran was rolled out. The first suborbital test flight of a scale-model (BOR-5) took place as early as July 1983. As the project progressed, five additional scale-model flights were performed. A test vehicle was constructed with four jet engines mounted at the rear; this vehicle is usually referred to as OK-GLI, or as the "Buran aerodynamic analogue". The jets were used to take off from a normal landing strip, and once it reached a designated point, the engines were cut and OK-GLI glided back to land. This provided invaluable information about the handling characteristics of the Buran design, and significantly differed from the carrier plane/air drop method used by the United States and the *Enterprise* test craft. Twenty-four test flights of OK-GLI were performed after which the shuttle was "worn out". The developers considered using a couple of Mil Mi-26 helicopters to "bundle" lift the Buran, but test flights with a mock-up showed how risky and impractical that was.[*][13] The VM-T ferried components[*][14] and the Antonov An-225 Mriya (the heaviest airplane ever) was designed and used to ferry the shuttle.[*][15][*][16]

The flight and ground-testing software also required research. In 1983 the Buran developers estimated that the software development would require several thousand programmers if done with their existing methodology (in assembly language), and they appealed to Keldysh Institute of Applied Mathematics for assistance. It was decided to develop a new high-level "problem-oriented" programming language. Researchers at Keldysh developed two languages: PROL2 (used for real-time programming of onboard systems) and DIPOL (used for the ground-based test systems), as well as the development and debugging environment SAPO PROLOGUE.[*][17] There was also an operating system known as Prolog Manager.[*][18] Work on these languages continued beyond the end of the Buran project, with PROL2 being extended into SIPROL,[*][19] and eventually all three languages developed into DRAKON which is still in use in the Russian space industry. A declassified May 1990 CIA report citing open-source intelligence material states that the software for the Buran spacecraft was written in "the French-developed programming language known as Prolog",[*][20] possibly due to confusion with the name PROLOGUE.

11.2.3   Flight crew preparation

Main article: List of human spaceflight programs

Until the end of the Soviet Union in 1991, seven cosmonauts were allocated to the Buran programme and trained on the OK-GLI ("Buran aerodynamic analogue") test vehicle. All had experience as test pilots. They were: Ivan Ivanovich Bachurin, Alexei Sergeyevich Borodai, Anatoli Semyonovich Levchenko, Aleksandr Vladimirovich Shchukin, Rimantas Antanas Stankevičius, Igor Petrovich Volk and Viktor Vasiliyevich Zabolotsky.

A rule, set in place for cosmonauts because of the failed Soyuz 25 of 1977, insisted that all Soviet space missions contain at least one crew member who has been to space before. In 1982, it was decided that all Buran commanders and their back-ups would occupy the third seat on a Soyuz mission, prior to their Buran spaceflight.[*][9] Several people had been selected to potentially be in the first Buran crew. By 1985, it was decided that at least one of the two crew members would be a test pilot trained at the Gromov Flight Research Institute (known as "LII"), and potential crew lists were drawn up.[*][9] Only two potential Buran crew members reached space: Igor Volk, who flew in Soyuz T-12 to the space station Salyut 7, and Anatoli Levchenko who visited Mir, launching with Soyuz TM-4 and landing with Soyuz TM-3.[*][9] Both of these spaceflights lasted about a week.

Levchenko died of a brain tumour the year after his orbital flight, Bachurin left the cosmonaut corps because of medical reasons, Shchukin was assigned to the back-up crew of Soyuz TM-4 and later died in a plane crash, Stankevičius was also killed in a plane crash, while Borodai and Zabolotsky remained unassigned to a Soyuz flight until the Buran programme ended.

Spaceflight of I.P. Volk

Main article: Soyuz T-12

Igor Volk was planned to be the commander of the first manned Buran flight. There were two purposes of the Soyuz T-12 mission, one of which was to give Volk spaceflight experience. The other purpose, seen as the more important factor, was to beat the United States and have the first spacewalk by a woman.[*][9] At the time of the Soyuz T-12 mission the Buran programme was still a state secret. The appearance of Volk as a crew member caused some, including the British Interplanetary Society magazine *Spaceflight*, to ask why a test pilot was occupying a Soyuz seat usually reserved for researchers or foreign cosmonauts.[*][21]

Spaceflight of A.S. Levchenko

Anatoli Levchenko was planned to be the back-up commander of the first manned Buran flight, and in March 1987 he began extensive training for his Soyuz spaceflight.[*][9] In December 1987, he occupied the third seat aboard Soyuz TM-4 to Mir, and returned to Earth about a week later on Soyuz TM-3. His mission is sometimes called *Mir LII-1*, after the Gromov Flight Research Institute shorthand.[*][22] When Levchenko died the following year, it left the back-up crew of the first Buran mission again without spaceflight experience. A Soyuz spaceflight for another potential back-up commander was sought by the Gromov Flight Research Institute, but never occurred.[*][9]

11.2.4   Ground facilities

Maintenance, launches and landings of the Buran-class orbiters were to take place at the Baikonur Cosmodrome in the Kazakh S.S.R. Several facilities at Baikonur were adapted or newly built for these purposes:

- Site 110 —Used for the launch of the Buran-class orbiters. Like the assembly and processing hall at Site 112, the launch complex was originally constructed for the Soviet lunar landing programme and later converted for the Energia-Buran programme.

- Site 112 —Used for orbiter maintenance and to mate the orbiters to their Energia launchers (thus fulfilling a role similar to the VAB at KSC). The main hangar at the site, called *MIK RN* or *MIK 112*, was originally built for the assembly of the N-1 moon rocket. After cancellation of the N-1 programme in 1974, the facilities at Site 112 were converted for the Energia-Buran programme. It was here that Orbiter K1 was stored after the end of the Buran programme and was destroyed when the hangar roof collapsed in 2002.[23][24]

- Site 251 —Used as Buran orbiter landing facility, also known as *Yubileyniy Airfield* (and fulfilling a role similar to the SLF at KSC). It features one runway, called 06/24, which is 4,500 metres (4,900 yd) long and 84 metres (92 yd) wide, paved with "Grade 600" high quality reinforced concrete. At the edge of the runway was a special mating-demating device, designed to lift an orbiter off its Antonov An-225 Mriya carrier aircraft and load it on a transporter, which would carry the orbiter to the processing building at Site 254. A purpose-built orbiter landing control facility, housed in a large multi-storey office building, was located near the runway. *Yubileyniy Airfield* was also used to receive heavy transport planes carrying elements of the Energia-Buran system. After the end of the Buran programme, Site 251 was abandoned but later reopened as a commercial cargo airport. Besides serving Baikonur, Kazakh authorities also use it for passenger and charter flights from Russia.[25][26]

- Site 254 —Built to service the Buran-class orbiters between flights (thus fulfilling a role similar to the OPF at KSC). Constructed in the 1980s as a special four-bay building, it also featured a large processing area flanked by several floors of test rooms. After cancellation of the Buran programme it was adapted for pre-launch operations of the Soyuz and Progress spacecraft.[27]

11.2.5   Missions

Following a series of atmospheric test flights using the jet-powered OK-GLI prototype, the first operational spacecraft (Orbiter K1) flew one test mission on 15 November 1988 at 03:00:02 UTC.[28] The spacecraft was launched unmanned from and landed at Baikonur Cosmodrome in the Kazakh S.S.R. and flew two orbits, travelling 83,707 kilometres (52,013 mi) in 3 hours and 25 minutes (0.14 flight days).[29] *Buran* never flew again; the programme was cancelled shortly after the dissolution of the Soviet Union.[30] In 2002, the collapse of the hangar in which it was stored destroyed the Buran orbiter.[31][32]

Atmospheric test flights

An aerodynamic testbed, OK-GLI, was constructed in 1984 to test the in-flight properties of the Buran design. Unlike the American prototype Space Shuttle *Enterprise*, OK-GLI had four AL-31 turbofan engines fitted, meaning it was able to fly under its own power.

Orbital flight of Orbiter K1 in 1988

The only orbital launch of the Orbiter K1 *Buran* (also known as ""OK-1K1" or "Shuttle 1.01") was at 3:00 UTC on 15 November 1988 from pad 110/37 in Baikonur. The unmanned craft was lifted into orbit by the specially designed Energia booster rocket. The life support system was not installed and no software was installed on the CRT displays.[2] The shuttle orbited the Earth twice in 206 minutes of flight. On its return, it performed an automated landing on the shuttle runway at Baikonur Cosmodrome.[35]

Planned flights

The planned flights for the shuttles in 1989, before the downsizing of the project and eventual cancellation, were:[36]

- 1991 —Orbiter K2 *Ptichka* unmanned first flight, duration 1–2 days.

- 1992 —Orbiter K2 *Ptichka* unmanned second flight, duration 7–8 days. Orbital maneuvers and space station approach test.

- 1993 —Orbiter K1 *Buran* unmanned second flight, duration 15–20 days.

- 1994 —Orbiter K3 *Baikal* first manned space test flight, duration of 24 hours. Craft equipped with life-support system and with two ejection seats. Crew would consist of two cosmonauts with Igor Volk as commander, and Aleksandr Ivanchenko as flight engineer.

- Second manned space test flight, crew would consist of two cosmonauts.

- Third manned space test flight, crew would consist of two cosmonauts.

- Fourth manned space test flight, crew would consist of two cosmonauts.

The planned unmanned second flight of Ptichka was changed in 1991 to the following:

- December 1991 —Orbiter K2 *Ptichka* unmanned second flight, with a duration of 7–8 days. Orbital maneuvers and space station approach test:

  - automatic docking with Mir's Kristall module
  - crew transfer from Mir to the shuttle, with testing of some of its systems in the course of twenty-four hours, including the remote manipulator
  - undocking and autonomous flight in orbit
  - docking of the manned Soyuz-TM 101 with the shuttle
  - crew transfer from the Soyuz to the shuttle and onboard work in the course of twenty-four hours
  - automatic undocking and landing

11.2.6   Cancellation of the programme 1993

After the first flight of a Buran shuttle, the project was suspended due to lack of funds and the political situation in the Soviet Union. The two subsequent orbiters, which were due in 1990 (informally *Ptichka*) and 1992 (informally *Baikal*) were never completed. The project was officially terminated on 30 June 1993, by President Boris Yeltsin. At the time of its cancellation, 20 billion rubles (roughly US$71,534,000) had been spent on the Buran programme.[37]

The programme was designed to boost national pride, carry out research, and meet technological objectives similar to those of the U.S. Space Shuttle programme, including resupply of the Mir space station, which was launched in 1986 and remained in service until 2001. When Mir was finally visited by a space shuttle, the visitor was a U.S. Shuttle, not Buran.

The Buran SO, a docking module that was to be used for rendezvous with the Mir space station, was refitted for use with the U.S. Space Shuttles during the Shuttle-Mir missions.[38]

Baikonur hangar collapse

On 12 May 2002, a hangar roof at the Baikonur Cosmodrome in Kazakhstan collapsed because of a structural failure due to poor maintenance. The collapse killed 7 workers and destroyed one of the Buran craft (Orbiter K1), as well as a mock-up of an Energia booster rocket. It was not clear to outsiders at the time which Buran programme craft was destroyed, and the BBC reported that it was just "a model" of the orbiter.[32] It occurred at the *MIK RN/MIK 112* building at Site 112 of the Baikonur Cosmodrome, 14 years after the first and only Buran flight. Work on the roof had begun for a maintenance project, whose equipment is thought to have contributed to the collapse. Also, preceding 12 May there had been several days of heavy rain.[9]

11.3   Fleet status and locations

Most of the geo-location below show the shuttle bodies on the ground; in some cases Google Earth's History facility is required to see the shuttle within the dates specified.[39][40]

Amusement rides and Buran shuttle test vehicle OK-7M/OK-TVA *at Gorky Park in Moscow.*

11.3.1   Related test vehicles and models

11.4   Possibilities for a revival of the Buran Programme

Over time, several scientists looked into trying to revive the *Buran* programme, especially after the Space Shuttle Columbia disaster.[*][47]

The 2003 grounding of the U.S. Space Shuttles caused many to wonder whether the Russian Energia launcher or Buran shuttle could be brought back into service.[*][48] By then, however, all of the equipment for both (including the vehicles themselves) had fallen into disrepair or been repurposed after falling into disuse with the collapse of the Soviet Union.

In 2010 the director of Moscow's Central Machine Building Institute said the *Buran* project would be reviewed in the hope of restarting a similar manned spacecraft design, with rocket test launches as soon as 2015.[*][49] Russia also continues work on the PPTS but has abandoned the Kliper program, due to differences in vision with its European partners.[*][50][*][51][*][52]

Due to the 2011 retirement of the American Space Shuttle and the need for STS-type craft in the meantime to complete the International Space Station, some American and Russian scientists had been mulling over plans to possibly revive the already-existing Buran shuttles in the Buran programme rather than spend money on an entirely new craft and wait for it to be fully developed[*][48][*][47] but the plans did not come to fruition.

On the 25th anniversary of the *Buran* flight in November 2013, Oleg Ostapenko, the new head of Roscosmos, the Russian Federal Space Agency, proposed that a new heavy lift launch vehicle be built for the Russian space program. The rocket would be intended to place a payload of 100 tonnes (220,000 lb) in a baseline low Earth orbit and is projected to be based on the Angara launch vehicle technology.[*][53]

Recently there have been new interests in renewing the programme temporarily while Russia struggles with the CSTS and Kliper design stages.[*][54]

11.5 Technical description

11.5.1 Specifications

First Spaceplanes

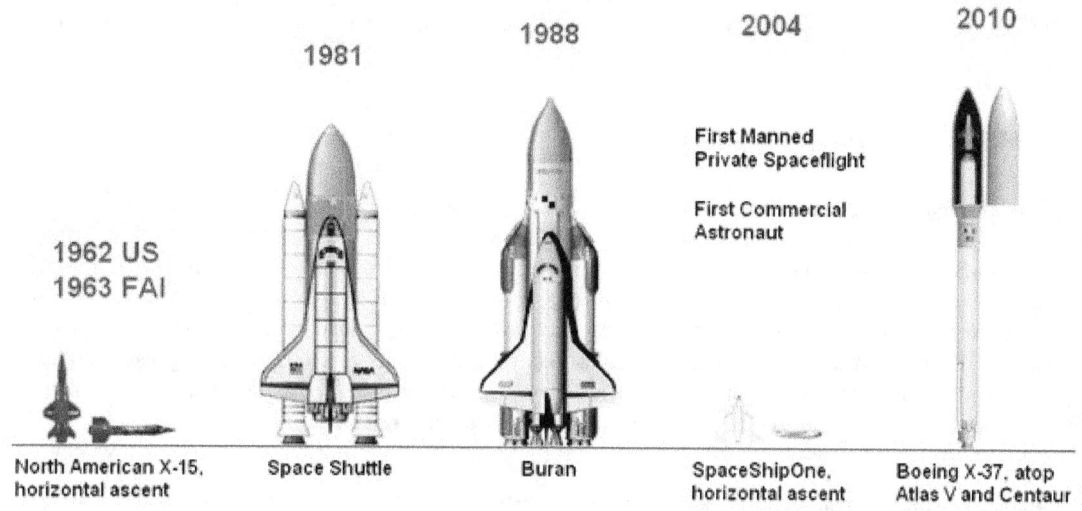

The Buran orbiter ranks among the world's first spaceplanes, with the North American X-15, the Space Shuttle, SpaceShipOne, and the Boeing X-37. Of these, only the Buran and X-37 spaceflights were unmanned.

Mass breakdown

- Mass of Total Structure / Landing Systems: 42,000 kg (93,000 lb)
- Mass of Functional Systems and Propulsion: 33,000 kg (73,000 lb)
- Maximum Payload: 30,000 kg (66,000 lb)
- Maximum liftoff weight: 105,000 kg (231,000 lb)

Dimensions

- Length: 36.37 m (119.3 ft)
- Wingspan: 23.92 m (78.5 ft)
- Height on Gear: 16.35 m (53.6 ft)
- Payload bay length: 18.55 m (60.9 ft)
- Payload bay diameter: 4.65 m (15.3 ft)
- Wing glove sweep: 78 degrees
- Wing sweep: 45 degrees

Propulsion

- Total orbital maneuvering engine thrust: 17,600 kgf (173,000 N; 39,000 lbf)
- Orbital Maneuvering Engine Specific Impulse: 362 seconds (3.55 km/s)
- Total Maneuvering Impulse: 5 kgf-sec (11 lbf-sec)

- Total Reaction Control System Thrust: 14,866 kgf (145,790 N; 32,770 lbf)

- Average RCS Specific Impulse: 275–295 seconds (2.70–2.89 km/s)

- Normal Maximum Propellant Load: 14,500 kg (32,000 lb)

Unlike the US Space Shuttle, which was propelled by a combination of solid boosters and the shuttle orbiter's own liquid-fuel engines fueled from a large fuel tank, the Soviet/Russian shuttle system used thrust from the rocket's four RD-170 liquid oxygen/kerosene engines developed by Valentin Glushko and another four RD-0120 liquid oxygen/liquid hydrogen engines.

11.5.2   Buran and the US Space Shuttle

Comparison to NASA's Space Shuttle

Because Buran's debut followed that of Space Shuttle *Columbia* 's, and because there were striking visual similarities between the two shuttle systems—a state of affairs which recalled the similarity between the Tupolev Tu-144 and Concorde supersonic airliners—many speculated that Cold War espionage played a role in the development of the Soviet shuttle. Despite remarkable external similarities, many key differences existed, which suggests that, had espionage been a factor in Buran's development, it would likely have been in the form of external photography or early airframe designs. One CIA commenter states that Buran was based on a rejected NASA design.*[55]

Key differences between Buran and NASA's Space Shuttle

- Buran had no main engines, and Energia's engines were expendable. The Space Shuttle main engines were part of the orbiter, and were reused for multiple flights.

- Energia could be configured for variety of payloads other than Buran, and was able to put up to 100 metric tons into orbit. The Space Shuttle orbiter was integral to its launch system and was the system's only payload.

- Energia's four boosters used liquid propellant (kerosene/oxygen). The Space Shuttle's two boosters used solid propellant.*[56]

- The liquid fueled booster rockets were not constructed in segments vulnerable to leakage through O-rings, which caused the destruction of *Challenger*.

- The Energia rocket was not covered in foam, the shedding of which from the large fuel tank led to the destruction of *Columbia*.

- Energia's four boosters were expended after each flight, though they were intended to eventually be recoverable. The Space Shuttle's boosters were recovered and reused.

- Buran's equivalent of the Space Shuttle Orbital Maneuvering System used GOX/LOX/Kerosene propellant, with lower toxicity and higher performance (a specific impulse of 362 seconds (3.55 km/s) using a turbopump system)*[57] than the Shuttle's pressure-fed monomethylhydrazine/dinitrogen tetroxide OMS engines.

- Buran was designed to be capable of both piloted and fully autonomous flight, including landing. The Space Shuttle was later retrofitted with automated landing capability, first flown on STS-121, but the system was intended to be used only in contingencies.*[58]

- The nose landing gear was located much farther down the fuselage rather than just under the mid-deck as with the NASA Space Shuttle.

- Buran could lift 30 metric tons into orbit in its standard configuration, comparable to the early Space Shuttle's original 27.8 metric tons*[59]*[60]

- Buran included a drag chute, the Space Shuttle did not originally but was later retrofitted to include one.

- The lift-to-drag ratio of Buran is cited as 6.5,*[61] compared to a subsonic L/D of 4.5 for the Space Shuttle.*[62]

- Buran and Energia were moved to the launch pad horizontally on a rail transporter, and then erected and fueled at the launch site.[*][63][*][64][*][65] The Space Shuttle was transported vertically on the crawler-transporter with loaded solid boosters.[*][66]

- The Buran was intended to carry a crew of up to ten, the Shuttle carried up to eight in regular operation and would have carried more only in a contingency.[*][60][*][67]

11.6 See also

- MAKS (spacecraft)

- Manned space missions

- Unmanned space missions

- Space exploration

- Space accidents and incidents

- Space Shuttle Program (United States)

- N1

11.7 References

[1] *Eight feared dead in Baikonur hangar collapse*, RSpaceflkight Now.

[2] "Shuttle Buran". *NASA.gov*. 12 November 1997. Archived from the original on 4 August 2006.

[3] http://www.sciencefirsthand.ru/gunko.pdf

[4] Harvey, Brian (2007). *The Rebirth of the Russian Space Program: 50 Years After Sputnik, New Frontiers*. Springer. p. 8. ISBN 0387713565.

[5] *Russian shuttle dream dashed by Soviet crash. YouTube.com* (Russia Today). 15 November 2007. Retrieved 16 July 2009.

[6] Chertok, Boris (2005); Rockets and People

[7] Paul Marks (2011-07-07). "Cosmonaut: Soviet space shuttle was safer than NASA's" .

[8] "Antonov An-225 Mryia (Cossack)". *The Aviation Zone*.

[9] Hendrickx & Vis 2007.

[10] Zak, Anatoly (20 November 2008). "Buran - the Soviet 'space shuttle'". *BBC News*. Retrieved 7 December 2008.

[11] Sparrow, Giles (2009). *Spaceflight: The Complete Story From Sputnik to Shuttle—and Beyond*. DK Publishers. p. 215. ISBN 9780756656416.

[12] Windrem, Robert (4 November 1997). "How the Soviets stole a space shuttle" . *NBC News*.

[13] Fedotov, V. A. "BURAN Orbital Spaceship Airframe Creation" . *Buran-Energia.com*. Retrieved 22 January 2013.

[14] Petrovitch, Vassili. "VM-T Atlant: Description" . *Buran-Energia.com*. Retrieved 22 January 2013.

[15] Goebel, Greg. "The Antonov Giants: An-22, An-124, & An-225 - Antonov An-225 Mriya ("Cossack")". *Airvectors.net*. Retrieved 21 August 2012.

[16] Goebel, Greg. "Postscript: The Other Shuttles - The Soviet Buran shuttle programme" . *Vectorsite.net*. Retrieved 21 August 2012.

[17] "Системное и прикладное программирование" [System and application programming]. 50th Anniversary of Institute for Applied Mathematics. Keldysh Institute of Applied Mathematics. 2004. Retrieved March 2015.

[18] "Отдел программных комплексов" [Department of software systems]. Keldysh Institute of Applied Mathematics. Retrieved March 2015.

[19] Kryukov, V.; Petrenko, A. (1996). *Интегрированный подход к разработке крупных программных систем управления реального времени* [*An integrated approach to the development of large software systems, real-time control*]. Индустрия программирования [Software industry]. Moscow.

[20] "Soviet Software Productivity: Isolated Gains in an Uphill Battle" (PDF). Central Intelligence Agency. May 1990. p. 7. SW 90-10029X. Archived from the original (PDF) on 11 June 2012.

[21] Hendrickx, Bart; Bert Vis (2007-10-04). *Energiya-Buran : The Soviet Space Shuttle*. Praxis. p. 526. ISBN 0-387-69848-5.

[22] Wade, Mark. "Mir LII-1". *Encyclopedia Astronautica*. Retrieved 15 November 2010.

[23] http://www.russianspaceweb.com/baikonur_energia_112.html

[24] http://www.buran-energia.com/bourane-buran/bourane-fin.php

[25] http://www.russianspaceweb.com/baikonur_energia_251.html

[26] http://ourairports.com/airports/UAON/pilot-info.html

[27] http://www.russianspaceweb.com/baikonur_energia_254.html

[28] Hendrickx & Vis 2007, p. 349.

[29] Hendrickx & Vis 2007, p. 356.

[30] *The New Book of Popular Science* **1**. Scholastic. 2008. p. 257. ISBN 9780717212262.

[31] Hendrickx & Vis 2007, p. 388.

[32] Whitehouse, David (13 May 2002). "Russia's space dreams abandoned". *BBC News*. Retrieved 14 November 2007.

[33] "Soviet shuttle". *The Christian Science Monitor*. 17 November 1988. p. 15. Retrieved 15 January 2013.

[34] Barringer, Felicity (16 November 1988). "Soviet Space Shuttle Orbits and Returns In Unmanned Debut". *The New York Times*. Retrieved 15 January 2013.

[35] Chertok, Boris E. (2005). Siddiqi, Asif A., ed. *Rockets and People* (PDF). NASA History Series **1**. National Aeronautics and Space Administration. p. 179. SP-2005-4110.

[36] Lukashevich, Vadim. Экипажи "Бурана": Несбывшиеся планы[The Crews of "Buran": Unfulfilled Plans]. *Buran.ru* (in Russian). Retrieved 5 August 2006.

[37] Wade, Mark. "Yeltsin cancels Buran project". *Encyclopedia Astronautica*. Retrieved 2 July 2006.

[38] Wade, Mark. "Mir-Shuttle Docking Module". *Encyclopedia Astronautica*. Retrieved 16 July 2009.

[39] Petrovitch, Vasili. "Buran-Energia". Retrieved 20 February 2015.

[40] Zak, Anatoliy. "Buran". Retrieved 20 February 2015.

[41] "Space shuttle Buran heat shield thermal black tile excellent condition".

[42] http://speyer.technik-museum.de/en/en/spaceshuttle-buran

[43] Zak, Anatoly. "Buran reusable shuttle". Russian Space WEB. Retrieved 22 February 2015.See the last line of the cronology.

[44] ru:OK-TBA

[45] "Buran: The Abandoned Russian Space Shuttle". Urban Ghost Media. 30 September 2010. Retrieved 21 August 2012.

[46] "Energia-Buran: Where are they now". *K26.com*. Retrieved 5 August 2006.

[47] Birch, Douglas (2003). "Russian space program is handed new responsibility" (url). Sun Foreign Staff. Retrieved 2008-10-17.

[48] Oberg, James (10 June 2005). "Russia ready to take lead on space station". *NBC News*. Retrieved 16 July 2009.

[49] "Russia To Review Its Space Shuttle Project". Space Daily. Xinhua. Retrieved 2010-07-28.

[50] "Soviet space shuttle could bail out NASA". Current.com. 2008-12-31. Retrieved 2009-07-15.

[51] "Soviet space shuttle could bail out NASA". Russia Today. Archived from the original on 1 December 2011. Retrieved 2009-07-15.

[52] "Russia, Europe abandon joint space project —Roscosmos". RIA Novosti. Retrieved 2009-01-29.

[53] "Russia starts ambitious super-heavy space rocket project". *Space Daily*. 2013-11-19. Retrieved 2013-12-13. *Buran* could stay in orbit for 30 days, while the American shuttle had a 15-day time limit. It could deliver into orbit 30 tonnes of cargo, compared to the US shuttle's 24 tonnes of cargo. It could carry a crew of 10 cosmonauts, while the American shuttle could carry seven astronauts. Preparation for the *Energia/Buran* launch at Baikonur Cosmodrome only took 15 days. However, it took one month of preparations before the US shuttle was launched from Cape Canaveral. The *Energia* rocket booster could be used to launch various payloads into orbit, whereas the American shuttle's booster was one-task. A year and a half before the *Buran* launch, *Energia* was launched with a full-scale mock-up of the Skif-DM orbital combat laser platform weighing 77 tonnes, measuring 37 meters long, and over four meters in diameter. Though the mock-up failed to reach the desired orbit and fell into the Pacific, the *Energia* booster did its job fine, delivering the huge space platform into intermediate orbit, 110 kilometers above the earth's surface. But the most important difference from the American model was that the Soviet spaceship could perform the flight and landing in totally automatic mode, which it brilliantly demonstrated on November 15, 1988. *Buran's* American counterpart used to land with switched-off engines, meaning it could make only one landing attempt. The Soviet spacecraft could take several tries if needed. When *Buran* approached Baikonur Cosmodrome and started landing in 1988, its sensors registered too strong side winds and the robotic system sent the huge machine for another rectangular traffic pattern approach, successfully landing the spacecraft on a second try. The *Buran* shuttle was designed to perform 100 flights to space, while its engines were ready to do 66 flights without replacement. During its flight, it lost just eight of its unique thermal-insulation tiles out of 38,800.

[54] "Soviet space shuttle could bail out NASA". *Russia Today*. 15 November 2008. Archived from the original on 30 November 2010.

[55] Weiss, Gus W. (1996). "The Farewell Dossier: Duping the Soviets - A Deception Operation". *Studies in Intelligence* (Central Intelligence Agency) **39** (5). Retrieved 8 August 2012.

[56] "Space Shuttle: Solid Rocket Boosters". *NASA.gov*. Retrieved 16 October 2010.

[57] Lukashevich, Vadim. "Объединенная двигательная установка (ОДУ)" [Joint Propulsion System (JPS)]. *Buran.ru*. Retrieved 21 November 2013.

[58] Malik, Tariq (29 June 2006). "Shuttle to Carry Tools for Repair and Remote-Control Landing". *Space.com*.

[59] Wade, Mark. "Shuttle". *Encyclopedia Astronautica*. Retrieved 20 September 2010.

[60] Scott, Jeff (5 February 2007). "Soviet Buran Space Shuttle". *Aerospaceweb.org*.

[61] ""Molniya" Research & Industrial Corporation". *Buran.ru*. Retrieved 20 September 2010.

[62] Chaffee, Norman, ed. (1985). *Space Shuttle Technical Conference, Part 1*. NASA. p. 258. N85-16889.

[63] "Buran, the First Russian Shuttle". *EnglishRussia.com*. 14 September 2006. Retrieved 21 August 2012.

[64] "Russian rockets". *The Mars Society*. Retrieved 21 August 2012.

[65] "6 Abandoned Mega-Machines: Jumbo Jets, Space Shuttle Transporters & More". *Urban Ghosts*. 26 January 2011. Retrieved 21 August 2012.

[66] Sands, Jason (May 2007). "NASA Diesel-Powered Shuttle Hauler - The Crawler". *Diesel Power*.

[67] Ceccacci, Anthony J.; Dye, Paul F. (12 July 2005). "Contingency Shuttle Crew Support (CSCS)/Rescue Flight Resource Book" (PDF). NASA. Retrieved September 2014.

11.8 Bibliography

- Hendrickx, Bart; Vis, Bert (2007). *Energiya-Buran: The Soviet Space Shuttle*. Springer-Praxis. doi:10.1007/978-0-387-73984-7. ISBN 0-387-69848-5.

11.9   External links

- Buran.ru, official website by NPO Molniya

- Buran at *Encyclopedia Astronautica*

- Buran and Energia at Buran-Energia.com

- Buran at RussianSpaceWeb.com

Buran's rear (1989)

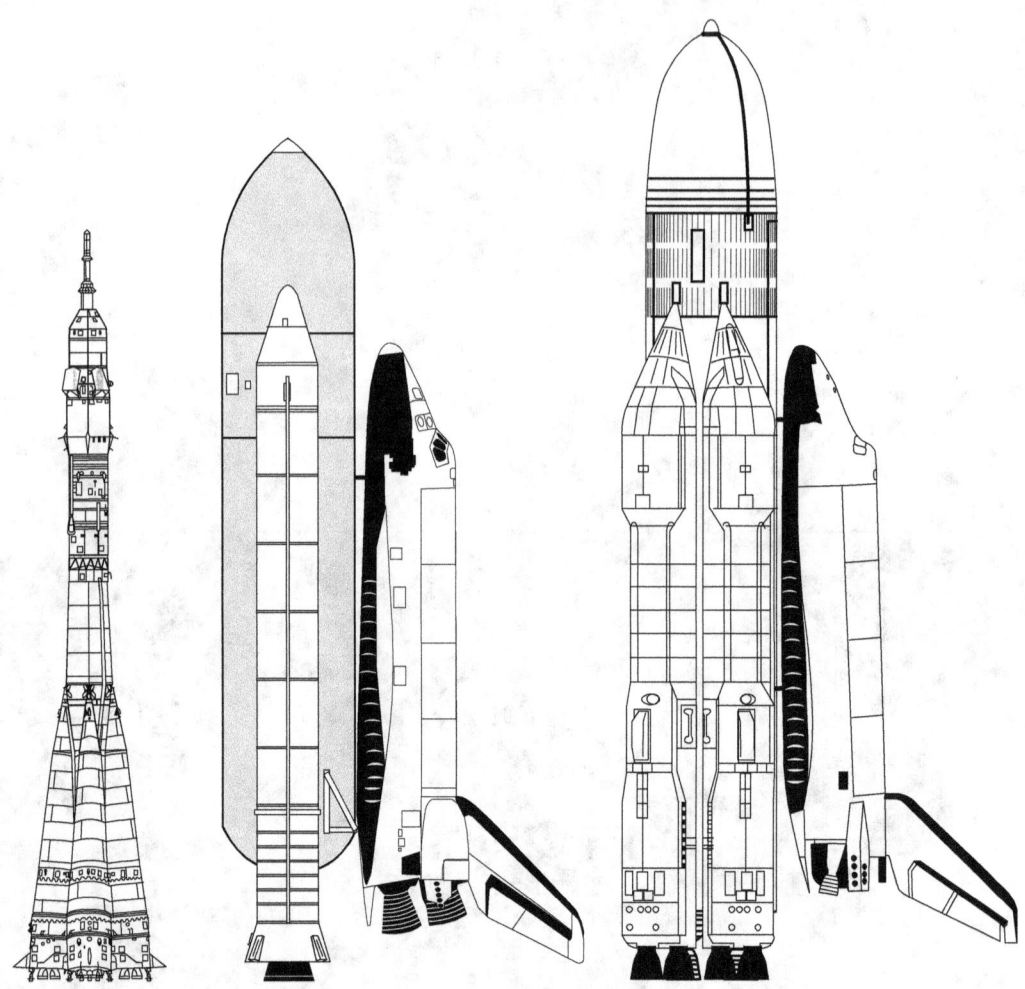

Soyuz, Space Shuttle, and Buran

Chapter 12

Dream Chaser

This article is about the Sierra Nevada Corporation (SNC) spaceplane. For other uses, see Dream Chasers.

The **Dream Chaser** is an American reusable crewed suborbital and orbital[6] lifting-body spaceplane being developed by Sierra Nevada Corporation (SNC) Space Systems. The Dream Chaser is designed to carry up to seven people to and from low Earth orbit. The vehicle would launch vertically on an Atlas V rocket and land horizontally automatically on conventional runways.[1]

12.1 Design

Dream Chaser is a reusable composite spacecraft designed to carry from two to seven people and/or cargo to orbital destinations such as the International Space Station.[7] It would have a built-in launch escape system[8] and could fly autonomously if needed.[4] It could use any suitable launch vehicle but is planned to be launched on a human-rated Atlas V 412 rocket.[4][9] The vehicle would be able to return from space by gliding (typically experiencing less than 1.5 g on re-entry) and landing on any airport runway that handles commercial air traffic.[10][11] Its reaction control system thrusters burn ethanol-based fuel,[4][10] which is not an explosively volatile material, allowing the Dream Chaser to be handled immediately after landing, unlike the Space Shuttle.[4] Its thermal protection system (TPS) is an ablative tile created by NASA's Ames Research Center that would be replaced as a large group rather than tile by tile, and would only need to be replaced after several flights.[4]

12.1.1 Rocket engines

On-orbit propulsion of the Dream Chaser is provided by twin hybrid rocket engines. The hybrid rocket motors are fueled with hydroxyl-terminated polybutadiene (HTPB) and nitrous oxide, or more simply put, "rubber and laughing gas".[8] These two substances are both non-toxic and easily stored, making them safer than liquid rocket fuels. Unlike solid rockets, Dream Chaser's hybrid fuel system would allow the motor to stop and start repeatedly, and be throttleable. SNC Space Systems was also developing a similar hybrid rocket, RocketMotor Two, for Virgin Galactic's SpaceShipTwo,[8] as a subcontractor to Scaled Composites. In May 2014 their involvement in the program ended, after Virgin Galactic elected to replace SNC's version of RocketMotorTwo, powered by HTBD rubber fuel, with its own internally developed hybrid motor using a polyamide plastic fuel, while continuing to use the same nitrous oxide oxidizer.[12]

12.1.2 Engine testing

Sierra Nevada completed an initial test phase on the Dream Chaser rocket engine in 2010, under the CCDev1 program, including three successful test firings on a single hybrid motor in a single day.[13]

A second phase of testing began in June 2013, with a motor firing and ignition test in order to validate the newly modified test stand, as a start to the Commercial Crew Integrated Capability (CCiCap) contract test phase.

12.1.3   Wind tunnel testing

In 2014, Sierra Nevada completed its wind tunnel testing as part of its CCiCAP Milestone 8. The Wind tunnel testing involved analyzing the flight dynamics characteristics that the vehicle will experience during orbital ascent and re-entry. Wind tunnel testing was also completed for the Dream Chaser Atlas V integrated launch system. These tests were completed at NASA Ames Research Center at Moffett Field, California, CALSPAN Transonic Wind Tunnel in New York, and at NASA Langley Research Center Unitary Plan Wind Tunnel in Hampton, Virginia.[14]

12.2   History

The historical antecedents of the Dream Chaser go back over 50 years in the US; with the 1957 X-20 Dyna-Soar concept and the 1966 Northrop M2-F2 and Martin X-23 PRIME lifting bodies.[15][16] Its design is derived from NASA's 1990 HL-20 lifting body design which was itself similar to the 1980s Soviet BOR-4, which in turn was considered by NASA engineers as influenced by the late 1960s HL-10,[17] and the Soviet Mikoyan-Gurevich MiG-105 military spaceplane concept,[18] a spaceplane studied as a means to develop a Soviet counterpart to the US's X-20 Dyna-Soar.[19]

The name "Dream Chaser" has been used for two separate space vehicles. One, planned to be an orbital vehicle based on the NASA HL-20, originated at SpaceDev when Jim Benson was still there. The second, a suborbital vehicle, was the result of Jim Benson having reused the name when he formed the Benson Space Company for the purposes of space tourism.[18]

12.2.1   SpaceDev Dream Chaser proposal

COTS

The Dream Chaser was publicly announced on 20 September 2004 as a candidate for NASA's Vision for Space Exploration and later Commercial Orbital Transportation Services Program (COTS).

When the Dream Chaser was not selected under Phase 1 of the COTS Program, SpaceDev founder Jim Benson stepped down as Chairman of SpaceDev and started Benson Space Company to pursue the development of the Dream Chaser.[20] In April 2007, SpaceDev announced that it had partnered with the United Launch Alliance to pursue the possibility of using the Atlas V booster rocket as the Dream Chaser's launch vehicle.[21] In June 2007, SpaceDev signed a Space Act agreement with NASA.[22]

CCDev

About two weeks after Benson's 10 October 2008 death, SpaceDev agreed to be acquired by Sierra Nevada Corporation, a privately owned company operated by Fatih Ozmen and Eren Ozmen, on 21 October 2008 for $38 million.[23] On 1 February 2010, Sierra Nevada Corporation was awarded $20 million in seed money under NASA's Commercial Crew Development (CCDev) phase 1 program for the development of the Dream Chaser.[24][25] Of the $50 million awarded by the CCDev program, Dream Chaser's award represented the largest share of the funds. SNC completed the four planned milestones on time which included program implementation plans, manufacturing readiness capability, hybrid rocket test fires, and the preliminary structure design.[26] Further initial Dream Chaser tests included the drop test of a 15% scaled version at the NASA Dryden Flight Research Center.[27] The 5-foot-long (1.52 meters) model was dropped from 14,000 feet (4,300 m) to test flight stability and collect aerodynamic data for flight control surfaces.[27]

For the CCDev phase 2 solicitation by NASA in October 2010, Sierra Nevada proposed extensions of Dream Chaser spaceplane technology. According to head of Sierra Nevada Space Systems Mark Sirangelo, the cost of completing the Dream Chaser should be less than $1 billion.[2][28]

On 18 April 2011, NASA awarded nearly $270 million in funding for CCDev 2, including $80 million to Sierra Nevada Corporation for Dream Chaser.[29] Since then, nearly a dozen further milestones have been completed under that Space Act Agreement. Some of these milestones included testing of the airfoil fin shape, integrated flight software and hardware, landing gear, and a full-scale captive carry flight test.[30][31]

An artist's impression of the X-20 Dyna-Soar being launched using a Titan booster, with large fins added to the Titan's first stage

CCiCap

On 3 August 2012, NASA announced the award of $212.5 million to Sierra Nevada Corporation to continue work on the Dream Chaser under the Commercial Crew Integrated Capability (CCiCAP) Program.*[32]

In December 2013, the German Aerospace Center (DLR) announced a funded study to investigate ways in which Europe might take advantage of the Dream Chaser crewed spaceplane technology. Named the DC4EU (Dream

Chaser for European Utilization), the project will study using it for sending crews and cargo to the ISS and on missions not involving the ISS, particularly in orbits of substantially greater altitude than the ISS can reach.[*][33]

In January 2014, the European Space Agency (ESA) agreed to be a partner on the DC4EU project, and will also investigate whether the Dream Chaser can use ESA avionics and docking mechanisms. ESA will also study launching options for the "Europeanized" Dream Chaser, particularly whether it can be launched within the Ariane 5's large aerodynamic cargo fairing – or, like the Atlas V, without it. In order to fit within the fairing, the Dream Chaser's wing length will have to be reduced slightly, which is thought to be easier than going through a full aerodynamic test program to evaluate and prove it along with the Ariane for flight without the fairing.[*][34]

In late January 2014, it was announced that the Dream Chaser orbital test vehicle was under contract to be launched on an initial orbital test flight, using an Atlas V rocket, from Kennedy Space Center in November 2016. This is a privately arranged commercial agreement, and is funded directly by Sierra Nevada and is not a part of any existing NASA contract.[*][35]

2014 CCtCap non-selection by NASA

After being involved with the NASA Commercial Crew Development program since 2009—and being selected as one of the contract award recipients in each prior phase of the program—NASA did not select the Dream Chaser for the next phase of the Commercial Crew Program announced 16 September 2014[*][36] due to lack of maturity.[*][37] Sierra Nevada filed a protest to the US Government Accountability Office (GAO) on 26 September. The GAO is investigating and will respond after a process that could take up to 100 days. Boeing and SpaceX were asked by NASA to "stop work" on the crewed spacecraft during the protest resolution.[*][38] However, on 22 October 2014, a Federal Judge ruled that NASA could proceed with contracts with Boeing and SpaceX to develop their "space taxis", while the GAO continues to consider Sierra Nevada's protest of NASA's original decision.

Two weeks after losing the Commercial Crew Transportation Capability (CCtCap) competition to SpaceX and Boeing on 16 September 2014,[*][39] Sierra Nevada Corporation announced it has designed a launch system that combines a scale version of the company's Dream Chaser space plane with the Stratolaunch Systems air launch system.[*][40] Earlier the same week, Sierra Nevada introduced new spaceflight opportunities to the world - coined the Dream Chaser Global Project"- which would provide customized access to low Earth orbit to global customers.[*][41]

Despite not being selected to continue forward under NASA's Commercial Crew transportation Capability (CCtCap) phase of the effort to send crews to orbit via private companies, SNC is still completing milestones under earlier phases of the CCP.[*][42] On 2 December 2014 SNC announced that it completed NASA's CCiCap Milestone 5a related to propulsion risk reduction for the Dream Chaser space system.[*][43]

By late December, details had emerged that "a high-ranking agency official"—"William Gerstenmaier, the agency's top human exploration official and the one who made the final decision"—"opted to rank Boeing's proposal higher than a previous panel of agency procurement experts." More specifically, Sierra Nevada asserted in their filings with the GAO that Gerstenmaier may have "overstepped his authority by unilaterally changing the scoring criteria."[*][44]

On 5 January 2015, the GAO denied Sierra Nevada's CCtCap challenge, stating that NASA made the proper decision when it decided to award Boeing $4.2 billion and SpaceX $2.6 billion to develop their vehicles. Ralph White, the GAO's managing associate counsel, announced that NASA "recognized Boeing's higher price but also considered Boeing's proposal to be the strongest of all three proposals in terms of technical approach, management approach and past performance, and to offer the crew transportation system with most utility and highest value to the government." Furthermore, the agency found "several favorable features" in SNC's proposal "but ultimately concluded that SpaceX's lower price made it a better value." [*][45]

12.2.2   Dream Chaser Global Project

In September 2014, SNC announced that it would, with global partners, use the Dream Chaser as the baseline space-craft for orbital access for a variety of programs, specializing the craft as needed.[*][46]

On 5 November 2014 during the Space Traffic Management Conference at Embry–Riddle Aeronautical University, SNC's Space Systems team presented the challenges and opportunities related to landing the Dream Chaser spacecraft at public-use airports.[*][47] According to the presentation, "Unlike the Space Shuttle, the Dream Chaser does not require any unique landing aids or specialized equipment as it uses all non-toxic propellants and industry standard subsystems." [*][48]

Stratolaunch+DreamChaser

In late November 2014, Vulcan Aerospace released the results of the SNC/Stratolaunch space transportation architecture, which indicated that the reduced-size Dream Chaser in conjunction with the Stratolaunch-based launch system mission capabilities. The system would have an outbound range of 1,900 kilometers; 1,200 miles (1,000 nmi) away from the airport where the aircraft departed. The launch vehicle would be a modified air-launched Orbital Sciences rocket that is approximately 37 m (120 ft) in length. The Dream Chaser payload would be a 75-percent sized version of the vehicle previously proposed to NASA—while maintaining the relative outer mold line—6.9 m (22.5 ft) in length with a wingspan of 5.5 m (18.2 ft), which could carry 2 to 3 crewmembers plus a variety of scientific and research payloads.*[49]

Dream Chaser for European Utilization

See also: Hermes (spacecraft) and Intermediate eXperimental Vehicle

In 2013 SNC and OHB entered into an agreement to study the feasibility of using SNC's Dream Chaser spacecraft for a variety of missions. The DC4EU study thoroughly reviewed applications for the Dream Chaser including crewed and uncrewed flights to low-Earth orbit (LEO) for missions such as microgravity science, satellite servicing and active debris removal (ADR).

On 3 February 2015, the Sierra Nevada Corporation's (SNC) Space Systems and OHB System AG (OHB) in Germany announced the completion of the initial Dream Chaser for European Utilization (DC4EU) study.*[50]

"The inherent design advantages of the Dream Chaser reusable lifting body spacecraft make it an ideal vehicle for a broad range of space applications," said Dr. Fritz Merkle, member of the Executive Management Board of OHB AG. "We partnered with SNC to study how the design of the Dream Chaser can be used to advance European interests in space. The study results confirm the viability of using the spacecraft for microgravity science and ADR. DC4EU can benefit the entire international space community with its unique capabilities. We look forward to further maturing our design with SNC as we expand our partnership." The cooperation was renewed in April 2015 for additional two years.*[51]

12.2.3 Dream Chaser Cargo System

The cargo variant of the SNC Dream Chaser is called the *Dream Chaser Cargo System*. Featuring an expendable cargo portion, containing solar panels, the cargo version of the spacecraft will be capable of taking 5000 kg back to Earth, undergoing re-entry forces of 1.5G. It has been proposed for the Phase II program for cargo resupply of the International Space Station.*[52]

CRS2

In December 2014, Sierra Nevada proposed *Dream Chaser* for CRS-2 consideration.*[53] It is in competition with the existing CRS-1 contract holders SpaceX Dragon capsule and Orbital Sciences Cygnus capsule, as well as fellow CCDev competitor Boeing CST-100.*[54]

To meet CRS2 guidelines, the cargo Dream Chaser will feature foldable wings, to fit within a 5m cargo fairing, unlike the passenger Dream Chaser, which did not use a cargo fairing. The ability to fit in a cargo fairing allows launches from Ariane 5 as well as Atlas 5 rocket launcher vehicles. To expand the cargo uplift capacity, an expendable cargo module is affixed aft, which will not support downlift, but can be used for disposal of up to 3250 kg of trash. Total uplift is planned for 5000 kg pressurized, 500 kg unpressurized, with downlift of 1750 kg wholly within the spaceplane.*[55]

12.3 Development progress

On 24 June 2011 SNC announced it had achieved two critical milestones for NASA's CCDev program. The first was a Systems Requirement Review (SRR), where SNC validated their requirements based on NASA's draft Commercial Crew Program Requirements. The SRR was successfully completed on 1 June 2011 with participation from NASA and SNC industry partners. The second milestone was a review of the improved airfoil fin shape for Dream Chaser

The completed craft on the day of its initial captive carry test

used to aid its control through the atmosphere. Testing in a wind tunnel and computational fluid dynamics analyses allowed the fin selection to pass the NASA milestone.[*][56]

As of October 2011, Sierra Nevada Corp had completed four of the 13 milestones set out in the CCDev Agreement.[*][57] The most recent milestones accomplished include: a System Requirements Review, a new cockpit simulator, finalizing the tip fin airfoil design and most recently,[*][58] a Vehicle Avionics Integration Laboratory (VAIL), which will be used to test Dream Chaser computers and electronics in simulated space mission scenarios.[*][57]

By February 2012, Sierra Nevada Corporation stated that it had completed the assembly and delivery of the primary structure of the first Dream Chaser flight test vehicle. With this, SNC completed all 11 of its CCDev milestones that were scheduled up to that point. SNC stated in a press release that it was "...on time and on budget." [*][59]

On 24 April 2012 Sierra Nevada Corporation announced the successful completion of wind tunnel testing of a scale model of the Dream Chaser vehicle.[*][60]

On 12 June 2012 SNC announced the commemoration of its fifth year as a NASA Langley partner in the design and development of Dream Chaser.[*][61] Together with ULA, the NASA/SNC team performed buffet tests on the Dream Chaser and Atlas V stack. To date, the Langley/SNC team has worked on aerodynamic and aerothermal analysis of Dream Chaser, as well as guidance, navigation and control systems.[*][61]

On 11 July 2012 SNC announced that they successfully completed testing of the nose landing gear for Dream Chaser.[*][62] This milestone evaluated the impact to the landing gear during simulated approach and landing tests as well as the impact of future orbital flights. The main landing gear was tested in a similar way in February 2012. The nose gear landing test was the last milestone to be completed before the free flight approach and landing tests scheduled for later in 2012.[*][62]

In August 2012, SNC completed CCiCap Milestone 1, or the 'Program Implementation Plan Review' . This included creating a plan for implementing design, development, testing, and evaluation activities through the duration of CCiCap funding.[*][63]

By October 2012 the "Integrated System Baseline Review" , or CCiCap Milestone 2, had been completed. This review demonstrated the maturity of the Dream Chaser Space System as well as the integration and support of the

Atlas V launch vehicle, mission systems, and ground systems.*[63]

On 30 January 2013 SNC announced a new partnership with Lockheed Martin. Under the agreement, SNC will pay Lockheed Martin $10 million to build the second airframe at its Michoud facility in New Orleans, Louisiana. This second airframe is slated to be the first orbital test vehicle, with orbital flight testing planned to begin within the next two years.*[64] In January 2014, SNC announced they had signed a launch contract to fly the first orbital test vehicle on a robotically controlled orbital test flight in November 2016.*[35]

In January 2013, Sierra Nevada also announced that the second captive carry and first unpowered drop test of Dream Chaser would take place at Edwards Air Force Base, California in March 2013. The spaceplane release would occur at 3,700 metres (12,000 ft) altitude and would be followed by an autonomous robotic landing.*[64]*[65]

On 13 March 2013, NASA announced that former space shuttle commander Lee Archambault was leaving the agency in order to join SNC. Archambault, a former combat pilot and 15-year NASA veteran who flew on Atlantis and Discovery, will work on the Dream Chaser program as a systems engineer and test pilot.*[66]*[67]

On 29 April 2013, Virgin Galactic's SpaceShipTwo sub-orbital vehicle was propelled on its first ever powered flight by SNC's Hybrid Rocket Motor. SNC manufactures the main oxidizer valve and the hybrid rocket motor, plus the nitrous oxide dump and pressurization system control valves. The hybrid rocket motor and oxidizer valve system are manufactured at an SNC facility in Poway, California, where motors for both Space Ship Two and Dream Chaser are produced.*[68]

On 1 August 2014, the first completed piece of the orbital test vehicle's composite airframe was unveiled at the Lockheed Martin Michoud Assembly Facility in Louisiana.*[69]

In October 2015, the thermal protection system was installed on the ETA for the next phase of atmospheric flight testing. The orbital cabin assembly of the FTA orbital test vehicle was also completed by contractor Lockheed Martin.*[70]

12.3.1 Flight test program

In May 2013, The Dream Chaser Engineering Test Article (ETA) was shipped to the Dryden Flight Research Center in California for a series of ground tests and aerodynamic flight tests.*[71] This move to Dryden came about a year after a captive carry test that was conducted near the Rocky Mountain Metropolitan Airport on 29 May 2012. During that test, an Erickson Skycrane was used to lift the Dream Chaser to better determine its aerodynamic properties.*[72] "The testing at Dryden will include tow, captive-carry and free-flight tests of the Dream Chaser. A truck will tow the vehicle down a runway to validate performance of the nose strut, brakes and tires. The captive-carry flights will further examine the loads the vehicle will encounter during flight and test the performance and flutter of the vehicle up to release from an Erickson Skycrane helicopter. The free-flight tests are designed to validate the Dream Chaser's aerodynamics as well as test the flight control surfaces to verify flight characteristics for approach, flare and landing." *[73] A second captive carry flight test was completed on 22 August 2013.*[74]

On 26 October 2013, the first free-flight occurred. The test vehicle was released from the "skycrane" helicopter, and flew the correct flightpath to touchdown less than a minute later. Just prior to landing, the left main landing gear failed to deploy resulting in a crash landing.*[75] In a press teleconference a short while later, Mark Sirangelo, corporate vice president of Sierra Nevada, told reporters that the view of the ETA was obscured by the dust as it skidded off the runway, but that the vehicle was found upright, with the crew compartment intact, and all systems inside still in working order. Sierra Nevada corporation engineers do not believe that the ETA flipped over.*[76]*[77]

The first two Dream Chasers —the ETA and the Flight Test Article (FTA) —have been given internal and external names, with some sources reporting that the ETA will be named **Eagle**.,*[71] while the FTA was originally named **Ascalon** before being changed to **Ascension**.*[78]

An initial orbital test flight of the Dream Chaser orbital test vehicle is planned for 1 November 2016, launching on an Atlas V rocket from Kennedy Space Center.*[35]

12.4 Technology partners

The following organizations have been named as technology partners:

- Boeing Phantom Works – construction of some early test articles*[11]

Dream Chaser model being tested at NASA Langley

- Charles Stark Draper Laboratory – Guidance, Navigation and Control[*][11]

- Aerojet – reaction control system technology[*][11]

- University of Colorado – human-rating[*][11]

- AdamWorks – composites[*][11]

- MDA – systems engineering[*][11]

- Lockheed Martin – airframe construction and human rating of the spaceplane[*][64][*][65]

12.5   See also

- BOR-4

- BOR-5

- Mikoyan-Gurevich MiG-105

- Boeing X-20 Dyna-Soar

- Boeing X-37

- HL-20 Personnel Launch System

- Commercial Crew Development

- Private spaceflight

- Skylon

- Spaceplane

- Dragon V2, a crew-carrying spacecraft being developed by SpaceX

- CST-100, a crew-carrying spacecraft being developed by Boeing

- Orion (spacecraft), a spacecraft being built for NASA by Lockheed Martin

12.6   References

[1] "Dream Chaser Model Drops in at NASA Dryden" (Press release). Dryden Flight Research Center: NASA. 2010-12-17. Archived from the original on 2014-01-07. Retrieved 2012-08-29.

[2] Chang, Kenneth (2011-02-01). "Businesses Take Flight, With Help From NASA" . *New York Times*. p. D1. Archived from the original on 2014-01-06. Retrieved 2012-08-29.

[3] Wade, Mark (2014). "Dream Chaser" . Encyclopedia Astronautix. Archived from the original on 2014-01-06. Retrieved 2012-08-29.

[4] Sirangelo, Mark (August 2011). "NewSpace 2011: Sierra Nevada Corporation" . Spacevidcast. Retrieved 2011-08-16. Sirangelo, Mark (24 August 2014). "Flight Plans and Crews for Commercial Dream Chaser's First Flights: One-on-One Interview With SNC VP Mark Sirangelo (Part 3)". AmericaSpace.

[5] Bayt, Rob (2011-07-26), *Commercial Crew Program: Key Driving Requirements Walkthrough* (Powerpoint), NASA, retrieved 2011-07-27

[6] Leonard, David (2011-02-07). "Private Spaceflight Innovators Attract NASA's Attention" . New York. Archived from the original on 2014-01-06. Retrieved 2014-01-06. Dream Chaser will become a fully capable suborbital vehicle on the way to reaching orbital capability.

[7] "NASA Deputy Administrator Lori Garver touts Colorado's role" . Youtube.com. 2011-02-05. Retrieved 2012-08-29.

[8] Klingler, Dave (2012-09-06). "50 years to orbit: Dream Chaser's crazy Cold War backstory: The reusable mini-spaceplane is back from the dead—again—and prepping for space" . *ars Technical* (Boston: Conde Nast). p. 3. Archived from the original on 2014-01-06. Retrieved 2012-09-07.

[9] "Moving Forward: Commercial Crew Development Building the Next Era in Spaceflight" (PDF). *Rendezvous: Where today meets tomorrow* **4** (2): 10–15. Summer 2010. Archived from the original on 2014-01-06. Retrieved 2014-01-06.

[10] "The Space Show : Mark Sirangelo interview" . David Livingston. 2012-01-04. Retrieved 2012-01-07.

[11] Frank Morring, Jr (2010-02-19). "Sierra Nevada Building On NASA Design" . Aviation Week.

[12] Doug Messier (24 May 2014). "Virgin Galactic Hails RocketMotorTwo Milestone" . ParabolicArc.

[13] "Sierra Nevada Corporation Begins Dream Chaser Main Hybrid Rocket Motor Testing" . *NewSpace Watch*. 2013-06-06. Retrieved 2013-06-11. (subscription required (help)).

[14] "Dream Chaser passes Wind Tunnel tests for CCiCap Milestone" . *NASASpaceflight.com*. 2014-05-19. Retrieved 2015-01-02.

[15] Eddy, Max (2012-04-02). "How the United States Will Return to Space" . *Geekosystem* (New York). Retrieved 2014-01-06.

[16] "Evolution of the Dream Chaser" .

[17] Hodges, Jim (Fall 2011). "The Dream Chaser: Back to the Future" . *ASK Magazine: The NASA Source for Project Management and Engineering Excellence* (44) (Washington, DC: Academy of Program/Project & Engineering Leadership NASA). Archived from the original on 2014-01-06. Retrieved 2013-11-16.

[18] Klingler, Dave (2012-09-06). "50 years to orbit: Dream Chaser's crazy Cold War backstory: The reusable mini-spaceplane is back from the dead—again—and prepping for space" . *ars Technical* (Boston: Conde Nast). p. 2. Archived from the original on 2014-01-06. Retrieved 2012-09-07.

[19] "Dream Chaser Builds on Decades of Experience" .

[20] Sirangelo, Mark (2006-09-26). "SpaceDev Announces Founder James Benson Steps Down as Chairman and CTO; Benson Starts Independent Space Company to Market SpaceDev's Dream Chaser" (Press release). Poway, California: SpaceDev. Archived from the original on 2007-06-17.

[21] "SpaceDev and United Launch Alliance to Explore Launching the Dream Chaser(TM) Space Vehicle on an Atlas V Launch Vehicle" (Press release). Poway, California: SpaceDev. Market Wire. 2007-04-10. Archived from the original on 2014-01-06. Retrieved 2014-01-06.

[22] "NASA Signs Commercial Space Transportation Agreements". NASA. 18 June 2007. Retrieved 16 August 2011.

[23] Fikes, Bradley J. (2008-10-21). "SpaceDev agrees to be acquired". *U-T San Diego*. Archived from the original on 2014-01-06. Retrieved 2014-01-06.

[24] "SNC receives largest award of NASA's CCDev Competitive Contract". SNC. 1 February 2010.

[25] "Text of Space Act Agreement" (PDF).

[26] "Commercial Crew: Sierra Nevada". NASA. Retrieved 25 July 2012.

[27] "Dream Chaser Model Drops in at NASA Dryden". NASA.

[28] "Sierra Nevada Space Systems Adds Key Former Nasa Leaders to Its Dream Chaser Orbital Space Vehicle Team" (Press release). Louisville, Colorado. 2011-07-05. Archived from the original on 2014-01-06. Retrieved 2014-01-06.

[29] Dean, James. "NASA awards $270 million for commercial crew efforts". space.com, 18 April 2011.

[30] "Sierra nevada corporation's dream chaser space system passes preliminary design review". *SNC Release*. 6 June 2012.

[31] "Sierra nevada corporation begins flight test program of the dream chaser orbital crew vehicle". *SNC Release*. 2012-05-30.

[32] "Boeing, SpaceX and Sierra Nevada Win CCiCAP Awards". spacenews.com, 3 August 2012.

[33] Messier, Doug (2013-12-16). "German Space Agency Funds Study on Uses of Sierra Nevada's Dream Chaser". *Parabolic Arc* (Mojave, California). Archived from the original on 2014-01-06. Retrieved 2013-12-16.

[34] Clark, Stephen (2014-01-08). "Europe eyes cooperation on Dream Chaser space plane". *Spaceflight Now*. Archived from the original on 2014-01-09. Retrieved 2014-01-09.

[35] Dream Chaser mini-shuttle given 2016 launch date. **BBC News**. (24 January 2014)

[36] Schierholz, Stephanie; Martin, Stephanie (16 September 2014). "NASA Chooses American Companies to Transport U.S. Astronauts to International Space Station". *www.nasa.gov*. Retrieved 17 September 2014.

[37] Norris, Guy. "Why NASA Rejected Sierra Nevada's Commercial Crew Vehicle" *Aviation Week & Space Technology*, 11 October 2014. Accessed: 13 October 2014. Archived on 13 October 2014

[38] Keeney, Laura (2014-10-03). "So Sierra Nevada protested NASA space-taxi contract, but what's next?". *Denver Post*. Retrieved 2014-10-05.

[39] "NASA Chooses American Companies to Transport U.S. Astronauts to International Space Station". NASA. 16 September 2014. Retrieved 18 September 2014.

[40] "Sierra Nevada and Stratolaunch Team Up on Dream Chaser Space Plane". NBC News. 1 October 2014. Retrieved 1 October 2014.

[41] "Sierra Nevada Corporation Introduces Dream Chaser Global Project Spaceflight Program Sept. 30". SpaceRef. 29 September 2014. Retrieved 29 September 2014.

[42] "Sierra Nevada completes Dream Chaser's milestone 15a for prior phase of Commercial Crew". Spaceflight Insider. 3 December 2014. Retrieved 3 December 2014.

[43] "SNC Tests Dream Chaser Propulsion System". NASA Blog blogs.nasa.gov. 2 December 2014. Retrieved 2 December 2014.

[44] Messier, Doug (23 December 2014). "Sierra Nevada Alleges Boeing Benefitted From Commercial Crew Criteria Changes". *Parabolic Arc*. Retrieved 25 December 2014.

[45] Davenport, Christian. "GAO denies Sierra Nevada's legal challenge to NASA space contract". *Washington Post*. Retrieved 5 January 2015.

[46] "Sierra Nevada Corporation to Introduce Dream Chaser® Global Project Spaceflight Program Sept. 30". SNC. 29 September 2014.

[47] "Sierra Nevada Corporation to Present Progress on Evaluating Dream Chaser Landing at Public Use Airports". WFXS FOX55 WAUSAU. 5 November 2014. Retrieved 1 November 2014.

[48] "Challenges and Opportunities Related to Landing the Dream Chaser® Commercial Reusable Space Vehicle at a Public-Use Airport". ERAU Scholarly Commons. 5 November 2014. Retrieved 1 November 2014.

[49] Gebhardt, Chris (2014-11-26). "SNC, Stratolaunch expand on proposed Dream Chaser flights". *NASASpaceFlight.com*. Retrieved 2014-11-27.

[50] Completion of the initial DC4EU study (2015-03-02). "http://www.sncorp.com/AboutUs/NewsDetails/749"

[51] de Selding, Peter B. (17 April 2015). "DLR Renews Cooperation with SNC on Dream Chaser". *Space News*. Retrieved 2015-04-21.

[52] Jeff Foust (13 March 2015). "Lockheed Martin Pitches Reusable Tug for Space Station Resupply". Space News.

[53] Christian Davenport (13 February 2015). "Grounded: Left behind in the contracting race to restore Americans to space". The Washington Post.

[54] Dan Leone (24 January 2015). "Weather Sat, CRS-2 Top U.S. Civil Space Procurement Agenda for 2015". Space-News.com.

[55] Jeff Foust (17 March 2015). "Sierra Nevada Hopes Dream Chaser Finds "Sweet Spot" of ISS Cargo Competition". Space News.

[56] Voss, Ed (2011-06-24). "Sierra Nevada Space Systems Successfully Completes Two Major Nasa Human Space Flight Development Milestones" (Press release). Poway, California: Sierra Nevada Space Systems. Archived from the original on 2011-10-11.

[57] "Commercial Crew Development Industry Partners Continue Progress" (PDF). Retrieved 29 August 2012.

[58] "Next Spacex Cargo Demo Flight" (PDF). Retrieved 29 August 2012.

[59] "Sierra Nevada Corporation's Space Systems Delivers the Dream Chaser® First Flight Test Vehicle Structure, Completing a Major Milestone for NASA's Commercial Crew Program" (Press release).

[60] "Sierra Nevada News & Press Releases". Sncorp.com. 24 April 2012. Retrieved 7 May 2012.

[61] "SNC and NASA Langley announce Five Years of Partnership".

[62] "Sierra Nevada Corporation Announces Successful Completion of Dream Chaser Cew Vehicle Nose Gear Landing Test". SNC. Retrieved 15 August 2012.

[63] "Sierra Nevada Completes Dream Chaser Safety Review". 10 May 2013. Retrieved 15 May 2013.

[64] Rosenberg, Zach (2013-01-30). "Lockheed to build second Dream Chaser airframe for Sierra Nevada". *Flightglobal* (Sutton, Surrey, UK). Archived from the original on 2014-01-07. Retrieved 2013-03-25.

[65] Dean, James (2013-01-30). "Sierra Nevada's Dream Chaser will get Lockheed Martin's help". *Florida Today* (Melbourne, Florida). Archived from the original on 2014-01-07. Retrieved 2013-02-11.

[66] Bolden, Jay (13 March 2013). "NASA Astronaut Lee Archambault Leaving Agency". NASA. Retrieved 25 March 2013.

[67] "NASA Astronaut Lee Archambault Joins Sierra Nevada as Test Pilot". 13 March 2013. Retrieved 25 March 2013.

[68] "SNC's Hybrid Rocket Engines Power SpaceShipTwo on its First Powered Flight Test". 29 April 2013. Retrieved 15 May 2013.

[69] First Piece of Private Dream Chaser Space Plane Unveiled. (6 August 2014)

[70] "Dream Chaser preps for 2nd free-flight test and first orbital test". Space Daily. 9 October 2015.

[71] Bergin, Chris (2013-05-12). "Dream Chaser ETA heads to Dryden for drop tests". *NasaSpaceFlight.com*. Archived from the original on 2014-01-06. Retrieved 2013-05-14.

[72] "Sierra Nevada's Dream Chaser spacecraft tested at Broomfield airport". dailycamera.com. 29 May 2012. Retrieved 29 May 2012.

[73] Lindsey, Clark (2013-05-14). "More about SNC preparations for drop tests of Dream Chaser prototype". *NewSpace Watch*. Retrieved 2013-05-14. (subscription required (help)).

[74] Wall, Mike (2013-08-26). "Dream Chaser space plane dangles from helicopter for second flight test". *NBC News* (New York). Archived from the original on 2014-01-06. Retrieved 2014-01-06.

[75] Bergin, Chris (2013-10-26). "Dream Chaser suffers landing gear failure after first flight". *NASA Spaceflight*. Archived from the original on 2014-01-06. Retrieved 2014-01-06.

[76] Harwood, William (2013-10-29). "Sierra Nevada investigates Dream Chaser landing mishap". *CBS News* (New York). Archived from the original on 2014-01-06. Retrieved 2014-01-06.

[77] David, Leonard (2013-10-29). "Private Dream Chaser Space Plane Skids Off Runway After Milestone Test Flight (Video)". *Space.com* (New York). Archived from the original on 2014-01-06. Retrieved 2014-01-06.

[78] Dream Chaser still fighting for her place in space *Nasaspaceflight*. (06 October 2015)

12.7   External links

- Sierra Nevada Corporation Space Systems web site

- SNC Space Systems' Dream Chaser page

- SpaceDev web site

- United Launch Alliance web site

- CG rendering of Dream Chaser servicing ISS

- Video animation —SpaceDev International Lunar Observatory Human Servicing Mission concept

Chapter 13

Dream Chaser

This article is about the Sierra Nevada Corporation (SNC) spaceplane. For other uses, see Dream Chasers.

The **Dream Chaser** is an American reusable crewed suborbital and orbital[6] lifting-body spaceplane being developed by Sierra Nevada Corporation (SNC) Space Systems. The Dream Chaser is designed to carry up to seven people to and from low Earth orbit. The vehicle would launch vertically on an Atlas V rocket and land horizontally automatically on conventional runways.[1]

13.1 Design

Dream Chaser is a reusable composite spacecraft designed to carry from two to seven people and/or cargo to orbital destinations such as the International Space Station.[7] It would have a built-in launch escape system[8] and could fly autonomously if needed.[4] It could use any suitable launch vehicle but is planned to be launched on a human-rated Atlas V 412 rocket.[4][9] The vehicle would be able to return from space by gliding (typically experiencing less than 1.5 g on re-entry) and landing on any airport runway that handles commercial air traffic.[10][11] Its reaction control system thrusters burn ethanol-based fuel,[4][10] which is not an explosively volatile material, allowing the Dream Chaser to be handled immediately after landing, unlike the Space Shuttle.[4] Its thermal protection system (TPS) is an ablative tile created by NASA's Ames Research Center that would be replaced as a large group rather than tile by tile, and would only need to be replaced after several flights.[4]

13.1.1 Rocket engines

On-orbit propulsion of the Dream Chaser is provided by twin hybrid rocket engines. The hybrid rocket motors are fueled with hydroxyl-terminated polybutadiene (HTPB) and nitrous oxide, or more simply put, "rubber and laughing gas".[8] These two substances are both non-toxic and easily stored, making them safer than liquid rocket fuels. Unlike solid rockets, Dream Chaser's hybrid fuel system would allow the motor to stop and start repeatedly, and be throttleable. SNC Space Systems was also developing a similar hybrid rocket, RocketMotor Two, for Virgin Galactic's SpaceShipTwo,[8] as a subcontractor to Scaled Composites. In May 2014 their involvement in the program ended, after Virgin Galactic elected to replace SNC's version of RocketMotorTwo, powered by HTBD rubber fuel, with its own internally developed hybrid motor using a polyamide plastic fuel, while continuing to use the same nitrous oxide oxidizer.[12]

13.1.2 Engine testing

Sierra Nevada completed an initial test phase on the Dream Chaser rocket engine in 2010, under the CCDev1 program, including three successful test firings on a single hybrid motor in a single day.[13]

A second phase of testing began in June 2013, with a motor firing and ignition test in order to validate the newly modified test stand, as a start to the Commercial Crew Integrated Capability (CCiCap) contract test phase.

13.1.3 Wind tunnel testing

In 2014, Sierra Nevada completed its wind tunnel testing as part of its CCiCAP Milestone 8. The Wind tunnel testing involved analyzing the flight dynamics characteristics that the vehicle will experience during orbital ascent and re-entry. Wind tunnel testing was also completed for the Dream Chaser Atlas V integrated launch system. These tests were completed at NASA Ames Research Center at Moffett Field, California, CALSPAN Transonic Wind Tunnel in New York, and at NASA Langley Research Center Unitary Plan Wind Tunnel in Hampton, Virginia.[*][14]

13.2 History

The historical antecedents of the Dream Chaser go back over 50 years in the US; with the 1957 X-20 Dyna-Soar concept and the 1966 Northrop M2-F2 and Martin X-23 PRIME lifting bodies.[*][15][*][16] Its design is derived from NASA's 1990 HL-20 lifting body design which was itself similar to the 1980s Soviet BOR-4, which in turn was considered by NASA engineers as influenced by the late 1960s HL-10,[*][17] and the Soviet Mikoyan-Gurevich MiG-105 military spaceplane concept,[*][18] a spaceplane studied as a means to develop a Soviet counterpart to the US's X-20 Dyna-Soar.[*][19]

The name "Dream Chaser" has been used for two separate space vehicles. One, planned to be an orbital vehicle based on the NASA HL-20, originated at SpaceDev when Jim Benson was still there. The second, a suborbital vehicle, was the result of Jim Benson having reused the name when he formed the Benson Space Company for the purposes of space tourism.[*][18]

13.2.1 SpaceDev Dream Chaser proposal

COTS

The Dream Chaser was publicly announced on 20 September 2004 as a candidate for NASA's Vision for Space Exploration and later Commercial Orbital Transportation Services Program (COTS).

When the Dream Chaser was not selected under Phase 1 of the COTS Program, SpaceDev founder Jim Benson stepped down as Chairman of SpaceDev and started Benson Space Company to pursue the development of the Dream Chaser.[*][20] In April 2007, SpaceDev announced that it had partnered with the United Launch Alliance to pursue the possibility of using the Atlas V booster rocket as the Dream Chaser's launch vehicle.[*][21] In June 2007, SpaceDev signed a Space Act agreement with NASA.[*][22]

CCDev

About two weeks after Benson's 10 October 2008 death, SpaceDev agreed to be acquired by Sierra Nevada Corporation, a privately owned company operated by Fatih Ozmen and Eren Ozmen, on 21 October 2008 for $38 million.[*][23] On 1 February 2010, Sierra Nevada Corporation was awarded $20 million in seed money under NASA's Commercial Crew Development (CCDev) phase 1 program for the development of the Dream Chaser.[*][24][*][25] Of the $50 million awarded by the CCDev program, Dream Chaser's award represented the largest share of the funds. SNC completed the four planned milestones on time which included program implementation plans, manufacturing readiness capability, hybrid rocket test fires, and the preliminary structure design.[*][26] Further initial Dream Chaser tests included the drop test of a 15% scaled version at the NASA Dryden Flight Research Center.[*][27] The 5-foot-long (1.52 meters) model was dropped from 14,000 feet (4,300 m) to test flight stability and collect aerodynamic data for flight control surfaces.[*][27]

For the CCDev phase 2 solicitation by NASA in October 2010, Sierra Nevada proposed extensions of Dream Chaser spaceplane technology. According to head of Sierra Nevada Space Systems Mark Sirangelo, the cost of completing the Dream Chaser should be less than $1 billion.[*][2][*][28]

On 18 April 2011, NASA awarded nearly $270 million in funding for CCDev 2, including $80 million to Sierra Nevada Corporation for Dream Chaser.[*][29] Since then, nearly a dozen further milestones have been completed under that Space Act Agreement. Some of these milestones included testing of the airfoil fin shape, integrated flight software and hardware, landing gear, and a full-scale captive carry flight test.[*][30][*][31]

An artist's impression of the X-20 Dyna-Soar being launched using a Titan booster, with large fins added to the Titan's first stage

CCiCap

On 3 August 2012, NASA announced the award of $212.5 million to Sierra Nevada Corporation to continue work on the Dream Chaser under the Commercial Crew Integrated Capability (CCiCAP) Program.[*][32]

In December 2013, the German Aerospace Center (DLR) announced a funded study to investigate ways in which Europe might take advantage of the Dream Chaser crewed spaceplane technology. Named the DC4EU (Dream

Chaser for European Utilization), the project will study using it for sending crews and cargo to the ISS and on missions not involving the ISS, particularly in orbits of substantially greater altitude than the ISS can reach.[*][33]

In January 2014, the European Space Agency (ESA) agreed to be a partner on the DC4EU project, and will also investigate whether the Dream Chaser can use ESA avionics and docking mechanisms. ESA will also study launching options for the "Europeanized" Dream Chaser, particularly whether it can be launched within the Ariane 5's large aerodynamic cargo fairing – or, like the Atlas V, without it. In order to fit within the fairing, the Dream Chaser's wing length will have to be reduced slightly, which is thought to be easier than going through a full aerodynamic test program to evaluate and prove it along with the Ariane for flight without the fairing.[*][34]

In late January 2014, it was announced that the Dream Chaser orbital test vehicle was under contract to be launched on an initial orbital test flight, using an Atlas V rocket, from Kennedy Space Center in November 2016. This is a privately arranged commercial agreement, and is funded directly by Sierra Nevada and is not a part of any existing NASA contract.[*][35]

2014 CCtCap non-selection by NASA

After being involved with the NASA Commercial Crew Development program since 2009—and being selected as one of the contract award recipients in each prior phase of the program—NASA did not select the Dream Chaser for the next phase of the Commercial Crew Program announced 16 September 2014[*][36] due to lack of maturity.[*][37] Sierra Nevada filed a protest to the US Government Accountability Office (GAO) on 26 September. The GAO is investigating and will respond after a process that could take up to 100 days. Boeing and SpaceX were asked by NASA to "stop work" on the crewed spacecraft during the protest resolution.[*][38] However, on 22 October 2014, a Federal Judge ruled that NASA could proceed with contracts with Boeing and SpaceX to develop their "space taxis", while the GAO continues to consider Sierra Nevada's protest of NASA's original decision.

Two weeks after losing the Commercial Crew Transportation Capability (CCtCap) competition to SpaceX and Boeing on 16 September 2014,[*][39] Sierra Nevada Corporation announced it has designed a launch system that combines a scale version of the company's Dream Chaser space plane with the Stratolaunch Systems air launch system.[*][40] Earlier the same week, Sierra Nevada introduced new spaceflight opportunities to the world - coined the Dream Chaser Global Project"- which would provide customized access to low Earth orbit to global customers.[*][41]

Despite not being selected to continue forward under NASA's Commercial Crew transportation Capability (CCtCap) phase of the effort to send crews to orbit via private companies, SNC is still completing milestones under earlier phases of the CCP.[*][42] On 2 December 2014 SNC announced that it completed NASA's CCiCap Milestone 5a related to propulsion risk reduction for the Dream Chaser space system.[*][43]

By late December, details had emerged that "a high-ranking agency official"—"William Gerstenmaier, the agency's top human exploration official and the one who made the final decision"—"opted to rank Boeing's proposal higher than a previous panel of agency procurement experts." More specifically, Sierra Nevada asserted in their filings with the GAO that Gerstenmaier may have "overstepped his authority by unilaterally changing the scoring criteria."[*][44]

On 5 January 2015, the GAO denied Sierra Nevada's CCtCap challenge, stating that NASA made the proper decision when it decided to award Boeing $4.2 billion and SpaceX $2.6 billion to develop their vehicles. Ralph White, the GAO's managing associate counsel, announced that NASA "recognized Boeing's higher price but also considered Boeing's proposal to be the strongest of all three proposals in terms of technical approach, management approach and past performance, and to offer the crew transportation system with most utility and highest value to the government." Furthermore, the agency found "several favorable features" in SNC's proposal "but ultimately concluded that SpaceX's lower price made it a better value."[*][45]

13.2.2   Dream Chaser Global Project

In September 2014, SNC announced that it would, with global partners, use the Dream Chaser as the baseline spacecraft for orbital access for a variety of programs, specializing the craft as needed.[*][46]

On 5 November 2014 during the Space Traffic Management Conference at Embry–Riddle Aeronautical University, SNC's Space Systems team presented the challenges and opportunities related to landing the Dream Chaser spacecraft at public-use airports.[*][47] According to the presentation, "Unlike the Space Shuttle, the Dream Chaser does not require any unique landing aids or specialized equipment as it uses all non-toxic propellants and industry standard subsystems."[*][48]

Stratolaunch+DreamChaser

In late November 2014, Vulcan Aerospace released the results of the SNC/Stratolaunch space transportation architecture, which indicated that the reduced-size Dream Chaser in conjunction with the Stratolaunch-based launch system mission capabilities. The system would have an outbound range of 1,900 kilometers; 1,200 miles (1,000 nmi) away from the airport where the aircraft departed. The launch vehicle would be a modified air-launched Orbital Sciences rocket that is approximately 37 m (120 ft) in length. The Dream Chaser payload would be a 75-percent sized version of the vehicle previously proposed to NASA—while maintaining the relative outer mold line—6.9 m (22.5 ft) in length with a wingspan of 5.5 m (18.2 ft), which could carry 2 to 3 crewmembers plus a variety of scientific and research payloads.[*][49]

Dream Chaser for European Utilization

See also: Hermes (spacecraft) and Intermediate eXperimental Vehicle

In 2013 SNC and OHB entered into an agreement to study the feasibility of using SNC's Dream Chaser spacecraft for a variety of missions. The DC4EU study thoroughly reviewed applications for the Dream Chaser including crewed and uncrewed flights to low-Earth orbit (LEO) for missions such as microgravity science, satellite servicing and active debris removal (ADR).

On 3 February 2015, the Sierra Nevada Corporation's (SNC) Space Systems and OHB System AG (OHB) in Germany announced the completion of the initial Dream Chaser for European Utilization (DC4EU) study.[*][50]

"The inherent design advantages of the Dream Chaser reusable lifting body spacecraft make it an ideal vehicle for a broad range of space applications," said Dr. Fritz Merkle, member of the Executive Management Board of OHB AG. "We partnered with SNC to study how the design of the Dream Chaser can be used to advance European interests in space. The study results confirm the viability of using the spacecraft for microgravity science and ADR. DC4EU can benefit the entire international space community with its unique capabilities. We look forward to further maturing our design with SNC as we expand our partnership." The cooperation was renewed in April 2015 for additional two years.[*][51]

13.2.3 Dream Chaser Cargo System

The cargo variant of the SNC Dream Chaser is called the *Dream Chaser Cargo System*. Featuring an expendable cargo portion, containing solar panels, the cargo version of the spacecraft will be capable of taking 5000 kg back to Earth, undergoing re-entry forces of 1.5G. It has been proposed for the Phase II program for cargo resupply of the International Space Station.[*][52]

CRS2

In December 2014, Sierra Nevada proposed *Dream Chaser* for CRS-2 consideration.[*][53] It is in competition with the existing CRS-1 contract holders SpaceX Dragon capsule and Orbital Sciences Cygnus capsule, as well as fellow CCDev competitor Boeing CST-100.[*][54]

To meet CRS2 guidelines, the cargo Dream Chaser will feature foldable wings, to fit within a 5m cargo fairing, unlike the passenger Dream Chaser, which did not use a cargo fairing. The ability to fit in a cargo fairing allows launches from Ariane 5 as well as Atlas 5 rocket launcher vehicles. To expand the cargo uplift capacity, an expendable cargo module is affixed aft, which will not support downlift, but can be used for disposal of up to 3250 kg of trash. Total uplift is planned for 5000 kg pressurized, 500 kg unpressurized, with downlift of 1750 kg wholly within the spaceplane.[*][55]

13.3 Development progress

On 24 June 2011 SNC announced it had achieved two critical milestones for NASA's CCDev program. The first was a Systems Requirement Review (SRR), where SNC validated their requirements based on NASA's draft Commercial Crew Program Requirements. The SRR was successfully completed on 1 June 2011 with participation from NASA and SNC industry partners. The second milestone was a review of the improved airfoil fin shape for Dream Chaser

The completed craft on the day of its initial captive carry test

used to aid its control through the atmosphere. Testing in a wind tunnel and computational fluid dynamics analyses allowed the fin selection to pass the NASA milestone.[56]

As of October 2011, Sierra Nevada Corp had completed four of the 13 milestones set out in the CCDev Agreement.[57] The most recent milestones accomplished include: a System Requirements Review, a new cockpit simulator, finalizing the tip fin airfoil design and most recently,[58] a Vehicle Avionics Integration Laboratory (VAIL), which will be used to test Dream Chaser computers and electronics in simulated space mission scenarios.[57]

By February 2012, Sierra Nevada Corporation stated that it had completed the assembly and delivery of the primary structure of the first Dream Chaser flight test vehicle. With this, SNC completed all 11 of its CCDev milestones that were scheduled up to that point. SNC stated in a press release that it was "...on time and on budget." [59]

On 24 April 2012 Sierra Nevada Corporation announced the successful completion of wind tunnel testing of a scale model of the Dream Chaser vehicle.[60]

On 12 June 2012 SNC announced the commemoration of its fifth year as a NASA Langley partner in the design and development of Dream Chaser.[61] Together with ULA, the NASA/SNC team performed buffet tests on the Dream Chaser and Atlas V stack. To date, the Langley/SNC team has worked on aerodynamic and aerothermal analysis of Dream Chaser, as well as guidance, navigation and control systems.[61]

On 11 July 2012 SNC announced that they successfully completed testing of the nose landing gear for Dream Chaser.[62] This milestone evaluated the impact to the landing gear during simulated approach and landing tests as well as the impact of future orbital flights. The main landing gear was tested in a similar way in February 2012. The nose gear landing test was the last milestone to be completed before the free flight approach and landing tests scheduled for later in 2012.[62]

In August 2012, SNC completed CCiCap Milestone 1, or the 'Program Implementation Plan Review'. This included creating a plan for implementing design, development, testing, and evaluation activities through the duration of CCiCap funding.[63]

By October 2012 the "Integrated System Baseline Review", or CCiCap Milestone 2, had been completed. This review demonstrated the maturity of the Dream Chaser Space System as well as the integration and support of the

Atlas V launch vehicle, mission systems, and ground systems.[*][63]

On 30 January 2013 SNC announced a new partnership with Lockheed Martin. Under the agreement, SNC will pay Lockheed Martin $10 million to build the second airframe at its Michoud facility in New Orleans, Louisiana. This second airframe is slated to be the first orbital test vehicle, with orbital flight testing planned to begin within the next two years.[*][64] In January 2014, SNC announced they had signed a launch contract to fly the first orbital test vehicle on a robotically controlled orbital test flight in November 2016.[*][35]

In January 2013, Sierra Nevada also announced that the second captive carry and first unpowered drop test of Dream Chaser would take place at Edwards Air Force Base, California in March 2013. The spaceplane release would occur at 3,700 metres (12,000 ft) altitude and would be followed by an autonomous robotic landing.[*][64][*][65]

On 13 March 2013, NASA announced that former space shuttle commander Lee Archambault was leaving the agency in order to join SNC. Archambault, a former combat pilot and 15-year NASA veteran who flew on Atlantis and Discovery, will work on the Dream Chaser program as a systems engineer and test pilot.[*][66][*][67]

On 29 April 2013, Virgin Galactic's SpaceShipTwo sub-orbital vehicle was propelled on its first ever powered flight by SNC's Hybrid Rocket Motor. SNC manufactures the main oxidizer valve and the hybrid rocket motor, plus the nitrous oxide dump and pressurization system control valves. The hybrid rocket motor and oxidizer valve system are manufactured at an SNC facility in Poway, California, where motors for both Space Ship Two and Dream Chaser are produced.[*][68]

On 1 August 2014, the first completed piece of the orbital test vehicle's composite airframe was unveiled at the Lockheed Martin Michoud Assembly Facility in Louisiana.[*][69]

In October 2015, the thermal protection system was installed on the ETA for the next phase of atmospheric flight testing. The orbital cabin assembly of the FTA orbital test vehicle was also completed by contractor Lockheed Martin.[*][70]

13.3.1   Flight test program

In May 2013, The Dream Chaser Engineering Test Article (ETA) was shipped to the Dryden Flight Research Center in California for a series of ground tests and aerodynamic flight tests.[*][71] This move to Dryden came about a year after a captive carry test that was conducted near the Rocky Mountain Metropolitan Airport on 29 May 2012. During that test, an Erickson Skycrane was used to lift the Dream Chaser to better determine its aerodynamic properties.[*][72] "The testing at Dryden will include tow, captive-carry and free-flight tests of the Dream Chaser. A truck will tow the vehicle down a runway to validate performance of the nose strut, brakes and tires. The captive-carry flights will further examine the loads the vehicle will encounter during flight and test the performance and flutter of the vehicle up to release from an Erickson Skycrane helicopter. The free-flight tests are designed to validate the Dream Chaser's aerodynamics as well as test the flight control surfaces to verify flight characteristics for approach, flare and landing." [*][73] A second captive carry flight test was completed on 22 August 2013.[*][74]

On 26 October 2013, the first free-flight occurred. The test vehicle was released from the "skycrane" helicopter, and flew the correct flightpath to touchdown less than a minute later. Just prior to landing, the left main landing gear failed to deploy resulting in a crash landing.[*][75] In a press teleconference a short while later, Mark Sirangelo, corporate vice president of Sierra Nevada, told reporters that the view of the ETA was obscured by the dust as it skidded off the runway, but that the vehicle was found upright, with the crew compartment intact, and all systems inside still in working order. Sierra Nevada corporation engineers do not believe that the ETA flipped over.[*][76][*][77]

The first two Dream Chasers —the ETA and the Flight Test Article (FTA) —have been given internal and external names, with some sources reporting that the ETA will be named **Eagle**.,[*][71] while the FTA was originally named **Ascalon** before being changed to **Ascension**.[*][78]

An initial orbital test flight of the Dream Chaser orbital test vehicle is planned for 1 November 2016, launching on an Atlas V rocket from Kennedy Space Center.[*][35]

13.4   Technology partners

The following organizations have been named as technology partners:

- Boeing Phantom Works – construction of some early test articles[*][11]

Dream Chaser model being tested at NASA Langley

- Charles Stark Draper Laboratory – Guidance, Navigation and Control*[11]

- Aerojet – reaction control system technology*[11]

- University of Colorado – human-rating*[11]

- AdamWorks – composites*[11]

- MDA – systems engineering*[11]

- Lockheed Martin – airframe construction and human rating of the spaceplane*[64]*[65]

13.5   See also

- BOR-4

- BOR-5

- Mikoyan-Gurevich MiG-105

- Boeing X-20 Dyna-Soar

- Boeing X-37

- HL-20 Personnel Launch System

- Commercial Crew Development

- Private spaceflight

- Skylon

- Spaceplane

- Dragon V2, a crew-carrying spacecraft being developed by SpaceX

- CST-100, a crew-carrying spacecraft being developed by Boeing

- Orion (spacecraft), a spacecraft being built for NASA by Lockheed Martin

13.6 References

[1] "Dream Chaser Model Drops in at NASA Dryden" (Press release). Dryden Flight Research Center: NASA. 2010-12-17. Archived from the original on 2014-01-07. Retrieved 2012-08-29.

[2] Chang, Kenneth (2011-02-01). "Businesses Take Flight, With Help From NASA" . *New York Times*. p. D1. Archived from the original on 2014-01-06. Retrieved 2012-08-29.

[3] Wade, Mark (2014). "Dream Chaser" . Encyclopedia Astronautix. Archived from the original on 2014-01-06. Retrieved 2012-08-29.

[4] Sirangelo, Mark (August 2011). "NewSpace 2011: Sierra Nevada Corporation" . Spacevidcast. Retrieved 2011-08-16. Sirangelo, Mark (24 August 2014). "Flight Plans and Crews for Commercial Dream Chaser's First Flights: One-on-One Interview With SNC VP Mark Sirangelo (Part 3)". AmericaSpace.

[5] Bayt, Rob (2011-07-26), *Commercial Crew Program: Key Driving Requirements Walkthrough* (Powerpoint), NASA, retrieved 2011-07-27

[6] Leonard, David (2011-02-07). "Private Spaceflight Innovators Attract NASA's Attention" . New York. Archived from the original on 2014-01-06. Retrieved 2014-01-06. Dream Chaser will become a fully capable suborbital vehicle on the way to reaching orbital capability.

[7] "NASA Deputy Administrator Lori Garver touts Colorado's role" . Youtube.com. 2011-02-05. Retrieved 2012-08-29.

[8] Klingler, Dave (2012-09-06). "50 years to orbit: Dream Chaser's crazy Cold War backstory: The reusable mini-spaceplane is back from the dead—again—and prepping for space" . *ars Technical* (Boston: Conde Nast). p. 3. Archived from the original on 2014-01-06. Retrieved 2012-09-07.

[9] "Moving Forward: Commercial Crew Development Building the Next Era in Spaceflight" (PDF). *Rendezvous: Where today meets tomorrow* **4** (2): 10–15. Summer 2010. Archived from the original on 2014-01-06. Retrieved 2014-01-06.

[10] "The Space Show : Mark Sirangelo interview" . David Livingston. 2012-01-04. Retrieved 2012-01-07.

[11] Frank Morring, Jr (2010-02-19). "Sierra Nevada Building On NASA Design" . Aviation Week.

[12] Doug Messier (24 May 2014). "Virgin Galactic Hails RocketMotorTwo Milestone" . ParabolicArc.

[13] "Sierra Nevada Corporation Begins Dream Chaser Main Hybrid Rocket Motor Testing" . *NewSpace Watch*. 2013-06-06. Retrieved 2013-06-11. (subscription required (help)).

[14] "Dream Chaser passes Wind Tunnel tests for CCiCap Milestone" . *NASASpaceflight.com*. 2014-05-19. Retrieved 2015-01-02.

[15] Eddy, Max (2012-04-02). "How the United States Will Return to Space" . *Geekosystem* (New York). Retrieved 2014-01-06.

[16] "Evolution of the Dream Chaser" .

[17] Hodges, Jim (Fall 2011). "The Dream Chaser: Back to the Future" . *ASK Magazine: The NASA Source for Project Management and Engineering Excellence* (44) (Washington, DC: Academy of Program/Project & Engineering Leadership NASA). Archived from the original on 2014-01-06. Retrieved 2013-11-16.

[18] Klingler, Dave (2012-09-06). "50 years to orbit: Dream Chaser's crazy Cold War backstory: The reusable mini-spaceplane is back from the dead—again—and prepping for space" . *ars Technical* (Boston: Conde Nast). p. 2. Archived from the original on 2014-01-06. Retrieved 2012-09-07.

[19] "Dream Chaser Builds on Decades of Experience" .

[20] Sirangelo, Mark (2006-09-26). "SpaceDev Announces Founder James Benson Steps Down as Chairman and CTO; Benson Starts Independent Space Company to Market SpaceDev's Dream Chaser" (Press release). Poway, California: SpaceDev. Archived from the original on 2007-06-17.

[21] "SpaceDev and United Launch Alliance to Explore Launching the Dream Chaser(TM) Space Vehicle on an Atlas V Launch Vehicle" (Press release). Poway, California: SpaceDev. Market Wire. 2007-04-10. Archived from the original on 2014-01-06. Retrieved 2014-01-06.

[22] "NASA Signs Commercial Space Transportation Agreements". NASA. 18 June 2007. Retrieved 16 August 2011.

[23] Fikes, Bradley J. (2008-10-21). "SpaceDev agrees to be acquired". *U-T San Diego*. Archived from the original on 2014-01-06. Retrieved 2014-01-06.

[24] "SNC receives largest award of NASA's CCDev Competitive Contract". SNC. 1 February 2010.

[25] "Text of Space Act Agreement" (PDF).

[26] "Commercial Crew: Sierra Nevada". NASA. Retrieved 25 July 2012.

[27] "Dream Chaser Model Drops in at NASA Dryden". NASA.

[28] "Sierra Nevada Space Systems Adds Key Former Nasa Leaders to Its Dream Chaser Orbital Space Vehicle Team" (Press release). Louisville, Colorado. 2011-07-05. Archived from the original on 2014-01-06. Retrieved 2014-01-06.

[29] Dean, James. "NASA awards $270 million for commercial crew efforts". space.com, 18 April 2011.

[30] "Sierra nevada corporation's dream chaser space system passes preliminary design review". *SNC Release*. 6 June 2012.

[31] "Sierra nevada corporation begins flight test program of the dream chaser orbital crew vehicle". *SNC Release*. 2012-05-30.

[32] "Boeing, SpaceX and Sierra Nevada Win CCiCAP Awards". spacenews.com, 3 August 2012.

[33] Messier, Doug (2013-12-16). "German Space Agency Funds Study on Uses of Sierra Nevada's Dream Chaser". *Parabolic Arc* (Mojave, California). Archived from the original on 2014-01-06. Retrieved 2013-12-16.

[34] Clark, Stephen (2014-01-08). "Europe eyes cooperation on Dream Chaser space plane". *Spaceflight Now*. Archived from the original on 2014-01-09. Retrieved 2014-01-09.

[35] Dream Chaser mini-shuttle given 2016 launch date. **BBC News**. (24 January 2014)

[36] Schierholz, Stephanie; Martin, Stephanie (16 September 2014). "NASA Chooses American Companies to Transport U.S. Astronauts to International Space Station". *www.nasa.gov*. Retrieved 17 September 2014.

[37] Norris, Guy. "Why NASA Rejected Sierra Nevada's Commercial Crew Vehicle" *Aviation Week & Space Technology*, 11 October 2014. Accessed: 13 October 2014. Archived on 13 October 2014

[38] Keeney, Laura (2014-10-03). "So Sierra Nevada protested NASA space-taxi contract, but what's next?". *Denver Post*. Retrieved 2014-10-05.

[39] "NASA Chooses American Companies to Transport U.S. Astronauts to International Space Station". NASA. 16 September 2014. Retrieved 18 September 2014.

[40] "Sierra Nevada and Stratolaunch Team Up on Dream Chaser Space Plane". NBC News. 1 October 2014. Retrieved 1 October 2014.

[41] "Sierra Nevada Corporation Introduces Dream Chaser Global Project Spaceflight Program Sept. 30". SpaceRef. 29 September 2014. Retrieved 29 September 2014.

[42] "Sierra Nevada completes Dream Chaser's milestone 15a for prior phase of Commercial Crew". Spaceflight Insider. 3 December 2014. Retrieved 3 December 2014.

[43] "SNC Tests Dream Chaser Propulsion System". NASA Blog blogs.nasa.gov. 2 December 2014. Retrieved 2 December 2014.

[44] Messier, Doug (23 December 2014). "Sierra Nevada Alleges Boeing Benefitted From Commercial Crew Criteria Changes". *Parabolic Arc*. Retrieved 25 December 2014.

[45] Davenport, Christian. "GAO denies Sierra Nevada's legal challenge to NASA space contract". *Washington Post*. Retrieved 5 January 2015.

[46] "Sierra Nevada Corporation to Introduce Dream Chaser® Global Project Spaceflight Program Sept. 30". SNC. 29 September 2014.

[47] "Sierra Nevada Corporation to Present Progress on Evaluating Dream Chaser Landing at Public Use Airports". WFXS FOX55 WAUSAU. 5 November 2014. Retrieved 1 November 2014.

[48] "Challenges and Opportunities Related to Landing the Dream Chaser® Commercial Reusable Space Vehicle at a Public-Use Airport". ERAU Scholarly Commons. 5 November 2014. Retrieved 1 November 2014.

[49] Gebhardt, Chris (2014-11-26). "SNC, Stratolaunch expand on proposed Dream Chaser flights". *NASASpaceFlight.com*. Retrieved 2014-11-27.

[50] Completion of the initial DC4EU study (2015-03-02). "http://www.sncorp.com/AboutUs/NewsDetails/749"

[51] de Selding, Peter B. (17 April 2015). "DLR Renews Cooperation with SNC on Dream Chaser". *Space News*. Retrieved 2015-04-21.

[52] Jeff Foust (13 March 2015). "Lockheed Martin Pitches Reusable Tug for Space Station Resupply". Space News.

[53] Christian Davenport (13 February 2015). "Grounded: Left behind in the contracting race to restore Americans to space". The Washington Post.

[54] Dan Leone (24 January 2015). "Weather Sat, CRS-2 Top U.S. Civil Space Procurement Agenda for 2015". Space-News.com.

[55] Jeff Foust (17 March 2015). "Sierra Nevada Hopes Dream Chaser Finds "Sweet Spot" of ISS Cargo Competition". Space News.

[56] Voss, Ed (2011-06-24). "Sierra Nevada Space Systems Successfully Completes Two Major Nasa Human Space Flight Development Milestones" (Press release). Poway, California: Sierra Nevada Space Systems. Archived from the original on 2011-10-11.

[57] "Commercial Crew Development Industry Partners Continue Progress" (PDF). Retrieved 29 August 2012.

[58] "Next Spacex Cargo Demo Flight" (PDF). Retrieved 29 August 2012.

[59] "Sierra Nevada Corporation's Space Systems Delivers the Dream Chaser® First Flight Test Vehicle Structure, Completing a Major Milestone for NASA's Commercial Crew Program" (Press release).

[60] "Sierra Nevada News & Press Releases". Sncorp.com. 24 April 2012. Retrieved 7 May 2012.

[61] "SNC and NASA Langley announce Five Years of Partnership".

[62] "Sierra Nevada Corporation Announces Successful Completion of Dream Chaser Cew Vehicle Nose Gear Landing Test". SNC. Retrieved 15 August 2012.

[63] "Sierra Nevada Completes Dream Chaser Safety Review". 10 May 2013. Retrieved 15 May 2013.

[64] Rosenberg, Zach (2013-01-30). "Lockheed to build second Dream Chaser airframe for Sierra Nevada". *Flightglobal* (Sutton, Surrey, UK). Archived from the original on 2014-01-07. Retrieved 2013-03-25.

[65] Dean, James (2013-01-30). "Sierra Nevada's Dream Chaser will get Lockheed Martin's help". *Florida Today* (Melbourne, Florida). Archived from the original on 2014-01-07. Retrieved 2013-02-11.

[66] Bolden, Jay (13 March 2013). "NASA Astronaut Lee Archambault Leaving Agency". NASA. Retrieved 25 March 2013.

[67] "NASA Astronaut Lee Archambault Joins Sierra Nevada as Test Pilot". 13 March 2013. Retrieved 25 March 2013.

[68] "SNC's Hybrid Rocket Engines Power SpaceShipTwo on its First Powered Flight Test". 29 April 2013. Retrieved 15 May 2013.

[69] First Piece of Private Dream Chaser Space Plane Unveiled. (6 August 2014)

[70] "Dream Chaser preps for 2nd free-flight test and first orbital test". Space Daily. 9 October 2015.

[71] Bergin, Chris (2013-05-12). "Dream Chaser ETA heads to Dryden for drop tests". *NasaSpaceFlight.com*. Archived from the original on 2014-01-06. Retrieved 2013-05-14.

[72] "Sierra Nevada's Dream Chaser spacecraft tested at Broomfield airport". dailycamera.com. 29 May 2012. Retrieved 29 May 2012.

[73] Lindsey, Clark (2013-05-14). "More about SNC preparations for drop tests of Dream Chaser prototype". *NewSpace Watch*. Retrieved 2013-05-14. (subscription required (help)).

[74] Wall, Mike (2013-08-26). "Dream Chaser space plane dangles from helicopter for second flight test". *NBC News* (New York). Archived from the original on 2014-01-06. Retrieved 2014-01-06.

[75] Bergin, Chris (2013-10-26). "Dream Chaser suffers landing gear failure after first flight". *NASA Spaceflight*. Archived from the original on 2014-01-06. Retrieved 2014-01-06.

[76] Harwood, William (2013-10-29). "Sierra Nevada investigates Dream Chaser landing mishap". *CBS News* (New York). Archived from the original on 2014-01-06. Retrieved 2014-01-06.

[77] David, Leonard (2013-10-29). "Private Dream Chaser Space Plane Skids Off Runway After Milestone Test Flight (Video)". *Space.com* (New York). Archived from the original on 2014-01-06. Retrieved 2014-01-06.

[78] Dream Chaser still fighting for her place in space *Nasaspaceflight.* (06 October 2015)

13.7   External links

- Sierra Nevada Corporation Space Systems web site

- SNC Space Systems' Dream Chaser page

- SpaceDev web site

- United Launch Alliance web site

- CG rendering of Dream Chaser servicing ISS

- Video animation —SpaceDev International Lunar Observatory Human Servicing Mission concept

Chapter 14

Falke (spacecraft)

Falke was a German program to fly a subscale model of the Space Shuttle orbiter in real conditions in order to obtain aerodynamic data in the frame of the preparation of the Hermes spaceplane. One flight test was performed in 1990.

14.1   Organization

The program was funded by the German federal Ministry of Research. The leadership of the program was DLR. The flight model was produced by the German company OHB-System

14.2   Flight model characteristics

The Space Shuttle, whose shape was used for Falke

- Length 6.8 m

- Wing span 4.4 m

- Height 2.6 m

- Mass 730 kg

The shape of Falke was the one of the Space Shuttle orbiter with a 1/5 scale factor. Falke had its own power, an autopilot and a computer to control the hydraulically actuated flight control surfaces of the spaceplane.

The sensor suite of Falke was measuring attitude, temperature, flux, pressure and acceleration.

CNES was tracking Falke by radar and telemetry.

14.3   Flight history

The only flight of Falke took place on September 6, 1990. French space agency CNES launched a stratospheric balloon from its Aire-sur-l'Adour center carrying Falke. After a 2 h 43 mn ascent, Falke was released at an altitude of 38.8 km.

At the end of the flight, a parachute was deployed at an altitude of 6 km and Falke landed in horizontal position on airbags.

14.4   Outcome

Three further flights were foreseen, but they were cancelled when the European Space Agency cancelled Hermes.

14.5   References

- *French Balloon Operational Activity - Overall view and two examples: FALKE, an aeronautic project using Stratospheric balloon, and Arctic long-duration flights during the ILAS campaign* - Pierre Faucon - CNES, Aire-sur-l'Adour - Paper ISTS 98 - j - 19V

Chapter 15

Goodyear Meteor Junior

The **Goodyear Meteor Junior** was a 1954 concept for a fully reusable spacecraft and launch system designed by Darrell C. Romick and two of his colleagues employed by Goodyear Aerospace, a subsidiary of the American Goodyear Tire and Rubber Company.[*][1] Darrell Romick originally estimated that the craft would cost about the same as an intercontinental B-52 bomber.[*][2]

15.1   Concept

The concept was introduced in 1954, before the dawn of human spaceflight, to be displayed at the annual conference of the American Rocket Society. The design called for a winged spacecraft piloted by a crew of three. The craft was to contain three stages, of which only the uppermost, containing the crew, would ultimately be propelled to orbit around Earth. All three of these stages were to contain landing gear, allowing the stages to glide to a landing to be reused in a future launch.[*][1]

The craft would have been 142 ft (43 m) long, and would weigh 500 tons.[*][2] The craft itself would have been impractically massive for the time period; however, the design ignited public interest in that it was an early concept for a reusable vehicle to transport crew and cargo to space and back,[*][1] and was the subject of an article in the December 1957 issue of Popular Science magazine.[*][2]

The original concept envisioned the craft launching from the White Sands Missile Range. At a downrange distance of 300 mi (480 km) and an altitude of 24 mi (39 km), the first stage would separate from the craft. At 1,000 mi (1,600 km) downrange, the second stage would separate at an altitude of 41 mi (66 km). After launch was completed, the third stage/crew compartment would orbit the Earth at an altitude of 500 mi (800 km) at a velocity of 16,660 miles per hour (26,810 km/h), where it would stay for approximately two months before returning to Earth.[*][2]

15.2   References

[1]   "Model, Space Shuttle, Goodyear Meteor Jr. 3-Stage Fully Reusable Concept". National Air and Space Museum. Retrieved 27 July 2011.

[2]   "Satellite with crew? By 1965, U.S. experts say". Popular Science. Retrieved 27 July 2011.

Chapter 16

Hermes (spacecraft)

Hermes was a proposed spaceplane designed by the French Centre National d'Études Spatiales (CNES) in 1975, and later by the European Space Agency. It was superficially similar to the US X-20. France proposed in January 1985 to go through with Hermes development under the auspices of the ESA.[*][1] Hermes was to have been part of a manned space flight program. It would have been launched using an Ariane 5. The project was approved in November 1987, with an initial pre-development phase from 1988 to 1990, with a green light for full-rate development depending on the outcome of the phase. The project suffered numerous delays and funding issues. It was canceled in 1992 since neither cost nor performance goals could be achieved. No Hermes shuttles were ever built.

16.1 Configuration

Hermes was to be launched on top of an Ariane 5 launcher and would consist of two parts: one part, a cone-shaped Resource Module which attached to the rear of the vehicle would be left behind before re-entry. Only the space plane itself would re-enter Earth's atmosphere and land.

In the configuration envisioned prior to project termination, Hermes would carry three astronauts and a 3,000 kg pressurized payload. The final launch weight would be 21,000 kg, which was seen as the upper limit of what an extended Ariane 5 could lift. The Hermes would be 19.00 m in length.

16.1.1 Landing sites

Possible landing sites for the shuttle would be:

- Guiana Space Centre

- Fort de France island (Martinique Aimé Césaire International Airport)

- Bermudas

- Istres (Istres-Le Tubé Air Base)

16.1.2 Typical missions

Four typical missions were proposed:

- Experiments are made on board, while in an equatorial 800 km orbit

- Flights to NASA's space station *Freedom* at a 28.5 degree orbit

- Flights to ESA's space station *Columbus* at a 60 degree orbit.

- Flights to ESA's unmanned remote sensing Polar Platform at a 98 degree 500 km orbit

16.2   Development

To ensure European autonomous access to space, in the mid-1980s the French space agency CNES pushed for a European shuttle mimicking those of the Soviet Union and the United States. The European Space Agency approved the project, dubbed Hermes, in November 1987. It was envisioned that Hermes would service a small space station, the Columbus Man-Tended Free Flyer (MTFF), built primarily by the German and Italian Industry in the frame of the Columbus program.

Hermes was to be developed in two phases.

16.2.1   *Phase 1: Study and pre-development.*

This phase was scheduled to end in 1990. Initially the plans called for a capacity to lift 6 astronauts and 4,550 kg of cargo, but after the Challenger disaster, it was felt necessary to include ejection capacity of some form to give astronauts at least a small chance of survival in case of catastrophe. Accordingly the six seats were now curtailed to only three regular ejection seats, which were chosen over an entirely ejecting crew capsule that would have given the crew an escape option at heights over 28 km. The cargo capacity was limited to 3,000 kg. Hermes would not be able to place objects into orbit as its cargo hold could not be opened; again this option was abandoned due to weight concerns.

Although Hermes was originally seen as being entirely reusable (up to 30 successive re-entries without major servicing), problems aligning the capacity of the Ariane 5 launcher with the design of Hermes itself forced it to leave behind its rear part, the Resource Module, before re-entry. A newly built resource module would then be attached to the Hermes space plane and the entire structure would be launched again.

Phase 1 was not completed until the end of 1991 and by then the political climate surrounding Hermes had changed considerably. The Iron Curtain had been lifted and the Cold War was ending. As a result, ESA decided to interject a year-long "reflection" period to examine if it still made sense for Europe to build its own space shuttle and space station or if new partners could be found to share cost and development.

Officially, Phase 1 completed at the end of 1992, after a year of reflection.

16.2.2   *Phase 2: Final development, manufacture & initial operations.*

This phase was never properly started, as ESA and the Russian Aviation and Space Agency (RKA) had agreed to cooperate on future launchers and a replacement space station for Mir. Economic concerns prevented RKA from properly participating in a future launcher program, but at this point most of ESA's crew transport capabilities had been reoriented towards a capsule type system (as opposed to the glider system that Hermes represented) which was what the joint Russian/European designs called for.

When both Russia and ESA joined up with NASA to build the International Space Station, the immediate need for a European crew transport system disappeared as both Russia and the USA had existing capabilities that did not need expansion. Accordingly ESA decided to abandon the Hermes project.

16.3   Partners

The primary companies involved in Hermes were Aérospatiale and Dassault-Breguet, both French. Arianespace built the Ariane 5 launcher and was seen as a strong candidate for running the Hermes infrastructure. No contract was ever signed, however.

16.4   Hardware and mock-ups

A full-scale mock-up (built by CNES in 1987) is stored in Paris – Le Bourget Airport for repairs and perhaps future display

- A 1/7 scale model (built by EADS) is on display at the Bordeaux-Mérignac Airport

Hermes 1:1 mock-up on display in 1992

16.5 See also

- Hopper, an unmanned, reusable satellite launch system that was developed and ultimately canceled by the ESA in the early 2000s.

- Intermediate eXperimental Vehicle, European lifting body spacecraft tested in February 2015

- E.S.S Mega, a 1991 space simulation game for MS-DOS, that feature the Hermes shuttle, along with other ESA vehicles.

- SOAR, Swiss private spacecraft derived from Hermes

16.6 References

[1] Martin Bayer, *Hermes: Learning from our mistakes*, Space Policy, Volume 11, Number 3, August 1995, pp. 171-180(10)

16.7 External links

- Hermes at Astronautix.com

- Hermes at Aerospaceguide.net

- 1:1 Mockup photos, including cockpit

- HERMES, l' avion spatial inachevé...

Chapter 17

HOPE-X

HOPE was a Japanese experimental spaceplane project designed by a partnership between NASDA and NAL (both now part of JAXA), started in the 1980s. It was positioned for most of its lifetime as one of the main Japanese contributions to the International Space Station, the other being the Japanese Experiment Module. The project was eventually cancelled in 2003, by which point test flights of a sub-scale testbed had flown successfully.

17.1 History

The original HOPE project called for the building of a sub-scale orbital prototype known as **HOPE-X**, for *H-2 Orbiting Plane, Experimental*. This would be used for flight testing and systems validation, before moving onto the larger HOPE, which used many of the same parts and general design in a 4-man 22-metric-ton (49,000 lb) design. As the name implies, both would be launched on Japan's new H-2 launcher, the full-scale HOPE requiring substantial upgrades in performance. At the time, Japan was an up-and-coming industrial powerhouse, and their space program was moving from success to success. There was little doubt, and a little trepidation, that HOPE would be successful.

As part of the overall Japanese space program, testing for technologies that would be used on HOPE and other projects was well advanced. In February 1994 the first test flight of the new H-2 launcher was used to also launch the experimental OREX ballistic re-entry vehicle, which tested various communications systems, heating profiles and heat shielding components. Another project, Hyflex, followed in February 1996. Hyflex was intended to test the carbon-carbon heat shielding tiles that were intended to be used on HOPE, as well as having the same body shaping in order to gather data on hypersonic lifting. Hyflex was successful, but sank in the Pacific after splashdown before it could be recovered.

In 1997, well into the study, it was decided that HOPE-X should be modified into an unmanned cargo vehicle with the addition of automated approach and docking systems, and a cargo bay with doors similar to the one on the U.S. Space Shuttle. It was believed this would result in a "quick and dirty" cargo supply system for ISS, which was suffering from continued delays due to problems with the Shuttle program. It was estimated that such a conversion could be completed for an additional US$292 million, less expensive than designing a completely new ballistic cargo vehicle for the H-2 launcher, and much less expensive that the estimated US$2.9 billion needed to complete the full-sized HOPE. Even the small HOPE-X launched on unmodified H-2A rockets would deliver a useful 3 metric tons (6,600 lb) to ISS, about the same as the Progress spacecraft's approximate 2,500 kilograms (5,500 lb). HOPE-X was about 15.2 metres (50 ft) long with a 9.7 metres (32 ft) wingspan,[1] and looked quite a bit like the U.S. X-20 Dyna-Soar.

In 1998, the H-2 suffered from a string of failures. A re-evaluation of the entire space program followed, and budget constraints later forced a reduction in overall funding by US$690 million to US$4.22 billion for the five-year spending period between 1998 and 2002. This would force a delay in the timeline for the HOPE-X, with its first flight in 2003. By this time NASDA had spent only US$305 million since the project was approved in 1988, reflecting the status as a research project. The next year the H-2 project was cancelled outright, proceeding with the simplified and lighter H-2A alone. Hughes pulled out of the H-2A project at about this time; they had initially purchased ten launches on the system and it was considered a major international success for NASDA.

HOPE continued to soldier on. In 2000, an agreement was signed to land the returning vehicle at Christmas Island's Aeon Airstrip. The High Speed Flight Demonstration project consisted of 25% scale models of HOPE-X to test navigation technologies and flight characteristics.[2] As the 2003 deadline approached a number of debates broke out about the launcher profile, with many arguing that the H-2 should be replaced with a jet-powered cargo aircraft for an air-start. The first flight was pushed back further to 2004. Before this milestone was reached a major re-organization of NASDA took place in order to address its obvious overcommitment in light of Japan's economic contraction, especially now that there were demands for a crash program to develop spy satellites in order to track North Korean nuclear efforts. JAXA was formed, and HOPE was cancelled during this process.

17.2 See also

- Fuji (Spacecraft)

- Space Shuttle program

- EADS Phoenix - successor to the cancelled Hermes program that was a contemporary of HOPE.

- Buran (spacecraft)

- X-37

- H-II Transfer Vehicle

17.3 References

[1] "HOPE_X 開発から将来宇宙輸送系に向けて" [HOPE_X Lessons Learned for Future Space Transportation Systems] (PDF). *Technical Review* (in Japanese). MHI. January 2002. Retrieved 2012-06-06.

[2] "High Speed Flight Demonstration "HSFD"". *Technical Review*. JAXA. July 2003. Retrieved 2014-06-10.

17.4 External links

- HOPE-X Program

- Drop test of a scaled model of the Hope-X vehicle

Chapter 18

Hopper (spacecraft)

Hopper was a proposed European Space Agency orbital and reusable launch vehicle. The shuttle prototype spaceplane was one of several proposals for a European reusable launch vehicle (RLV) planned to cheaply ferry satellites into orbit by 2015.[*][1] The 'Phoenix' was a German-European project for a one-seventh scale model of the Hopper concept vehicle.[*][2]

18.1 Concept

The suborbital Hopper was a FESTIP (Future European Space Transportation Investigations Programme) system study design[*][3] selected for further analysis during Future Launchers Preparatory Programme. It would be composed of a single stage reusable vehicle which would not reach orbital velocity. This vehicle was to be launched on a 4 km magnetic horizontal track which would accelerate it to launch speed. At a height of 130 km, the vehicle would fire an expendable rocket powered upper stage which would reach orbital speed and place its satellite payload into orbit. Finally, the vehicle would then glide down to an island in the Atlantic Ocean and be taken back to French Guiana by ship.[*][1] An EADS spokesperson stated that a reusable launch vehicle like Hopper could halve the cost of sending a satellite into orbit, which now stands at $15,000 USD per kilogram of payload.[*][1]

18.2 Contractors

EADS was responsible for the project management and for the entire software equipment of the system. Other partner companies were also involved in the development. The European Space Agency (ESA) and EADS hoped to complete development of Hopper between 2015 and 2020. After the first glide test on May 2004, no updates were forthcoming and eventually the project was canceled.

18.3 Prototype

The *Phoenix RLV* launcher, the prototype of Hopper launcher, was part of the German national program ASTRA, a €40 million project founded by the German Federal Government, EADS Astrium Space Transportation and the state Free Hanseatic City of Bremen with one third each. Both EADS and the State of Bremen invested at least €8.2 million and €4.3 million, respectively. Another contribution of €16 million came from partner companies such as the Bremen-based OHB-System, DLR and the Federal Ministry for Education and Research.

The *Phoenix RLV* prototype was 6.9 meters long (23 ft), had a weight of 1,200 kilograms (2,640 lb), and a wingspan of 3.9 meters (13 ft). The prototype, at one sixth the size of the planned vehicle, was last in the *alpha stage* of development at Bremen labs of EADS.

On Saturday 8 May 2004, the prototype was dropped from 2.4 kilometers (8,000 ft) by a helicopter and landed precisely and without incident after a GPS-guided 90 second glide.[*][4] The test was conducted at the North European Aerospace Test range in Kiruna, 1,240 km (770 mi) north of Stockholm, Sweden. The primary aim of the test was to assess the glider potential of the craft. No subsequent tests were reported.

The final version of the vehicle was expected to be able to support the reentry forces, generated heat, and be able to glide from an altitude of 129 kilometers (80 mi).

18.4   See also

- Hermes

- IXV

- Maglev - magnetic rails

- Rocket sled launch

- Liquid Fly-back Booster

18.5   References

[1] Europe's space shuttle passes early test I 10 May 2004

[2] Launching the next generation of rockets - BBC News, 2004.

[3] Possible Future European Launchers, A Process of Convergence I ESA Bulletin Number 97 I March 1999

[4] Phoenix Flight Day I Swedish Space Corporation. I May 8, 2004

18.6   External links

- European Space Shuttle Glides To Success

- Glide test images: Zarm.uni-bremen.de, Spacetec.zarm.uni-bremen.de

Chapter 19

HOTOL

For the generic concept of *horizontal takeoff and landing*, see HTOL.

HOTOL, for **Horizontal Take-Off and Landing**, was a British design for an Airbreathing jet engine spaceplane

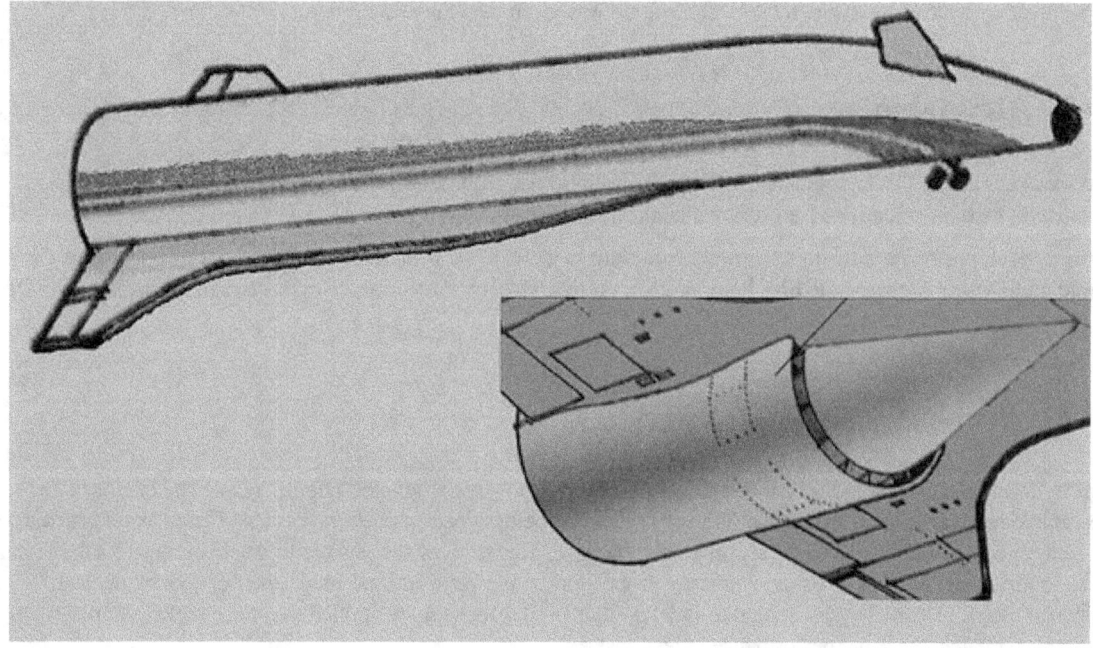

HOTOL

by Rolls-Royce and British Aerospace.

Designed as a single-stage-to-orbit (SSTO) reusable winged launch vehicle, it was to be fitted with a unique air-breathing engine, the RB545 or Swallow, to be developed by the Rolls-Royce company. The engine was technically a liquid hydrogen/liquid oxygen design, but dramatically reduced the amount of oxidizer needed to be carried on board by utilising atmospheric oxygen as the spacecraft climbed through the lower atmosphere.

Since propellant typically represents the majority of the takeoff weight of a rocket, HOTOL was to be considerably smaller than normal pure-rocket designs, roughly the size of a medium-haul airliner such as the McDonnell Douglas DC-9/MD-80. Ultimately, comparison with a rocket vehicle using similar construction techniques failed to show much advantage, and funding for the vehicle ceased.

19.1 Description

HOTOL would have been 63 metres long, 12.8 metres high, 7 metres in diameter and with a wingspan of 28.3 metres. The unmanned craft was intended to put a payload of around 7 to 8 tonnes in orbit, at 300 km altitude. It was intended to take off from a runway, mounted on the back of a large rocket-boosted trolley that would help get the craft up to "working speed". The engine was intended to switch from jet propulsion to pure rocket propulsion at 26–32 km high, by which time the craft would be travelling at Mach 5 to 7. After reaching orbit, HOTOL was intended to re-enter the atmosphere and glide down to land on a conventional runway (approx 1,500 metres minimum). HOTOL was designed for automatic, unmanned flights, although later stages would reintroduce a pilot. The internal landing gear would have been too small to carry the weight of the fully fueled rocket, so emergency landings would have required the fuel to be dumped.[1]

19.1.1   Engine

The RB545 was an air-breathing rocket engine. The exact details of this engine were covered by the UK's Official Secrets Act, and there is consequently little public information about its operation, although it has apparently been declassified.[2]

Within the atmosphere, hydrogen fuel would be used in a heat exchanger to precool air entering a high overall pressure ratio turbojet-like engine cycle, before being burnt with the air in a rocket motor, to produce a very high velocity propulsive jet.

Once out of the atmosphere, the RB545 would be capable of burning the hydrogen with on-board liquid oxygen as a high-efficiency hydrogen/oxygen rocket. The engine was given the Rolls Royce name "Swallow".[3]

19.2   Development

The ideas behind HOTOL originated from work done by Alan Bond for precooled jet engines which he had done specifically with the intention of powering a launch system.[4]

Formal development began with government funding in 1982. The design team was a joint effort between Rolls-Royce and British Aerospace led by John Scott-Scott and Dr Bob Parkinson. About the same time, the Rockwell X-30 scramjet programme was announced in America.

19.3   Problems

During development, it was found that the comparatively heavy rear-mounted engine moved the center of mass of the vehicle rearwards. This meant that the vehicle had to be designed to push the center of drag as far rearward as possible to ensure stability during the entire flight regime. Redesign of the vehicle to do this required a large mass of hydraulic systems, which cost a significant proportion of the payload, and made the economics unclear.[5] In particular, some of the analysis seemed to indicate that similar technology applied to a pure rocket approach would give approximately the same performance at less cost.

19.4   Shutdown

In 1988 the government withdrew further funding. The project was almost at the end of its design phase but the plans were still speculative and dogged with aerodynamic problems and operational disadvantages.

19.5   Successors

A cheaper redesign, **Interim HOTOL** or **HOTOL 2**, to be launched from the back of a modified Antonov An-225 *Mriya* transport aircraft, was offered by BAe in 1991 but that too was rejected. Interim HOTOL was to have dispensed with an air-breathing engine cycle and was designed to use more conventional LOX and liquid hydrogen.

In 1989, HOTOL co-creator Alan Bond formed Reaction Engines Limited (REL) which has since been working on the Skylon vehicle intended to solve the problems of HOTOL. In November 2012, REL conducted tests on an engine observed by the European Space Agency and declared the tests a success and that a major technical obstacle had been removed.[6] In July 2013 the UK government announced a £60m investment in REL.[7]

19.6 See also

- NASP - a scramjet vehicle with which HOTOL would have competed

- Reaction Engines Skylon - a follow on design that attempts to avoid HOTOL's shortcomings

- Reaction Engines A2 - a design for a hypersonic antipodal airliner

- Liquid air cycle engine - a related engine cycle that liquifies the air

19.7 References

[1] *Flight international* 1 March 1986

[2] Dr Bob Parkinson discusses HOTOL in an oral history interview recorded for the National Life Stories project Oral History of British Science at the British Library

[3] Flight Global: secret files reveal US interest in UK HOTOL spaceplane.

[4] BBC Four: The Three Rocketeers, retrieved 14 Sept 2012

[5]

[6] http://www.bbc.co.uk/news/science-environment-20510112

[7] http://www.bbc.co.uk/news/science-environment-23332592

19.8 Further reading

- Julian Moxon, Hotol: where next?, *Flight International*, 1 March 1986. (Two page discussion with technical description).

- Alan Postlethwaite, Hotol fights for life, *Flight International*, 25 March 1989. (With detailed cutaway)

- Rob Coppinger, Secret files reveal US interest in UK HOTOL spaceplane, *Flight International*, 23 February 2009. (Government thinking in 1984-5).

19.9 External links

- Cutaway drawing of the HOTOL

- HOTOL-related patent on jetisonable control surfaces

- Listen to Dr Bob Parkinson discuss the HOTOL in an oral history interview recorded for the National Life Stories project Oral History of British Science at the British Library

Chapter 20

Hyflex

Hyflex (*Hypersonic Flight Experiment*) was a NASDA reentry demonstrator prototype which was launched in 1996 on the only flight of the J-I launcher. It was a successor of **OREX** and was a precursor for the Japanese space Shuttle Hope.

Hyflex tested the carbon-carbon heat shielding tiles that were intended to be used on HOPE, as well as having the same body shaping in order to gather data on hypersonic lifting. Hyflex flew in space at 110 km altitude and succeeded in re-entry, but sank in the Pacific after splashdown before it could be recovered.

20.1 See also

- 1996 in spaceflight

20.2 External links

- Description of Hyflex on NASDA web page

Chapter 21

Intermediate eXperimental Vehicle

"IXV" redirects here. For the IATA code, see Along Airport.

The **Intermediate eXperimental Vehicle** (**IXV**) is an European Space Agency (ESA) experimental suborbital re-entry vehicle (spaceplane prototype) to validate European reusable launchers, evaluated in the frame of the Future Launchers Preparatory Programme (FLPP),[7] and developed under the leadership of the NGL Prime SpA company.[6] The IXV successfully completed its 100-minute[8] mission on 11 February 2015[9] being the first ever lifting body to perform full atmospheric reentry from orbital speed[10]:23

It inherited the principles of previous studies such as CNES's **Pre-X** and ESA's **AREV** (**Atmospheric Reentry Experimental Vehicle**), and the successful Atmospheric Reentry Demonstrator (ARD) flown in 1998. The successor of IXV will be the PRIDE spaceplane (Programme for Reusable In-orbit Demonstrator in Europe).[11]

21.1 Design

IXV used a lifting body arrangement with no wings of any sort, using two movable flaps for re-entry flight control. Re-entry was accomplished in a nose-high attitude like the Space Shuttle, with manoeuvring accomplished by rolling out-of-plane and then lifting in that direction, like an aircraft. Landing was accomplished by parachutes ejected through the top of the vehicle. The airframe was based on a traditional hot-structure/cold-structure arrangement, and was supported on-orbit by a separate manoeuvring and support module similar to the Resource Module intended for the Hermes. The avionics were controlled by a LEON2-FT microprocessor, and interconnected by a MIL-STD-1553B serial bus.[12]

On 18 December 2009, ESA announced a contract with Thales Alenia Space valued at €39.4 million to cover 18 months of preliminary IXV work.[5][13] The total estimated cost for the project is €150 million.[4]

21.2 Pre-launch testing

The IXV's subsonic parachute system was tested at the Yuma Proving Ground in Arizona in late 2012.[14] Water impact tests were conducted at Consiglio Nazionale delle Ricerche's INSEAN research tank near Rome.[15]

On 21 June 2013 an IXV test vehicle was dropped from an altitude of 3 km (1.9 mi) in the Salto di Quirra range off Sardinia. The test was to validate the water landing system including the subsonic parachute, flotation balloons, and beacon deployment. A small anomaly was encountered when inflating the balloons, but the other systems performed as expected. After the test the vehicle was taken for further analysis.[16]

On 23 June 2014 the recovery ship *Nos Aries* conducted a training exercise with an IXV test article off the coast of Tuscany.[17]

In June 2014 the IXV test vehicle arrived at the ESTEC Technical Centre in Noordwijk, the Netherlands, to undergo a rigorous test campaign to confirm its flight readiness in anticipation of a flight on a Vega rocket in November.[18]

21.3   Flight

Originally planned to make its flight in 2013,[19] it was scheduled to perform the launch on 18 November 2014,[20] however this initial launch window was missed due to unresolved range safety concerns.[21] Finally the IXV was launched on 11 February 2015 by a Vega rocket[22] as part of the VV04 mission.[1] Spacecraft launched at 08:40 local time,[23] separated from Vega launch vehicle at 333 km altitude and ascended to 412 km, after which it descended to begin reentry at 120 km altitude with a speed of 7.5 km/s, the same as for Low Earth Orbit spacecraft reentry. Following that IXV glided over Pacific ocean before opening parachutes to slow down its descent, flying 7300m from the beginning of the reentry.[10]:25–26 The ESA's small launcher, descended to the Pacific Ocean for later recovery by the *Nos Aries* ship, and analysis of the spacecraft and recorded mission data.[24][25]

21.4   Follow up mission

After a flawless test flight, ESA officials decided to plan an additional test flight for 2019 or 2020. This time the IXV will land on the ground instead of a splashdown by either installing a parafoil or landing gear. Planning was to begin in March 2015, with design work starting in mid 2015.[26]

21.5   Specifications

This is the drop-test model of the IXV with the flotation balloons inflated, as displayed in ESA ESTEC. Note that the flaps in this model cannot move.

Data from ESA,[4] Space.com,[19] Gunter's Space Page[6]

General characteristics

- **Crew:** None

- **Capacity:** None

- **Length:** 5 m (16.4 ft)

- **Wingspan:** 2.2 m (7.2 ft)

- **Height:** 1.5 m (4.9 ft)

- **Empty weight:** 480 kg (1,058 lb)

- **Loaded weight:** 1,900 kg with propulsion module (4,188 lb)

- **Power:** Batteries

Performance

- **Maximum speed:** 7700 m/s (27,720 km/h) 17,224 mph

- **Range:** 7,500 km (4,660 mi)

- **Max altitude:** 412 km (256 mi) suborbital flight

21.6 See also

- 2015 in spaceflight

- Atmospheric Reentry Demonstrator (ARD) - ESA reentry testbed flown in 1998

- European eXPErimental Re-entry Testbed (EXPERT) - research programme developing materials used in IXV

- Future Launchers Preparatory Programme - parent programme for IXV

- Programme for Reusable In-orbit Demonstrator in Europe (PRIDE) - future robotic spaceplane concept which IXV is developed for[26]

- Italian Aerospace Research Centre

- Hopper - an earlier ESA project on developing manned spaceplane, cancelled

- Hyflex (Hypersonic Flight Experiment) - equivalent Japanese spaceplane demonstrator developed and flown by NASDA in 1996

- Avatar - Indian spaceplane demonstrator, in development by ISRO

21.7 References

[1] "European space plane set for February launch" . *News.com.au* (News Corp Australia). 22 November 2014. Retrieved 20 January 2015.

[2] "European space plane set for February launch: firm" . 21 November 2014. Retrieved 24 November 2014.

[3] "Worldwide launch schedule" . 18 November 2014. Retrieved 19 November 2014.

[4] "IXV e-book" (PDF within a ZIP). European Space Agency. 2011. Retrieved 4 November 2011.

[5] de Selding, Peter B. (18 December 2009). "ESA Spending Freeze Ends with Deals for Sentinel Satellites, Ariane 5 Upgrade" . Space News. Retrieved 4 November 2011. The contract is valued at 39.4 million euros to cover preliminary IXV work for 18 months, Fabrizi said.

[6] Krebs, Gunter Dirk. "IXV" . Retrieved 4 November 2011.

[7]   "New milestone in IXV development". ESA. 15 September 2010. Retrieved 4 November 2011. The Intermediate eXperimental Vehicle (IXV), under ESA's Future Launchers Preparatory Programme (FLPP), is the step forward from the successful Atmospheric Reentry Demonstrator flight in 1998, establishing Europe's role in this field.

[8]   "IXV Mission Timeline". ESA. 9 February 2015. Retrieved 11 February 2015. It will navigate through the atmosphere within its reentry corridor before descending, slowed by a multistage parachute, for a safe splashdown in the Pacific Ocean some 100 minutes after liftoff.

[9]   "ESA experimental spaceplane completes research flight". ESA. 11 February 2015. Retrieved 12 February 2015. ESA's Intermediate eXperimental Vehicle flew a flawless reentry and splashed down in the Pacific Ocean just west of the Galapagos islands.

[10]   "ESA Bulletin 161 (1st quarter 2015)" (PDF). ESA. 2015. ISSN 0376-4265. Retrieved 30 May 2015.

[11]   Chris Bergin (2015-02-24). "IXV's Pride: Europe's spaceplane homecoming prelude to future goals". NASA SpaceFlight. Retrieved 2015-02-25. ESA leaders have already authorised development to begin on the IXV's successor: the Programme for Reusable In-orbit Demonstrator in Europe, or PRIDE.

[12]   Rodríguez, Enrique; Giménez, Pablo; de Miguel, Ignacio; Fernández, Vicente (25 September 2012). *SCOE for IXV GNC*. Simulation & EGSE Facilities for Space Programmes (SESP 2012). European Space Agency.

[13]   "ESA and Thales Alenia Space establish agreement for development of Intermediate eXperimental Vehicle (IXV)". ESA. 19 June 2009. Retrieved 4 November 2011.

[14]   "ESA's IXV Reentry Vehicle Prepares for Soft Landing". ESA. 9 November 2012. Retrieved 2 July 2013.

[15]   Iafrati, A. "Water impact and hydrodynamic loads". Consiglio Nazionale delle Ricerche. Retrieved 22 March 2014.

[16]   "Safe splashdown for IXV". ESA. 21 June 2013. Retrieved 2 July 2013.

[17]   "Bringing back our Spaceplane". ESA. 24 June 2014. Retrieved 30 June 2014. Yesterday, the ship and crew aiming to recover Europe's unmanned IXV spacecraft in November had a practice run off the coast of Tuscany, Italy.

[18]   "Unboxing IXV". ESA. 2 July 2014. Retrieved 2 July 2014. The moment when ESA's IXV Intermediate eXperimental Vehicle is removed from its protective container, safely inside the cleanroom environment of the Agency's Technical Centre.

[19]   Coppinger, Rob (13 June 2011). "Europe Aims to Launch Robotic Mini-Shuttle By 2020". Space.com. Retrieved 16 June 2011. In 2013, a Vega rocket will carry ESA's Intermediate eXperimental Vehicle into space.

[20]   "Europe's IXV atmospheric reentry demonstrator ready for final tests". space-travel.com. 24 March 2014. Retrieved 26 March 2014. The launch, using Europe's new Vega light launcher, is scheduled for October 2014.

[21]   de Selding, Peter B. (30 October 2014). "Thales Alenia Officials Flabbergasted as Safety Concerns Sideline IXV". *SpaceNews*. Retrieved 3 February 2015.

[22]   "Vega to fly ESA experimental reentry vehicle". ESA. 16 December 2011. Retrieved 16 December 2011. The launch of ESA's IXV Intermediate eXperimental Vehicle on Europe's new Vega rocket is now in detailed planning, a major step towards the craft's flight in 2014.

[23]   "European Mini-Space Shuttle Aces 1st Test Flight". Space.com. 11 February 2015. Retrieved 30 May 2015.

[24]   "ESA's IXV reentry vehicle prepares for soft landing". ESA. 9 November 2012. Retrieved 16 November 2012. it will fly the experimental hypersonic phase over the Pacific Ocean, descend by parachute and land in the ocean to await recovery and analysis.

[25]   "Europe's mini-space shuttle returns". *BBC News*. 11 February 2015. Retrieved 12 February 2015.

[26]   Howell, Elizabeth (23 February 2015). "Europe's Newly-Tested Space Plane Aims for Next Launch in 2019". *Space.com*. Retrieved 23 February 2015.

21.8 Further reading

- Tumino, Giorgio; Gerard, Yves (November 2006). "IXV: the Intermediate eXperimental Vehicle" (PDF). *ESA bulletin* (128): 62–67. External link in |work= (help)

- Tumino, Giorgio; Angelino, Enrico; Leleu, Frederic; Angelini, Roberto; Plotard, Patrice; Sommer, Josef (15 October 2008). *The IXV project: the ESA re-entry system and technolologies demonstrator paving the way to European autonomous space transportation and exploration endeavours* (PDF). 3rd FLPP Industrial Workshop. European Space Agency. IAC-08-D2.6.01.

- Baiocca, Paolo (June 2007). *Pre-X experimental re-entry lifting body: Design of flight test experiments for critical aerothermal phenomena* (PDF). RTO-EN-AVT-130 —Flight Experiments for Hypersonic Vehicle Development (von Karman Institute, 24–27 October 2005). NATO Research and Technology Organisation. pp. 11–1–11–18. ISBN 978-92-837-0079-1.

- Baiocca, Paolo; Guedron, Sylvain; Plotard, Patrice; Moulin, Jacques (October 2006). *Proceedings of the 57th IAF Congress.* International Astronautical Congress (IAF). *The Pre-X atmospheric re-entry experimental lifting body: Program status and system synthesis*: 459–474. doi:10.1016/j.actaastro.2007.01.053.

- Gawehn, T.; Neeb, D.; Tarfeld, F.; Gülhan, A.; Dormieux, M.; Binetti, P.; Walloschek, T. (2011). "Experimental investigation of the influence of the flow structure on the aerodynamic coefficients of the IXV vehicle". *Shock Waves* **21** (3): 253–266. Bibcode:2011ShWav..21..253G. doi:10.1007/s00193-011-0326-y.

21.9 External links

- Official IXV website

- IXV twitter profile

- *Full replay from liftoff to splashdown for IXV reentry mission*, ESA Multimedia Gallery (11 February 2015)

- *IXV first results press conference*, ESA Space in Videos (16 June 2015)

- *ESA' s IXV reentry vehicle mission*, ESA Multimedia Gallery (2012 animation)

- *IXV: learning to come back from Space*, IXV Video News Release VNR

- *ESA' s Intermediate eXperimental Vehicle*, ESA Multimedia Gallery (2008 animation)

- ESA Euronews: "Splashdown —the re-entry test" (2013-08-22). Video on YouTube

- CNES reusable atmospheric re-entry vehicle: PRE-X

Chapter 22

Keldysh bomber

The **Keldysh bomber** was a Soviet design for a rocket-powered sub-orbital bomber spacecraft which drew heavily upon work carried out by Eugen Sänger and Irene Bredt for the German Silbervogel project.

22.1 Development

During the closing weeks of World War II, the German work at Peenemünde was investigated by Soviet intelligence, amongst whom was rocket motor constructor Alexey Isayev, who found a copy of Sänger and Bredt's report.*[1] A translation was soon circulating among Soviet rocket designers, and a condensed version made its way to Stalin himself.*[2]

In November 1946 the NII-1 NKAP research institute was formed with mathematician Mstislav Vsevolodovich Keldysh as its head to investigate and develop the German Sänger-Bredt design. In 1947, studies indicated that the high fuel consumption of Sänger's rocket-based design rendered the concept impracticable in the short term. Using engines which were considered to be available in a reasonable timespan, 95% of the vehicle's initial mass would have to be propellant. However, use of ramjets during the acceleration phase would give the craft a more reasonable 22% dead weight and still achieve the 5 km/s velocity required for a 12,000 km intercontinental range.

It was estimated that it would take until the mid-1950s before a draft project of a feasible design could be prepared, and by that time the design had been made obsolete by more advanced designs. However, the work carried out would lead to the EKR, MKR, Buran, and Burya ramjet cruise missiles.*[3]

22.2 Proposed mission profile

- The 100 tonne craft would be accelerated to 500 m/s using a sled running along a 3 km track and powered by five or six RKDS-100 rocket engines of 600 tonnes total thrust. Separation velocity would be reached 11 seconds after ignition.

- After separation from the sled, the craft would climb using its main RKDS-100 rocket engine and two wingtip-mounted ramjets which would accelerate it to an altitude of 20 km and a speed of over Mach 3.

- The rocket would continue working after the ramjets had flamed out at high altitude; it had a specific impulse of 285 seconds, a thrust of 100 tonnes, and used Liquid oxygen/Kerosene propellants.

22.3 Specifications

22.3.1 General characteristics

- **Function**: Sub-orbital bomber
- **Launch mass**: 100,000 kg

- **Total length**: 28 m
- **Launch platform**: Rocket sled
- **Status**: Canceled

22.3.2   Launch sled (stage 0)

- **Engine**: 5/6 × RKDS-100 rocket engines
- **Length**: 14 m (45 ft)
- **Diameter**: 3.6 m (11.8 ft)
- **Thrust**: 5,880 kN (1,321,870 lbf)
- **Oxidizer**: LOx
- **Combustible**: Kerosene

22.3.3   Keldysh bomber (stage 1)

- **Engine**: 1 × RKDS-100 rocket, 2 x ramjets
- **Speed** : Mach 3
- **Range**: 12,000 km
- **Flight altitude**:
- **Warhead**:
- **Length**: 28.0 m
- **Diameter**: 3.6 m
- **Wing span**: 15 m
- **Wing area**: 126 m^2

22.4   See also

- Silbervogel
- Spacecraft propulsion
- X-20 Dynasoar

22.5   References

[1] Sänger, Eugen; Irene Sänger-Bredt (August 1944). "A Rocket Drive For Long Range Bombers" (PDF). Astronautix.com. Retrieved 2008-01-17.

[2] Westman, Juhani (2006). "Global Bounce". Retrieved 2008-01-17.

[3] Wade, Mark. "Keldysh Bomber". Astronautix.com. Retrieved 2008-01-17.

22.6   External links

- Astronautix.com

Chapter 23

Kliper

Kliper (**Клипер**, English: **Clipper**) was a proposed partly reusable manned spacecraft by RSC Energia. Due to lack of funding from the ESA and RSA, the project has been indefinitely postponed as of 2006.

Designed primarily to replace the Soyuz spacecraft, Kliper was proposed in two versions: as a pure lifting body design and as spaceplane with small wings. In either case, the craft would have been able to glide into the atmosphere at an angle that produces much less stress on the human occupants than the current Soyuz. Kliper was intended to be designed to be able to carry up to six people and to perform ferry services between Earth and the International Space Station.

23.1 Development

23.1.1 Announcement of the program

Soyuz TMA-6 spacecraft approaching the International Space Station - the Soyuz spacecraft would have been replaced by Kliper

In February 2004 Nikolai Moiseyev, the deputy director of Russian Federal Space Agency (FSA) told journalists that the Kliper project had been included in the Russian federal space program for 2005-15. At that point he announced that if the program is implemented successfully the first launch may even take place in five years' time. Kliper had been developed since 2000 and reportedly relied heavily on research studies as well as proposals for a small Russian lifting body spacecraft from the 1990s. Externally its design was comparable to the cancelled European minishuttle Hermes or the NASA study X-38. It was planned to be the successor to the veteran spacecraft Soyuz, which has been built in various modifications since 1961.

23.1.2 Early search for support

In 2005 Kliper was displayed in several air shows around Europe and Asia, in order to reach out to international partners who would be interested to co-fund and co-develop the spacecraft. The Russian Space Agency especially looked to Europe as the European Space Agency (ESA) had become its major partner in space activities during the last years. In May 2005 rumours started in the press that Europe would join the Kliper project in a specially funded venture that would be part of the Aurora Programme. These rumours turned out to be correct when both Russian and European space officials announced their cooperation to build Kliper during the Paris Air Show in Le Bourget on June 10, 2005.[1][2]

Vladimir Taneev, the leading designer of the Kliper system, speculated on the contribution of Europe to the project in the following way:

> The European companies will likely contribute avionics, materials, and cabin systems. Many different options are on the table, and in the near future we expect to form Russian-European working groups specialized in different subsystems and fields of design.

The Russian Space Agency as well as ESA announced that they would continue to look for other international partners such as Japan to invest in Kliper. A substantive cooperation with NASA was unlikely, due to the parallel development of America's own next-generation manned launch vehicle, the Crew Exploration Vehicle (CEV).

A further element of this process was made public on October 12, 2005, when various press agencies revealed that JAXA, the Japanese space agency, had been officially approached by Russia to participate in the project. JAXA has made it clear that they are more likely to join the project if ESA does so first, which is in doubt after ESA members rejected a study for Europe's involvement in the Kliper project in December 2005. The addition of Japan would make Kliper a truly multinational project, potentially combining the rugged reliability of Russian launchers with Japanese computer technology. A greater pan-national consensus would have allowed for a lighter funding burden on each participant as well.

23.1.3 Estimated costs

Announcements and speculations following the February 2004 press conference suggested a development budget of PP 10 billion (about US$400 million). However, in looking at today's costs for human space travel it was clear that the 10 billion rubles figure was a rather low estimate. In May 2005 The Guardian reported that costs are estimated to be roughly US$3 billion (for development and construction of Kliper until 2015) of which the bulk of US$1.8 billion was speculated to come from Europe.[3] Different sources in 2005 have reported that the money needed for the program would be €1.5 billion (about US$1.8 billion)[4] and on December 12, 2005 an article stated it would be €1 billion (solely in relation to development costs).[5]

On July 14, 2005 the Russian government approved the national space program for 2006 to 2015 with a budget of PP 305 billion (about US$11 billion). The whole budget for the 10-year period will be PP 425 billion (about US$15 billion).[6] The budget included the needed funding for the Kliper program.[7] Thus in face of Europe's denial to fund a €50 million feasibility study for the Kliper project at the European space summit in December 2005, Russian space officials have announced that Russia would fund Kliper even without any European contribution.[5]

The most recent article on Kliper stated that the project would have incurred PP 16 billion (about US$600 million) in development costs, PP 11 billion of which will be financed by the government and PP 5 billion by contractors.[8]

23.1.4   First launch and target for regular flights

In 2004 it was announced that it was likely that Kliper would make its first launch as early as 2010 or 2011, the same time the Space Shuttle was scheduled to be retired. However, it was reported by BBC News on September 27, 2005, that the first flight tests were not planned until 2011, with the first manned flights in 2012 and the Soyuz being phased out over time until 2014. An article on December 3, 2005[9] cited the president of the Energia Rocket and Space Corporation Nikolai Sevastyanov that "the first regular lift-off is scheduled for 2012, while a complete transport system will be in place by 2015." After the termination of the Russian Space Agency's tender for a new spacecraft, Energia announced that this would push its Kliper proposal's first flight —if developed at all —back further.

23.1.5   ESA's part in Kliper —uncertainty over European cooperation

Kliper was planned to be Russia's and even Europe's primary access route to the International Space Station

On September 28, 2005 the BBC reported that Alan Thirkettle, head of ESA's Human Spaceflight Development Department, stated that Kliper would be used: "For future exploration, when we have the objective of going to the Moon, it is important to have several possibilities to go there, and within this framework of cooperation to have our own access to orbit around the Moon." In the same context, Alain Fournier-Sicre, head of the ESA permanent mission in the Russian Federation, also stated that: "The objective is to have a vehicle which is more comfortable than the Soyuz capsule which will be used with pilots and four passengers···It is meant to service the space station and to go between Earth and an orbit around the Moon with six crew members."

Although there seemed to be a lot of enthusiasm for Kliper within Alan Thirkettle's team at ESA (as outlined in the above paragraph), on December 7, 2005, the European space summit of governmental officials of ESA member states declined to approve a 50-million-euro two-year study focusing on ESA's potential involvement in the Kliper project. In denying funding for the study ESA members stated that, among other factors that seemed unfavourable, under the current Russian proposal Europe would not share control over the design of the program and would be limited to being a small industrial contributor.

Jean-Jacques Dordain, ESA's Director General, put the refusal to fund the study into context: "It is not a question of member states for and member states against. I think the decision could not be taken for reasons that are not linked to Clipper itself. The decision could not be taken because of budgetary restraints." Dordain concluded that he was convinced that European support for Kliper was vital for ESA's future involvement in space transport and that a favourable decision can be achieved until June 2006. In concluding "We need two transportation systems in the world", *[10] Dordain also outlined shortly after the European Space Summit that the primary requirement of Europe's involvement in the Kliper project was to rely on two separate systems to support the ISS as had been proven vital after the Columbia Space Shuttle disaster in 2003.

Dordain's remarks were echoed by Daniel Sacotte, ESA's director of human spaceflight, microgravity and exploration, in saying simply that "The Russians are not going to finance it, we will finance it from our side", despite adding a cautionary note that "We needed the support from at least two states out of France, Italy and Germany. We didn't get it." What this means in practical terms remains to be seen; however, what is clear is that ESA officials are still pushing for Europe's involvement in the Kliper project.

Very negative comments relative to Kliper were brought by the various national delegations at the December meeting, in particular by the French Minister of Research François Goulard. In short, there remain for the time being member states strongly committed to Kliper, and others just as strongly opposed. The long-term view remains uncertain.

In 2006, Jean-Jacques Dordain explained that money allocated to space transportation development, which ESA currently funds in the amount of 300 million for the next 3 years, could be used for Europe's involvement in the project. Given the February 2006 statement that 5 billion rubles (~$200 million) of the development costs will come from "contractors", a limited involvement of ESA in Kliper might have been forthcoming.

23.1.6   Russian Space Agency's tender for Kliper

At the end of 2005, Roskosmos announced that a tender for Kliper would be held in January 2006 between RKK Energia, Khrunichev and Molniya with a selection date of February 3, 2006. However concerns about the bids led to a delay in the process, with a resubmittal deadline of March, 2006 and selection was rescheduled for April 2006. Following further delays, the tender was cancelled on 18 July 2006.*[11]

In late July 2006, the Russian Space Agency and the European Space Agency agreed to collaborate on a different project to develop a new spacecraft. They decided to fund a study under a program labelled Crew Space Transportation System (CSTS) which started in September 2006 and evaluate a capsule type concept, derived from Soyuz. While this program is the follow-on project of the RSA's and ESA's collaboration on a new spacevehicle, this program is no longer connected to Energia's winged Kliper design.

RSC Energia continued to pursue the project without Russian government support and announced that it would seek private investment for the craft.*[12] News reports in Russia indicated that Kliper was still expected to be ready for Russian Space Agency test flights around the year 2012,*[13] as part of Russian spacecraft upgrade program.*[14]*[15] The project has been officially halted in June 2007, after the major proponent of the project, Nikolai Sevastyanov, was dismissed from the position of the president of RSC Energia. *[16] The newly appointed president of RSC Energia, Vitaly Lopota, confirmed that Kliper would not be displayed on the 2007 MAKS aviation and space show. He said that Energia would spend more time on the project analysis, perform additional dynamic modeling, revise the design and appearance and then would come up with new proposals for Roscosmos.*[17]

In 2008 Vitaly Lopota shared his vision about new Russian spacecraft. He mentioned two possible options: a space capsule, which better suits missions to the Moon and Mars, and a lifted body design for low Earth orbit missions.*[18]*[19] According to new plans, instead of Kliper the new PPTS (Rus) will be developed since 2009 to 2017-2018.

23.2   Design

Given the Russian Space Agency's preference for Energia's lifting body proposal this part of the article concentrates entirely on Energia's design for Kliper.

23.2.1   Overview

Kliper's design was another attempt to solve the geometric problems of spacecraft. Soyuz has an Orbital Module, a hollow sphere, to be used for eating and hygiene, and an airlock located above the Reentry module (the capsule), with the docking mechanism at the top. In the event of an emergency, it would be lifted away from the rocket along with the reentry module, and the fairing over the spacecraft was designed to successfully split apart either circumferentially just below the reentry module in such an emergency or longitudinally if the flight should be successful. Kliper was designed with the Orbital Module below its reentry module, and the docking mechanism below that. This was made possible by constructing a reentry module broader than the orbital module, so that a pair of rocket nozzles for orbital maneuvering could have been fitted alongside it, as the later Salyut space stations had.

In connection with this new design, Kliper will feature a launch escape system that will enable it to detach from the carrier rocket if an abort of the mission during orbital ascent is required. An abort will be possible during every phase of the launch with the limitation of the first seconds after launch.[*][20]

23.2.2   Lifting body design

On return from space, Kliper's lifting body design would not only allow a smoother descent into Earth's atmosphere than the capsule design, such as Soyuz; but also permit control. RKK Energia claimed that the craft would be able to land in a predetermined one-square-kilometre area. Artistic impressions showed that the Kliper would have resembled a cylinder topped by a cone. Originally, landing proposals involved both a landing by parachute and as an alternative, in a modified version, a landing on a runway similar to an aircraft, or the Space Shuttle. However, leading designer Vladimir Daneev commented on this issue in June 2005:

> *We are 99% sure that it will be a spaceship with upturned little wings, enabling the Kliper to land on any*
> *class-one military airfield with a runway from three to three and a half kilometres in length.*[*][21]

Kliper, as a vehicle alone, would have been primarily a manned spaceship, carrying six cosmonauts and payloads of up to 700 kilograms (mostly experiments and other equipment used for carrying through experiments in orbit) and was planned to stay in orbit for approximately 15 days independently and for up to 360 days if docked to the International Space Station. This highlighted both the Russian/European and the American change in space transportation philosophy. Rather than focusing on the lifting of cargo and a crew, in the same way as the Space Shuttle or Buran, the Russian space agency adopted a 'people first' philosophy with the aim of 'bolting' extra capabilities for more advanced missions onto Kliper at a later date. Each orbiter was intended to make 25 flights prior to retirement.

23.2.3   Using a space tug

During autumn of 2005 Kliper's design was changed again. In order to fit the Kliper on the planned upgraded version of the Soyuz-2 rocket, labeled the Soyuz-2-3, Kliper would be 'split up' into two spacecraft, the Kliper crew vehicle and Parom (rus. *"ferry"*), a space tug. Parom would have been a permanent orbital spacecraft awaiting Kliper in orbit, docking with it and then providing orbital manoeuvering and boosting Kliper to higher orbits in order to dock with the International Space Station. The Parom was planned to be indefinitely reusable, refueling itself via the cargo container, space station, or spacecraft that it is attached to.[*][22][*][23]

23.2.4   Final version of Energia's proposal

The version of Kliper presented during the bid in January 2006 differs again from the original design. It showed a lifting body with larger wings, that, according to Energia officials, could be folded around the core crew module and unfold after atmospheric re-entry in order to provide cross-range and better landing accuracy for the spacecraft. The light Kliper version proposed was stripped down to 7 tons and uses the 'split-up'-option with Parom as a spacetug.

23.3   Missions

The Kliper program was proposed as the Russian-European counterpart to the American Orion Spacecraft and was therefore designed (similar to the Orion) to be part of a modular system that enabled it to be both a LEO-shuttle type vehicle as well as part of a spacecraft able to go beyond Earth orbit to the Moon and even Mars (there were outline suggestions of lunar applications in September 2005). The modular design would have included the Kliper crew module and - depending on the mission - a mission module or propulsion module. Although far-fetched, this corresponds to announcements by the Russian Space Agency that according to a lunar mission study, using the Soyuz, a landing on the Moon could be achieved within the next decade.[*][24]

Information on Kliper's beyond LEO mission capabilities were expanded further by RSC Energia, with a picture released in December 2005 of what a possible Kliper interplanetary configuration might have looked like. The design was entirely theoretical but made for a view of where RSC Energia saw the Kliper operating, and how it might have done so. This configuration was unlike anything seen so far for a manned space vehicle, with the solar arrays needed for electrical power vastly bigger than the habitable volume at the centre. It was also unclear what the mode of propulsion was. The very large solar array suggested an ion propulsion system might have been contemplated for such a mission, though it might also simply be that there was another reason for such a large array, such as increased power for better telemetry transmission rates over large distances.

23.4 Carrier rockets

The present Soyuz rocket would not be able to lift Kliper into low earth orbit, because the spacecraft (the version designed without Parom) was expected to weigh between 13 and 14.5 metric tons (with payload and crew) whereas Soyuz only has a lifting capacity of around 8 metric tons. It was originally planned to heavily enhance the Soyuz rocket - a project that was labelled the Onega rocket or Soyuz-3. Until fall of 2005 it was much more likely that Kliper would have used an Angara-A3 rocket, which was scheduled to make its first launch in 2012 (however the Angara program has been delayed and Angara-A3 may not be developed in light of the funding of the development of Soyuz 2-3) or possibly a Zenit rocket that is built in Ukraine.

At the end of 2005, Kliper's design was changed again (as outlined above) and the most likely solution for a carrier rocket became the Soyuz 2-3, an upgraded Soyuz 2 rocket. This enhanced Soyuz should have been able to launch Kliper into space because of weight reduction resulting in the use of the Parom as a space tug.*[25]

With regard to launch sites for Kliper, further information became available as of October 2005, with a planning-stage declaration from Nikolai Moiseev, Deputy Director of the Russian Space Agency that Kliper could have been launched from ESA's Guiana Space Centre in French Guiana. Though this aim had already been suggested, the comment was made in the context of facility upgrades for Kourou that are already under way since 2003 and are expected to be finished in 2007 with the first launch of a Soyuz rocket from French Guiana in 2008. It had been suggested that Kliper could have been launched from both Baikonur and Kourou, by Alan Thirkettle, head of ESA's human spaceflight, microgravity and exploration directorate, in December 2005.*[25]*[26]

23.5 See also

- Parom

- Prospective Piloted Transport System (PPTS, Rus)

- Crew Space Transportation System (CSTS) - European counterpart program with Russian participation

- Crew Exploration Vehicle - the American counterpart program

- Shuttle Derived Launch Vehicle

- Spaceplane

23.6 References

[1] Phsorg: ESA to join Russia's Clipper Program (retrieved Dec 27, 2006)

[2] BBC News: Plans for Euro-Russian Spaceplane (retrieved Dec 27, 2006)

[3] Guardian: Europe to hitch Spaceride on Russia's rocket (retrieved Dec 27, 2006)

[4] Deutsche Welle: Europeans take on NASA (retrieved Dec 27, 2006)

[5] Deutsche Welle: Europe Keen to join Russia in new spaceship project (retrieved Dec 27, 2006)

[6] "Russian govt agrees 12.5 bln eur 10-yr space programme" . Forbes. July 15, 2005. Archived from the original on 2007-05-01.

[7] RBCNews: Government approves space program for 2006-2015 (retrieved Dec 27, 2006)

[8] Flight International: Kliper choice delayed (retrieved Dec 27, 2006)

[9] RIA Novosti: Russian technologies can put cosmonauts on Moon

[10] "News" , *Europe unites over space bugdet*, Nature, 2005-05-12.

[11] www.flightglobal.com

[12] Flightglobal: Farnborough Air Show - Energia's Klipper work continues

[13] RIA Novosti: Russia remains leader in spacecraft launches

[14] RIA Novosti: Russia implements ambitious space program

[15] RSC Energia: Concept of Russian Manned Space Navigation

[16] Energia Chief Dismissed Amid Differences with Roskosmos

[17] Stumbling Kliper (in Russian)

[18] What spacecraft will replace the Soyuz? (in Russian)

[19] Energia has returned to the Kliper(in Russian)

[20] RIA Novosti - Science & Technologies - Russia's new spaceship to save crew in possible LV accident

[21] http://en.rian.ru

[22] Kliper (Clipper) spacecraft

[23] Lighter Kliper could make towed trip to ISS-01/11/2005-Flight International

[24] Mosnews.com

[25] Flight International

[26] SPACE.com - Russia's Next Spaceship: Alternative to NASA's CEV

23.7   External links

- Images and information at Russian Space Web

- Images and information at Astronautix

- Kliper section at www.buran.ru (in Russian) (Also see buran.ru author's comments about this section at Новости Космонавтики forum: 1, 2).

- Detailed website about the Kliper shuttle, contains lots of schems and explanations

- Winged Kliper at MAKS-2005 Air Show (August 2005)

- Kliper mock-up photo gallery (May 2005). Also see the thread at Новости Космонавтики forum about the presentation on November 30, 2004 (in Russian).

- ESA on its permanent cooperation with Russia in space and Putin's commitment to the Kliper project - 2004

- Europe envisages cooperation on new Russian space plane July 1, 2005

23.8   In the news

- December 20, 2006 - RIAN Opinion & analysis: Russia set to implement ambitious space program (Part II)

- July, 2006 - RSC Energia: Concept of national manned space navigation

- April, 2006 - Russian Space - Manned Spaceflight To Be Cost-Efficient

- February 14, 2006 - Kliper choice delayed

- January 31, 2006 - Energia holds lead in Kliper contest

- January 18, 2006 - Tender to build new-generation spaceship has started. (in Russian)

- January 17, 2006 - Clipper spacecraft constructor to be announced in February 2006

- December 9, 2005 - ESA vows to clinch cash for shuttle

- December 9, 2005 - Europe keen to join Russia in new spaceship project

- December 6, 2005 - Europeans Unlikely to Back Russia's Manned Space Vehicle

- November 1, 2005 - A lighter version of Kliper (which would work in combination with the Parom space tug) is under consideration by RSA and ESA

- October 12, 2005 - Japanese Space Agency confirms its invitation to participate in the project

- October 6, 2005 - Short note concerning the RSA's outline intention to use the Kourou launch site for Kliper

- September 28, 2005 - ESA chiefs release more information on the collaboration

- August 21, 2005 -- ESA and Russia collaborate on Kliper

- July 15, 2005 -- Europeans take on NASA

- June, 2005 -- Report on the Paris Air Show, partly about Kliper (in Russian)

- June 17, 2005 -- Spectrum article on Kliper at Le Bourget

- June 15, 2005 -- Europe and Russia team up for the Kliper spacecraft

- June 15, 2005 -- Article in the Pravda over the European-Russian partnership with Kliper (English)

- June 10, 2005 -- Information on Kliper's launch escape system

- February 15, 2005 -- Article in the Pravda on planned Kliper exposition at Le Bourget'2005 airshow

- September 2004 -- Mosnews - Russia Prepares Launch of New Space Shuttle -- (with image of barebone Kliper)

- April 2004—Energija's Nikolaj Brjuchanov and ESA's Joerg Feustel-Buechl on Kliper in ARD (German TV) (in German)

Chapter 24

Lockheed L-301

Lockheed L-301 (sometimes called the **X-24C**, though this designation was never officially assigned) was an experimental air-breathing hypersonic aircraft project. It was developed by the NASA and United States Air Force (USAF) organization National Hypersonic Flight Research Facility[*][1] (NHFRF or NHRF[*][2]), with Skunk Works as the prime contractor. In January 1977, the program was "tentatively scheduled to operate two vehicles for eight years and to conduct 100 flights per vehicle." [*][3] NASA discontinued work on L-301 and NHRF in September 1977 due to budget constraints and lack of need.[*][1]

24.1 Development

The L-301 HGV was intended to be a follow-on to the X-15 and X-24 (specifically the X-24B) programs, to take lessons learned from both and integrate them into an airframe capable of at least reaching Mach 8 and engaging in hypersonic skip-glide maneuvers for long range missions. While the NASA program, one of several to use the tentative X-24C designator, was ostensibly canceled in 1977, it was only canceled at the time because of USAF disclosures of duplicate black programs with the same contractors for similar vehicles. The vehicle used both air breathing ram or scramjet propulsion as well as a rocket engine, carrying both RP-1 and LH2 propellant as well as on board stores of LOX.

It is not known whether the black program ever resulted in flight tests, however wind tunnel models are well documented online by both Lockheed and USAF websites,[*][4] while Lockheed drawings have appeared on the web,[*][5] particularly on the sites of modelers producing models of this vehicle. Aviation historian Rene Francillion believes Lockheed did fly a testbed aircraft in 1982.

24.2 Design

24.2.1 Propulsion

Originally intended to carry the same XLR-99 engine used by the X-15, the primary engine was changed to the LR-105, which was the sustainer engine used on the Atlas launcher. This rocket engine, burning RP-1 and LOX, was intended to accelerate the X-24C to hypersonic speeds in order to ignite the hydrogen fueled, air breathing ram/scramjet mounted in the belly of the airframe with which it would attain cruise speeds of at least Mach 6 and peak velocities of Mach 8+ at altitudes of 90,000 feet or more.

As such, this vehicle was plainly not intended to reach orbit, but may have served as a technology testbed for development of later black orbiter programs, perhaps even the purported Blackstar project. It may also have served as an intermediate stage for an expendable upper stage capable of putting a small payload in orbit.

24.2.2 Airframe

Design of the aircraft in various wind tunnel models and contractor drawings seems to follow variations of the FDL-5 and FDL-8 lifting body shapes originally developed by the USAF Flight Dynamics Laboratory in the 1950s, which were used in the earlier X-23 and X-24A/B programs. With a radically swept delta wing, and 2, 3, or 4 vertical stabilizers, as well as several body flaps (depending on the model), the vehicle did not lack for control surfaces. The vehicle measured 74 feet 10 inches long, 24 ft, 2 in wingspan, and 20 ft, 7 in height.

Various drawings show a payload bay twelve feet long and perhaps five feet diameter. This would certainly have been sufficient for delivering military ordnance on a transcontinental skip-glide strike mission. It may also have been large enough to carry an upper stage and a small satellite for a surprise orbiting which would eliminate the problem spy satellites have of having their ephemerides predicted and used by enemy nations to hide sensitive observation targets.

24.3 References

[1] http://books.google.com/books?id=DUkl5bH6k6EC&pg=PA98&lpg=PA98&dq=National+Hypersonic+Research+Facility+ x-15+x-24c&source=bl&ots=Ubvm4kazo1&sig=16Y5rv1y8HLZX8fy8mHZGmDBev4&hl=en&ei=7exrSujtNpLWM4a5tPkG& sa=X&oi=book_result&ct=result&resnum=3 [Lockheed Secret Projects by Dennis R. Jenkins]

[2] http://www.darpa.mil/tto/solicit/BAA08-53/VULCAN_Industry_Day_Presentations.pdf ["X-24C NHRF"]

[3] http://ntrs.nasa.gov/archive/nasa/casi.ntrs.nasa.gov/19790008668_1979008668.pdf [CONFIGURATION DEVELOPMENT STUDY OF THE X-24C HYPERSONIC RESEARCH AIRPLANE - PHASE II]

[4] Arnold Air Force Base - Library

[5] Lockheed's X-24C (L-301)

24.4 Further reading

- Miller, Jay. The X-Planes: X-1 to X-45. Hinckley, UK: Midland, 2001.

- Rose, Bill, 2008. Secret Projects: Military Space Technology. Hinckley, England: Midland Publishing.

24.5 External links

- Encyclopedia Astronautica

- Publicly archived Air Force picture

- Nasa Archives- X-24C Phase 3

Chapter 25

Lockheed Martin X-33

The **Lockheed Martin X-33** was an unmanned, sub-scale technology demonstrator suborbital spaceplane developed in the 1990s under the U.S. government-funded Space Launch Initiative program. The X-33 was a technology demonstrator for the VentureStar orbital spaceplane, which was planned to be a next-generation, commercially operated reusable launch vehicle. The X-33 would flight-test a range of technologies that NASA believed it needed for single-stage-to-orbit reusable launch vehicles (SSTO RLVs), such as metallic thermal protection systems, composite cryogenic fuel tanks for liquid hydrogen, the aerospike engine, autonomous (unmanned) flight control, rapid flight turn-around times through streamlined operations, and its lifting body aerodynamics.

Failures of its 21-meter wingspan and multi-lobed, composite material fuel cells during pressure testing ultimately led to the withdrawal of federal support for the program in early 2001. Lockheed Martin has conducted unrelated testing, and has had a single success after a string of failures as recently as 2009 using a 2-meter scale model.*[3]

25.1   Design and development

Through the use of the lifting body shape, composite multi-lobed liquid fuel tanks, and the aerospike engine, NASA and Lockheed Martin hoped to test fly a craft that would demonstrate the viability of a single-stage-to-orbit (SSTO) design. A spacecraft capable of reaching orbit in a single stage would not require external fuel tanks or boosters to reach low-earth orbit. Doing away with the need for "staging" with launch vehicles, such as with the Shuttle and the Apollo rockets, would lead to an inherently more reliable and safer space launch vehicle. While the X-33 would not approach airplane-like safety, the X-33 would attempt to demonstrate 0.997 reliability, or 3 mishaps out of 1,000 launches, which would be an order of magnitude more reliable than the Space Shuttle. The 15 planned experimental X-33 flights could only begin this statistical evaluation.

X-33 launch facility already completed at Edwards Air Force Base.

The unmanned craft would have been launched vertically from a specially designed facility constructed on Edwards

Air Force Base,[4] and landed horizontally (VTHL) on a runway at the end of its mission. Initial sub-orbital test flights were planned from Edwards AFB to Dugway Proving Grounds southwest of Salt Lake City, Utah. Once those test flights were completed, further flight tests were to be conducted from Edwards AFB to Malmstrom AFB in Great Falls, Montana, to gather more complete data on aircraft heating and engine performance at higher speeds and altitudes.

On July 2, 1996, NASA selected Lockheed Martin Skunk Works of Palmdale, California, to design, build, and test the X-33 experimental vehicle for the RLV program. Lockheed Martin's design concept for the X-33 was selected over competing designs from Boeing and McDonnell Douglas. Boeing featured a Space Shuttle-derived design, and McDonnell Douglas featured a design based on its vertical takeoff and landing (VTVL) DC-XA test vehicle.

The X-33 was never intended to fly higher than an altitude of 100 km, nor faster than one-half of orbital velocity. Had any successful tests occurred, extrapolation would have been necessary to apply the results to a proposed orbital vehicle.[5]

25.1.1 Commercial spaceflight

Based on the X-33 experience shared with NASA, Lockheed Martin hoped to make the business case for a full-scale SSTO RLV, called VentureStar, that would be developed and operated through commercial means. The intention was that rather than operate space transport systems as it has with the Space Shuttle, NASA would instead look to private industry to operate the reusable launch vehicle and NASA would purchase launch services from the commercial launch provider. Thus, the X-33 was not only about honing space flight technologies, but also about successfully demonstrating the technology required to make a commercial reusable launch vehicle possible.

The VentureStar was to be the first commercial aircraft to fly into space. The unmanned X-33 was slated to fly 15 suborbital hops to near 75.8 km altitude.[5] It was to be launched upright like a rocket and rather than having a straight flight path it would fly diagonally up for half the flight, reaching extremely high altitudes, and then back down for the rest of the flight. The VentureStar was intended for long inter-continental flights and supposed to be in service by 2012, but this project was never funded or begun.

The decision to design and build the X-33 grew out of an internal NASA study titled "Access to Space".[6] Unlike other space transport studies, "Access to Space" was to result in the design and construction of a vehicle.

25.1.2 NASA cancellation

Construction of the prototype was some 85% assembled with 96% of the parts and the launch facility 100%[4] complete when the program was canceled by NASA in 2001, after a long series of technical difficulties including flight instability and excess weight.

In particular, the composite liquid hydrogen fuel tank failed during testing in November 1999. The tank was constructed of honeycomb composite walls and internal structures to reduce its weight. A lighter tank was needed for the craft to demonstrate necessary technologies for single-stage-to-orbit operations. A hydrogen fueled SSTO craft's mass fraction requires that the weight of the vehicle without fuel be 10% of the fully fueled weight. This would allow for a vehicle to fly to low earth orbit without the need for the sort of external boosters and fuel tanks used by the Space Shuttle. But, after the composite tank failed on the test stand during fueling and pressure tests, NASA came to the conclusion that the technology of the time was simply not advanced enough for such a design. While the composite tank walls themselves were lighter, the odd hydrogen tank shape resulted in complex joints increasing the total mass of the composite tank to above that of an aluminum-based tank.[7]

NASA had invested $922 million in the project before cancellation and Lockheed Martin a further $357 million. Due to changes in the space launch business—including the challenges faced by companies such as Globalstar, Teledesic, and Iridium and the resulting drop in the number of anticipated commercial satellite launches per year—Lockheed Martin deemed that continuing development of the X-33 privately without government support would not be profitable.

25.1.3 Stats

Length: 69 ft

Width: 77 ft

Aerospike engine test at Stennis Space Center, August 6, 2001

Takeoff weight: 285,000 lbs

Fuel: LH2/LO2

Fuel weight: 210,000 lbs

Main Propulsion: 2 J-2S Linear Aerospikes

Take-off thrust: 410,000 lbs

Maximum speed: Mach 13+

Payload to Low Earth Orbit: N/A

25.1.4   Continued research

After the cancellation in 2001, engineers were able to make a working liquid oxygen tank out of carbon fiber composite.

On September 7, 2004, Northrop Grumman and NASA engineers unveiled a liquid hydrogen tank made of carbon fiber composite material that had demonstrated the ability for repeated fuelings and simulated launch cycles.*[8] Northrop Grumman concluded that these successful tests have enabled the development and refinement of new manufacturing processes that will allow the company to build large composite tanks without an autoclave; and design and engineering development of conformal fuel tanks appropriate for use on a single-stage-to-orbit vehicle.*[9]

25.2   See also

- Spaceplane

Figure E.2.2-1. Failed LH$_2$ Tank

Microcracking problem discovered in the Liquid Hydrogen (LH2) Multi-Lobed tank core by NASA scientists at Goddard Space Flight Center ultimately caused NASA to cancel the X-33 program

- Skylon (spacecraft)*[10]
- Blue Origin
- Blue Origin New Shepard
- McDonnell Douglas DC-X
- VentureStar
- Masten Space Systems
- Armadillo Aerospace
- Interorbital Systems
- Quad (rocket)
- Zarya
- Kankoh-maru
- Lunar Lander Challenge
- Reusable Vehicle Testing program by JAXA
- Bristol Spaceplanes

25.3   References

[1] Mark Wade. "X-33". Encyclopedia Astronautica. Retrieved 25 February 2015.

[2] Wikisource:X-33 Advanced Technology Demonstrator

[3] David, Leonard (15 Oct 2009). "Reusable rocket plane soars in test flight". MSNBC. Retrieved 27 Oct 2009.

[4] "X-33 Launch Complex (Area 1-54)" (PDF). USAF. Retrieved 2011-06-30.

[5] "Environmental Impact Statement, Notice of Intent 96-118". NASA. October 7, 1996. Flight tests would involve speeds of up to Mach 15 and altitudes up to approximately 75,800 meters... The test program is currently baselined for a combined total of 15 flights.

[6] "The Policy Origins of the X-33". NASA. September 23, 1998.

[7] Bergin, Chris (January 4, 2006). "X-33/VentureStar —What really happened". NASA Space Flight.

[8] Northrop Grumman. "Northrop Grumman, NASA Complete Testing of Prototype Composite Cryogenic Fuel Tank", *News Releases*, September 7, 2004, accessed April 27, 2011.

[9] Black, Sara (November 2005). "An update on composite tanks for cryogens". *High-Performance Composites*.

[10] *X-33* (news background), US: Nasa.

25.4   External links

- *X-33* (History), US: NASA.

- "X-33", *X planes*, Federation of American Scientists.

- *Status of the X-33 Reusable Launch Vehicle Program* (PDF), US: GAO, August 1999.

- *X-33 cancellation* (press release), NASA, March 1, 2001.

- *X-33 Launch Complex (Area 1-54)* (PDF), WP AFB: Air force

Chapter 26

XCOR Lynx

The **XCOR Lynx** is a suborbital horizontal-takeoff, horizontal-landing (HTHL), rocket-powered spaceplane under development by the California-based company XCOR Aerospace to compete in the emerging suborbital spaceflight market. The Lynx is projected to carry one pilot, a ticketed passenger, and/or a payload above 100 km altitude. As of March 2014, the passenger ticket was projected to cost $95,000.[1][2]

The concept has been under development since 2003, when a two-person suborbital spaceplane was announced under the name Xerus. According to a September 2015 report, the first flight of the Lynx spaceplane is likely to be in the second quarter of 2016 from Midland, Texas.[3]

26.1 History

26.1.1 Xerus

In 2003, XCOR proposed the **Xerus** (pronunciation: zEr'us) suborbital spaceplane concept. It was to be capable of transporting one pilot and one passenger as well as some science experiments and it would even be capable of carrying an upper stage which would launch near apogee and therefore would potentially be able to carry satellites into low-Earth orbit.[4] As late as 2007, XCOR continued to refer to their future two-person spaceplane concept as *Xerus*.[5]

26.1.2 Lynx

The Lynx was initially announced on March 26, 2008, with plans for an operational vehicle within two years.[6] In December of that year a ticket price of $95,000 per seat was announced.[7] The build of the Lynx Mark I flight article did not commence until mid 2013 and, as of October 2014 XCOR claims that the first flight would take place in 2015.[8][9] In July 2015 ticket prices increased by 50% to $150,000.[10]

Passengers hoping to make flights in the Lynx include the winners from the Axe Apollo Space Academy worldwide perfume contest, and Justin Dowd of Worcester, Massachusetts, who won[11] Metro International's Race for Space newspaper contest.[12]

26.2 Description

The Lynx will have four liquid rocket engines at the rear of the fuselage burning a mixture of LOX-Kerosene, each engine producing 2,900 pounds-force (13,000 N) of thrust.[13]

26.2.1 Mark I Prototype

- Maximum Altitude: 62 km (203,000 ft)[2]

- Primary Internal Payload: 120 kg (260 lb)*[14]

- Secondary payload spaces include a small area inside the cockpit behind the pilot or outside the vehicle in two areas in the aft fuselage fairing.*[14]

- Composite LOX tank*[15]

- Mach 2 (1,522 mph) speed of ascent*[16]

- 4G re-entry loading*[16]

26.2.2   Mark II Production Model

- Maximum Altitude: 107 km (351,000 ft)*[2]

- Primary Internal Payload: 120 kg (260 lb)*[17]*[18]

- Secondary payload spaces include the same as the Mark I.*[19]

- Non-toxic (non-hydrazine) reaction control system (RCS) thrusters,*[20] type 3N22*[21]

- Nonburnite LOX composite tank*[22]

26.2.3   Mark III

The Lynx Mark III is the same vehicle as the Mark II with External Dorsal Mounted Pod: 650 kg (1,430 lb) and is large enough to hold a two-stage carrier to launch a microsatellite or multiple nanosatellites into low-Earth orbit.*[23]

26.2.4   Lynx XR-5K18 engine

The XR-5K18 is a piston pump fed LOX/RP-1 engine using an expander cycle.*[23] The engine chamber and regenerative nozzle are cooled by RP-1*[22]*[23]

The development program of the XCOR Lynx 5K18 LOX/kerosene engine reached a major milestone in March 2011. Integrated test firings of the engine/nozzle combination demonstrated the ability of the aluminum nozzle to withstand the high temperatures of rocket-engine exhaust.*[24]

In March 2011, United Launch Alliance (ULA) announced they had entered into a joint-development contract with XCOR for a flight-ready, 25,000 to 30,000 pounds-force (110,000–130,000 N) cryogenic LH2/LOX upper-stage rocket engine (see XCOR/ULA liquid-hydrogen, upper-stage engine development project). The Lynx 5K18 effort to develop a new aluminum alloy engine nozzle using new manufacturing techniques removes several hundred pounds of weight from the large engine leading to significantly lower-cost and more-capable commercial and US government space flights.*[25]

26.2.5   Airframe

It was reported in 2010 that the Mark I airframe could use a carbon/epoxy ester composite, and the Mark II a carbon/cyanate with a nickel alloy for the nose and leading-edge thermal protection.*[26]

26.3   Mark I build

The flight article Lynx Mark I is claimed as being fabricated and assembled in Mojave beginning in mid 2013.*[27] The cockpit of the Lynx (made of carbon fibre and designed by AdamWorks, Colorado) was reported as being one of the items that held up the assembly.*[14]

At the start of October 2014, the cockpit was attached the to fuselage.*[28] The rear carry-through spar was attached to the fuselage shortly after Thanksgiving 2014.*[29] At the beginning of May 2015, the strakes were attached to the airframe.*[30] The last major component, the wings, are expected to be delivered in late 2015.*[31]

26.4 Test program

Tests of the XR-5K18 main engine began in 2008.[*][32]

As of February 2011, engine tests were largely complete[*][20] and the vehicle aerodynamic design had completed two rounds of wind tunnel testing. A third and final round of tests was completed in late 2011 using a "1/60-scale supersonic wind tunnel model of Lynx." [*][16][*][20]

As of October 2014, XCOR claimed that flight tests of the Mark I prototype would start in 2015.[*][8][*][9][*][33]

26.5 Operations

26.5.1 NASA sRLV program

As of March 2011, XCOR has submitted the Lynx as a reusable launch vehicle for carrying research payloads in response to NASA's suborbital reusable launch vehicle (sRLV) solicitation, which is a part of NASA's Flight Opportunities Program. XCOR projects 110 km (68 mi) altitude in flights of 30 to 45 minutes duration, while carrying up to 140 kg (310 lb) internal—or 650 kg (1,430 lb) external—of research payload. Flights will provide up to three minutes of microgravity below 0.01 g.[*][34]

26.5.2 Commercial operations

According to XCOR, the Lynx will fly four or more times a day, and will also have the capacity to deliver payloads into space. A Lynx prototype called Mark I was expected to perform its first test flight in 2015,[*][1][*][9][*][35] followed with a flight of the Mark II production model twelve to eighteen months after.[*][9] XCOR currently plans to have the Lynx's initial flights from the Mojave Air and Spaceport in Mojave, California[*][36] or any licensed spaceport with a 2,400 meter (7900 ft) runway. Towards the end of 2015[*][37] or in 2016[*][1] the Lynx is expected to begin flying suborbital space tourism flights and scientific research missions from a new spaceport on the Caribbean island of Curaçao.[*][38][*][39][*][40]

Because it lacks any propulsion system other than its rocket engines, the Lynx will have to be towed to the end of the runway. Once positioned on the runway, the pilot will ignite the four rocket engines and begin a steep climb. The engines will be shut off at approximately 138,000 feet (42 km) and Mach 2. The spaceplane will then continue to climb, unpowered until it reaches an apogee of approximately 200,000 feet (61 km). The spacecraft will experience a little over four minutes of weightlessness before re-entering the Earth's atmosphere. The occupants of the Lynx may experience up to four times normal gravity during re-entry. Once it has completed re-entry, the Lynx will then glide down and perform an unpowered landing. The total flight time is projected to last about 30 minutes.[*][23] The Lynx is expected to be able to perform 40 flights before maintenance is required.

As of March 2011, Orbital Outfitters was designing pressure suits for XCOR use.[*][41]

As of 2012, the successor to the Mark II may be a two-stage fully reusable orbital vehicle that takes off and lands horizontally.[*][42]

26.5.3 Development costs

Mark I production is planned to cost $10 million,[*][43][*][44] and Mark II around $12 million.[*][45]

26.6 See also

- Private spaceflight

- EADS Astrium Space Tourism Project

- Rocketplane XP

- SpaceShipTwo

- Blue Origin New Shepard

- XCOR EZ-Rocket

- XCOR Mark-I X-Racer

26.7 References

[1] Woollaston, Victoria (14 March 2014) 'Budget' XCOR space trip set to launch in 2016 will let you pilot the ship for £57,000 Daily Mail, Retrieved 26 October 2014

[2] Belfiore, Michael. "XCOR Lynx: Don't Sleep on the Space Corvette". Popular Mechanics. Retrieved 2 October 2012.

[3] "MSDC president: Lynx will launch from Midland this fiscal year". *MRT.com*. Retrieved 31 October 2015.

[4] Space.com: *XCOR Zeroes in on Xerus, space.com*, 2003-05-19, accessed 2011-01-04.

[5] David, Leonard (2007-04-23). "XCOR Pursues Dream a Step at a Time". *Space.com*. Retrieved 2013-10-21.

[6] "XCOR AEROSPACE SUBORBITAL VEHICLE TO FLY WITHIN TWO YEARS". XCOR. 26 March 2008. Retrieved 4 December 2013.

[7] "Rocket company offers $95,000 trips to space". New Space/Reuters. 2 December 2008. Retrieved 21 July 2015.

[8] Schilling, Govert (September 16, 2013). "Lynx Space Plane Taking Off: Q&A with XCOR Aerospace CEO Jeff Greason". Space.com. Retrieved October 23, 2013.

[9] Norris, Guy (8 October 2014) XCOR Lynx Moves Into Final Assembly Aviation Week, Retrieved 20 January 2015

[10] . Parabolic Arc. July 2015 http://www.parabolicarc.com/2015/07/17/xcor-hike-ticket-prices-50-percent/. Retrieved 21 July 2015. Missing or empty |title= (help)

[11] METRO WORLD NEWS (13 October 2014). "America's Justin Dowd wins Metro's Race for Space". *Metro*. Retrieved 28 June 2015.

[12] "Race for Space". *metroinspace.com*. Archived from the original on 13 June 2013. Retrieved 31 October 2015.

[13] "Rocket Test Paves Way For XCOR Lynx Flights". Aviation Week. 28 March 2013. Retrieved 8 November 2013.

[14] Belfore. Michael (November 2013) "The Lynx's Leap, Can a suborbital spaceship help XCOR reach orbit?" *Air & Space Magazine*, Smithsonian, Retrieved 14 October 2013

[15] "The Private Space Race". Composites World. 31 August 2010. Retrieved 8 November 2013.

[16] Joiner, Stephen (2011-05-01). "The Mojave Launch Lab". *Air & Space Smithsonian*. Retrieved 2013-11-03.

[17] Spark, Joel (2 March 2012). "XCOR, Southwest Research Institute Move Up Suborbital Payload Testing". Space Safety Magazine. Retrieved 4 December 2013.

[18] Spark, Joel (2 March 2012). "XCOR, Southwest Research Institute Move Up Suborbital Payload Testing". Space Safety Magazine. Retrieved 4 December 2013.

[19] "XCOR Aerospace Lynx". Zap 16. 2011-12-30. Retrieved 2014-06-27.

[20] Foust, Jeff (2011-02-28). "Suborbital back out of the shadows". *The Space Review*. Retrieved 2011-02-28. *the 5K18 engine, four of which will power the Lynx ... the last few technical milestones for the engine are largely complete. ... non-toxic reaction control system (RCS) thrusters, a project that Greason said was more challenging in some respects than the larger main engine, but critical to the company's vision of rapid turnaround times that would not be possible if conventional hydrazine RCS systems are used. The Lynx design has been through two rounds of wind tunnel tests, with a final round planned for later this year for some final tweaks*

[21] "A Spaceplane Is Born". MoonandBack. 2013-09-30. Retrieved 2014-05-01.

[22] "Lee Valentine on How XCOR Will Open Up Space". parabolicarc.com. March 19, 2012.

[23] "XCOR Aerospace's multi-talented Lynx spaceplane set for KSC". nasaspaceflight.com. August 27, 2012. Retrieved May 1, 2014.

[24] "Demo'd is a revolutionary rocket engine nozzle and a new engine development partnership". Satnews. 2011-03-22. Retrieved 2014-06-27.

[25] Morring, Frank, Jr. (2011-03-23). "ULA, XCOR to Develop Upper-Stage Engine". *Aviation Week*. Retrieved 2011-03-25. *United Launch Alliance (ULA) and XCOR Aerospace are planning a joint effort to develop a low-cost upper-stage engine in the same class as the venerable RL-10, using technology XCOR is developing for its planned Lynx suborbital spaceplane. The two companies have been testing actively cooled aluminum nozzles XCOR is developing for its liquid oxygen/kerosene 5K18 engine for the Lynx, a reusable two-seat piloted vehicle the company plans to use for commercial research and tourist flights.*

[26] "The Private Space Race". compositesworld.com. 2010-08-31.

[27] Messier, Doug (2013-09-19). "XCOR Follow the Build Looks at Subsonic Wind Tunnel Testing". *Parabolic Arc*. Retrieved 2013-10-12.

[28] James Dean (10 October 2014). "XCOR installs cockpit into Lynx space place". Florida Today.

[29] "XCOR Lynx suborbital spacecraft nears final assembly". Composites World. 23 December 2014.

[30] "XCOR Aerospace Announces Strakes Bonded to Lynx Mark I Spacecraft". SpaceRef. 8 May 2015.

[31] Jeff Foust (22 July 2015). "XCOR To Raise Ticket Prices for Suborbital Flights". Space News.

[32] Keith Cowing (December 17, 2008). "Successful First Test Fire of Engine for Lynx Suborbital Launch Vehicle". NASA Watch.

[33] "The Age of Space Flights is about to begin". flyfighterjet.com. 2013-10-12.

[34] "sRLV platforms compared". NASA. 2011-03-07. Retrieved 2011-03-10. Lynx: Type: HTHL/Piloted

[35] Belfiore, Michael, (9 January 2013) Lynx Rocket Plane Readying for Summer Flight Moon and Back, Retrieved 5 April 2013

[36] "XCOR Unveils New Suborbital Rocketship". SPACE.com.

[37] Nilsson, Eric and Zhangyu, Deng (25 October 2014) "The Final Frontier" Daily Telegraph supplement "China Watch" Page 1

[38] (2012) SXC - Buying your tickets into space! SXC web page, Retrieved 5 April 2013

[39] Staff writers (October 6, 2010). "Space Expedition Corporation Announces Wet Lease of XCOR Lynx Suborbital". *Space Media Network Promotions*. Space-Travel.com. Retrieved 2010-10-06.

[40] "Space Experience Curacao". *Home*. Space Experience Curacao. 2009–2010. Retrieved 2010-10-06.

[41] "Commercial Spacesuit Companies Compete for Market Share". Parabolic Arc. March 21, 2011. Retrieved October 30, 2013.

[42] "The Next Frontier: An Interview with Lee Valentine". 30 November 2012. Retrieved 29 June 2015.

[43] ANDY PASZTOR (March 26, 2008). "Economy Fare ($100,000) Lifts Space-Tourism Race". wsj.com.

[44] John Antczak (2008-03-27). "New rocket aims for space tourism market". msnbc.msn.com.

[45] Jeff Foust (March 31, 2008). "One size may not fit all". thespacereview.com.

26.8   External links

- Lynx Suborbital Spacecraft Page

Chapter 27

MAKS (spacecraft)

The **MAKS (Multipurpose aerospace system)** (Russian: МАКС (Многоцелевая авиационно-космическая система)) is a cancelled Soviet air-launched with orbiter Reusable launch system project that was proposed in 1988, but cancelled in 1991. The orbiter was supposed to reduce the cost of transporting materials to Earth orbit by a factor of ten. The reusable orbiter and its external non-reusable fuel tank, was to have been launched by an Antonov AN-225 airplane, developed by Antonov ASTC (Kyiv, Ukraine). Had it been built, the system would have weighed 275 metric tons (271 long tons; 303 short tons), and would have been capable of carrying a 7-metric-ton (6.9-long-ton; 7.7-short-ton) payload.*[1]

Three variants of the MAKS system were conceived: MAKS-OS, the standard configuration; MAKS-T, with upgraded payload capability; and MAKS-M, a version that included its fuel tank within the envelope of the orbiter.*[2]

As of June 2010, Russia is considering reviving the MAKS program.*[3] In Ukraine this project has developed into other air-launched orbiter projects, such as Svityaz and Oril.

27.1   See also

- Air launch to orbit

- Buran programme

- State Space Agency of Ukraine#Svityaz project

- RD-701 - main engine

27.2   References

[1] "Maks Air Launch System". *Aerospaceguide.net*. 11 November 2010. Retrieved 22 December 2010.

[2] Lukashevich, Vadim (2005). "Multipurpose Aerospace System (MAKS)". Retrieved 22 December 2010.

[3] Hsu, Jeremy (3 June 2010). "High-Tech Space Planes Taking Shape in Italy, Russia". *Space.com*. Retrieved 22 December 2010.

Chapter 28

Martin X-23 PRIME

The Martin **X-23A PRIME** (Precision Reentry Including Maneuvering reEntry) was a small lifting body re-entry vehicle tested by the United States Air Force in the mid-1960s. Unlike ASSET, primarily used for structural and heating research, the X-23 PRIME was developed to study the effects of maneuvering during re-entry of Earth's atmosphere, including cross-range maneuvers up to 710 statute miles (1143 km) off of the ballistic track.

28.1   Design

Each X-23 was constructed from titanium, beryllium, stainless steel, and aluminium. The craft consisted of two sections —the aft main structure and a removable forward "glove section." The structure was completely covered with a Martin-developed ablative heat shield 20 to 70 mm (¾ to 2¾ inches) thick, and the nose cap was constructed of carbon phenolic material.

Aerodynamic control was provided by a pair of 12-inch (30 cm) square lower flaps, and fixed upper flaps and rudders. A nitrogen gas reaction control system was used outside the atmosphere. At Mach 2 a drogue ballute deployed and slowed the vehicle's descent. As it deployed, its cable sliced the upper structure of the main equipment bay, allowing a 47-foot (16.4 m) recovery chute to deploy. It would then be recovered in midair by a specially-equipped JC-130B Hercules aircraft.

28.2   Flight testing

The first PRIME vehicle was launched from Vandenberg AFB on 21 December 1966 atop an Atlas launch vehicle. This mission simulated a low Earth orbit reentry with a zero cross-range. The ballute deployed at 99,850 feet (30.43 km), though the recovery parachute failed to completely deploy. The vehicle crashed into the Pacific Ocean.

The second vehicle was launched on 5 March 1967. This flight simulated a 654-mile (1053-kilometre) cross-range reentry, and banking at hypersonic speeds. The recovery parachute deployed properly, and was located by two of the deployed recovery aircraft. During an inspection fly-by of the descending parachute system it was seen that reefing cutters had failed to actuate. These cutters are on the harness suspending the vehicle from the parachute to insure stability of the vehicle behind the JC-130B recovery aircraft during reel-in, and permit safely boarding the vehicle. Perforce, the parachute and vehicle were allowed to descend to the sea. Subsequently, the vehicle separated from its flotation "balloon" in the rough seas and, with the parachute, sank before a nearby range ship could arrive to retrieve it from the ocean.

The final PRIME mission was flown on 19 April 1967, and simulated reentry from low Earth orbit with a 710-mile (1143-kilometre) cross-range. This time, all systems performed perfectly, and the X-23 was successfully recovered. An inspection by a USAF-Martin team reported the craft "ready to fly again," although no later missions were carried out. The third X-23 is now on display at the National Museum of the United States Air Force at Wright-Patterson Air Force Base in Ohio.

28.3   Specifications (X-23)

General characteristics

- **Crew:** None
- **Length:** 6 ft 9 in (2.07 m)
- **Wingspan:** 3 ft 10 in (1.16 m)
- **Height:** 2 ft 1 in (0.64 m)
- **Loaded weight:** 890 lb (405 kg)
- **Powerplant:** × Nitrogen gas reaction control thrusters

Performance

- **Maximum speed:** Mach 25
- **Range:** 710 miles (1,143 km)

- **Hypersonic L/D Ratio:** 1:1

28.4   See also

Aircraft of comparable role, configuration and era

Molniya BOR-4
ASSET

28.5   References

- Dennis R. Jenkins, Tony Landis, Jay Miller: *AMERICAN X-VEHICLES An Inventory—X-1 to X-50*, Monographs in Aerospace History No. 31, SP-2003-4531, June 2003 (PDF)

28.6   External links

- Encyclopedia Astronautica

Chapter 29

Mikoyan-Gurevich MiG-105

The **Mikoyan-Gurevich MiG-105** part of a program known as the **Spiral (aerospace system)** was a manned test vehicle to explore low-speed handling and landing.[*][2] It was a visible result of a Soviet project to create an orbital spaceplane. This was originally conceived in response to the American Boeing X-20 Dyna-Soar military space project and may have been influenced by contemporary manned lifting body research being conducted by NASA's Flight Research Center in California. The MiG 105 was nicknamed "Lapot" Russian: лапоть, or bast shoe (the word is also used as a slang for "shoe") for the shape of its nose.

29.1 Development

The program was also known as EPOS (Russian acronym for Experimental Passenger Orbital Aircraft). Work on this project finally began in 1965, two years after Dyna-Soar's cancellation. The project was halted in 1969, to be briefly resurrected in 1974 in response to the U.S. Space Shuttle Program. The test vehicle made its first subsonic free-flight test in 1976, taking off under its own power from an old airstrip near Moscow. It was flown by pilot Aviard G. Fastovets to the Zhukovskii flight test center, a distance of 19 miles. Flight tests, totaling eight in all, continued sporadically until 1978. The actual space plane project was cancelled when the decision was made to instead proceed with the Buran project. The MiG test vehicle itself still exists and is currently on display at the Monino Air Force Museum in Russia.

Gleb Lozino-Lozinskiy was the leader of the Spiral development programme.

29.2 Differences between Dyna-Soar and Spiral

Although having basically the same mission, Dyna-Soar and Spiral were radically different vehicles. For example:

- While the X-20 Dyna-Soar was designed for launch atop a conventional expendable rocket such as the Titan III-C or Saturn I, Soviet engineers opted for a midair launch scheme for Spiral. Known as "50 / 50", the idea was that the spaceplane and a liquid fuel booster stage would be launched at high altitude from the back of a large, airbreathing mothership travelling at hypersonic speeds. The mothership was to have been built by the Tupolev Design Bureau (OKB-156) and utilize many of the same technologies developed for the Tu-144 supersonic transport and the Sukhoi T-4 Mach 3 bomber. It was never built.

- Dyna-Soar was designed as a lifting body, while Spiral was a conventional delta that featured an innovative variable-dihedral wing. During launch and reentry, these were folded upward at 60 degrees. After dropping to subsonic speeds post-reentry, the pilot lowered the wings into the horizontal position, giving the spaceplane better re-entry and flight characteristics.

- Spiral was built to allow for a powered landing and go-around maneuver in case of a missed landing approach. An air intake for a single Kolesov turbojet was mounted beneath the central vertical stabilizer. This was protected during launch and re-entry by a clamshell door which opened at subsonic speeds. By comparison, Dyna-Soar was designed primarily for a once-off, unpowered deadstick landing.

- High temperature superalloy metals such as niobium, molybdenum, tungsten and rene 41 were to have been used in the heatshield structure of the X-20. Spiral was to have been protected by what Soviet engineers termed "scale-plate armour": steel plates mounted on articulated ceramic bearings to allow for thermal expansion during reentry. Several BOR (Russian acronym for Unpiloted Orbital Rocketplane) craft were flown to test this concept.

- In the event of a booster explosion or in-flight emergency, the crew compartment of Spiral was designed to separate from the rest of the vehicle and parachute to earth like a conventional ballistic capsule; this could be done at any point in the flight. Such an escape crew capsule was also considered for Dyna-Soar, but American engineers eventually opted for a solid-fuel escape rocket that would kick the spaceplane away from an exploding booster, hopefully saving both pilot and spacecraft.

- Much like the Space Shuttle, Dyna-Soar was designed with a small payload bay behind the pressurized crew module. This could be used for lofting small satellites, carrying surveillance equipment, weapons or even an extra crewmember in a pop-in cockpit. Spiral, on the other hand, was intended to carry only its pilot.

- Both Dyna-Soar and Spiral were designed to land on skids. The landing skids on Dyna-Soar were designed to deploy from insulated doors on the underside of the vehicle, like a conventional aircraft. Soviet engineers designed the landing skids on Spiral to deploy from a set of doors on the sides of the fuselage just above and ahead of the wings.

29.3   Pilots

A cosmonaut training group for pilots assigned to fly this vehicle was formed in the early 1960s. It went through many changes and was eventually dissolved entirely. Known members included:

- Gherman Titov, the second man to orbit the Earth (see Vostok 2 mission).

- Vasily Lazarev, Cosmonaut who would later fly the first Soyuz 7K-T mission (see Soyuz 12)

- Aviard G. Fastovets, who piloted the vehicle during the majority of its atmospheric tests.

29.4   BOR

The *БОР* (Russian: Беспилотный Орбитальный Ракетоплан, *Bespilotnyi Orbital'nyi Raketoplan*, "Unpiloted Orbital Rocketplane"). Another spacecraft to use the Spiral design was the BOR series, unmanned subscale reentry test vehicles. American analogs X-23 PRIME and ASSET. Several of these craft have been preserved in aerospace museums around the world.

29.5   Operators

Soviet Union

- Soviet Air Force

29.6   Specifications (MiG 105-11)

Data from Soviet X-planes; Yefim Gordon, Bill Gunston

General characteristics

- **Crew:** 1

- **Length:** 10.6m (including instrument boom) (34 ft 9 in)

Gherman Titov, the second cosmonaut and the main test pilot of the MiG-105.

- **Wingspan:** 6.7m (21 ft 12 in)

- **Height:** m (ft in)

- **Wing area:** 24m^2 (258ft^2)

- **Empty weight:** 3,500 Kg (7,716 lb)

- **Loaded weight:** 4,220 Kg (9,303 lb)

- **Useful load:** Kg (lb)

- **Max. takeoff weight:** Kg (lb)

- 500 Kg Fuel

Performance

- **Never exceed speed:** km/h (knots, mph)

- **Maximum speed:** 800km/h (432 knots, 500 mph)

- **Cruise speed:** km/h (knots, mph)

- **Stall speed:** km/h (knots, mph)

- **Range:** km (nm, mi)

- **Service ceiling:** m (ft)

- **Rate of climb:** m/s (ft/min)

- **Wing loading:** 175 kg/m^2 (lb/ft^2)

Landing speed 250-270 km/h (155-168 mph)

29.7   See also

- BOR-4

- Buran Shuttle

- Dream Chaser

- Space Shuttle

Aircraft of comparable role, configuration and era

- X-20 Dyna-Soar

- X-24

- NASA X-38 Crew Return Vehicle

29.8   References

[1] Soviet X-planes; Yefim Gordon, Bill Gunston

[2] Soviet X-planes; Yefim Gordon, Bill Gunston

29.9   External links

- Predecessor of Shuttle and Buran: Spiral Orbital Aircraft Programme

- Spiral and EPOS project

- Spiral OS

- Spiral 50-50

- Spiral, MIG 105, Uragan

Chapter 30

MUSTARD

For other uses, see Mustard (disambiguation).

The **Multi-Unit Space Transport And Recovery Device** or **MUSTARD** was a concept explored by the British Aircraft Corporation (BAC) around 1968 for launching payloads weighing as much as 5,000 lb. into orbit.

30.1 History

The project started life as the English Electric Mustard. For one year, collaborative work was done at Edwards Air Base. Once this collaborative work was over, three American prototype similar-looking aircraft appeared at Edwards Air Base. The Space Shuttle would later have a comparative design, and function.

30.2 Design

MUSTARD was a delta-winged three-stage reusable vehicle which used the triamese concept.[*][1][*][2] The design team was led by Tom Smith, Chief of the Aerospace Department at BAC.[*][3]

30.3 Rocket stages

The three components of the design were three largely identical lifting bodies (each similar to the Northrop HL-10), stacked back-to-belly.

The units would be stacked for launch, and two of them would act as boosters to launch the third into Earth orbit. The booster units would feed any excess fuel to the unit which was to be the spacecraft. At 150,000 to 200,000 ft. (45,750 to 60,960 m), at around 30 nautical miles, the booster units would separate and land like aircraft.

The spacecraft would place its payload into orbit at around 1000 nautical miles, after 10 minutes from launch, and then return in a like manner.

30.4 See also

- List of space launch system designs

30.5 References

[1] Britain in Space

[2] Flight International 10 March 1966

[3] Flight Global 24 March 1966

30.6   External links

- Encyclopedia Astronautica - Mustard

- Unreal Aircraft - Weird Wings - BAC MUSTARD

- MUSTARD scale model (white one in display case backside)

- ^ same as above ^ (See also Martin Marietta Spacemaster)

- Economist, June 21, 2013, Thunderbirds are gone: A British defence firm opens its archives to reveal flights of fancy that never flew

Chapter 31

NASA X-38

The **X-38 Program,** under leadership of the NASA-Johnson Space Center, was focused on developing the technology for a prototype emergency Crew Return Vehicle for the International Space Station (ISS). The project also intended to develop a crew return vehicle design that could be modified for other uses, such as a possible joint U.S. and international human spacecraft that could be launched on the French Ariane 5 booster.[1]

The program would eventually develop a total of three test prototype flight demonstrators for the proposed Crew Return Vehicle, each having incremental improvements on its predecessor. All three were wingless lifting body vehicles used in drop tests. The X-38 program was cancelled in 2002 due to budget cuts.[2]

31.1   History

The maximum crew size for the ISS is dependent on crew rescue capacity. Since it is imperative that the crew members be able to return to Earth in case of an unexpected emergency, a Crew Return Vehicle able to hold up to seven crew members was initially planned by the ISS program leadership. This would have allowed the full complement of seven astronauts to live and work on the ISS.

During the early years of ISS on-orbit construction, the crew was limited to three, corresponding to a single Russian Soyuz TMA vehicle that could be docked to the station at any given time. Later in May 2009 provisions were added for a total of two docked Soyuz vehicles simultaneously and the ISS crew was increased to 6 members. NASA has designed several crew return vehicles over the years with varying levels of detail.[3]

A small, in-house development study of the X-38 concept first began at JSC in early 1995, however several types of emergency scenarios were recognized by NASA as early as 1992 that drove the need for crew return from the International Space Station: [4]
1. A serious illness or injury to a station astronaut
2. A fire or collision with space debris
3. Grounding of the space shuttle so that it could not deliver life-sustaining supplies.

In early 1996, a contract was awarded to Scaled Composites, Inc., of Mojave, Calif., for the construction of three full-scale atmospheric test airframes. The first vehicle airframe was delivered to JSC in September 1996.[5]

31.2   Development

In an unusual move for an X-plane, the program involved the European Space Agency and the German Space Agency DLR. It was originally called **X-35**. The program manager was John Muratore, while the Flight Test Engineer was future NASA astronaut Michael E. Fossum.

The X-38 design used a wingless lifting body concept originally developed by the U.S. Air Force in the mid-1960s during the X-24 program.

The X-38 program used unmanned mockups to test the CRV design. Flight models were indicated with the letter V for "Vehicle" followed by a number.

The X-38 V-132 research vehicle drops away from NASA's B-52 mothership immediately after being released from the wing pylon

- X-38 V-131

- X-38 V-132

- X-38 V-131-R, which was the V-131 prototype reworked with a modified shell

- X-38 V-201, which was an orbital prototype to be launched by the Space Shuttle

- X-38 V-121, V-133 and V-301 were also foreseen, but were never built.

The X-38 V-131 and V-132 shared the aerodynamic shape of the X-24A. This shape had to be enlarged for the Crew Return Vehicle needs (crew of seven astronauts) and redesigned, especially in the rear part, which became thicker.

The X-38 V-131-R was designed at 80 percent of the size of a CRV [24.5 ft long (7.5 m), 11.6 ft wide (3.5 m), 8.4 ft high (2.6 m)], and featured the final redesigned shape (Two later versions, V-133 and V-201, were planned at 100 percent of the CRV size). The 80% scale versions were flown at 15,000 to 24,000 pound weight. The X-38 V-201 orbital prototype was 90 percent complete, but never flown.

In drop tests the V-131, V-132 and V-131-R were dropped by a B-52 from altitudes of up to 45,000 ft (13,700 m), gliding at near transonic speeds before deploying a drogue parachute to slow them to 60 miles per hour (97 km/h). The later prototypes had their descent continue under a 7,500-square-foot (700 m^2) parafoil wing, the largest ever made.[*][6] Flight control was mostly autonomous, backed up by a ground-based pilot.

31.3 Design

The X-38 was intended to be semi-permanently docked to the ISS. If the crew became sick or injured during the course of their mission, they would enter the rescue vehicle through a hatched docking mechanism. With execution of a short procedure, the crew return vehicles would automatically fly the crew members safely to Earth. Once undocked,

The X-38 CRV prototype makes a gentle lakebed landing at the end of a July 1999 test flight at the Dryden Flight Research Center with a fully deployed parafoil.

the vehicle would be deorbited using a deorbital propulsion system (DPS). The eight-thruster DPS would adjust the spacecraft's attitude and retrofire to slow the X-38 down, allowing gravitational attraction to pull it back into Earth's atmosphere. A DPS module was developed by Aerojet and delivered to Johnson Space Center in 2002 for V-201.

Following the jettison of the DPS, the X-38 would have glided from orbit and used a steerable parafoil for its final descent and landing. The high speeds at which lifting body aircraft operate make them dangerous to land. The parafoil would have been used to slow the vehicle and make landing safer. The landing gear consisted of skids rather than wheels: the skids worked like sleds so the vehicle would have slid to a stop on the ground.

Both the shape and size of the X-38 were different from that of the Space Shuttle. The Crew Return Vehicle would have fit into the payload bay of the shuttle. This does not, however, mean that it would have been small. The X-38 weighed 10,660 kg and was 9.1 meters long. The battery system, lasting nine hours, was to be used for power and life support. If the Crew Return Vehicle was needed, it would only take two to three hours for it to reach Earth.

The parafoil parachute, employed for landing, was derived from technology developed by the U.S. Army. This massive parafoil deploys in 5 stages for optimum performance. A drag chute would have been released from the rear of the X-38. This drag chute would have been used to stabilize and slow the vehicle down. The parafoil (area of 687 square meters) was then released. It would open in five steps (a process called staging). While the staging process only takes 45 seconds, it is important for a successful chute deployment. Staging prevents high-speed winds from tearing the parafoil.

The spacecraft's landing was to be completely automated. Mission Control would have sent coordinates to the onboard computer system. This system would also have used wind sensors and the Global Positioning System (a satellite-based coordinate system) to coordinate a safe trip home. Since the Crew Return Vehicle was designed with medical emergencies in mind, it made sense that the vehicle could find its way home automatically in the event that crew members were incapacitated or injured. If there was a need, the crew would have the capability to operate the vehicle by switching to the backup systems. In addition, seven high altitude low opening (HALO) parachute packs were included in the crew cabin, a measure designed to provide for the ability to bail out of the craft.

An Advanced Docking Berthing System (ADBS) was designed for the X-38 and the work on it led to the Low Impact

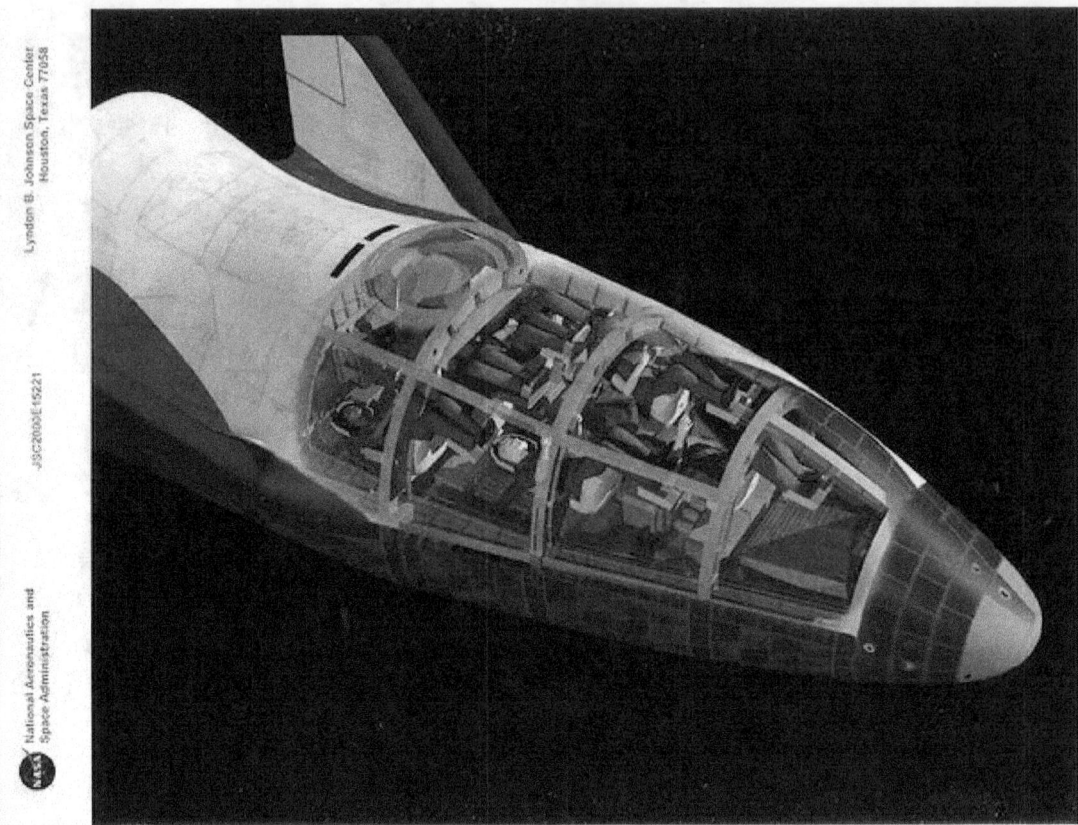

Graphical rendering of the X-38, with vehicle cutaway revealing 7-member crew's position during re-entry.

Docking System the Johnson Space Center later created for the planned vehicles in Project Constellation.

The X-38 vehicle was also known as the X-35 (but that designation was already allocated by the USAF to another vehicle) and X-CRV (experimental - Crew Return Vehicle)

31.4   Cancellation and Vehicle Redeployment

Severe cost overruns plagued the ISS program during its development and construction during the late 1990s and early 2000s. To bring costs under control the International Space Station Management and Cost Evaluation (IMCE) Task Force was created. The task force introduced a new concept known as "American Core Complete," whereby the U.S. would unilaterally reduce the previously agreed-upon American contributions to the ISS while retaining its role as the controlling member of the International partners. Core Complete (as opposed to the originally planned "Station Complete") deleted the American Habitation Module, the American CRV, and Node-3 from the ISS design without any negotiations with international partners. NASA Administrator, Sean O'Keefe, appointed by President George W. Bush, stated in December 2001 that he intended to adhere to the recommendations of the IMCE, including the implementation of Core Complete. The X-38 project cancellation was announced on April 29, 2002*[2] as a cost-cutting measure in accordance with the IMCE's recommendations.

The Core Complete concept was roundly criticized by many experts at the time since a majority of development work on the X-38 had been completed. The prototype space vehicle was approximately 90% complete at the time it was canceled.

The X-38 V-132 is now on permanent loan from NASA to the Strategic Air and Space Museum at Ashland Nebraska. As of October 2015 the 90% complete X-38 V-201, having been moved out of Building 220 at Johnson Space Center is now sitting outside Building 49 wrapped in construction webbing at Johnson Space Center, Houston, Texas. As of November 2010, the X-38 V-131R is on loan from NASA to the Evergreen Aviation Museum in McMinnville, Oregon.

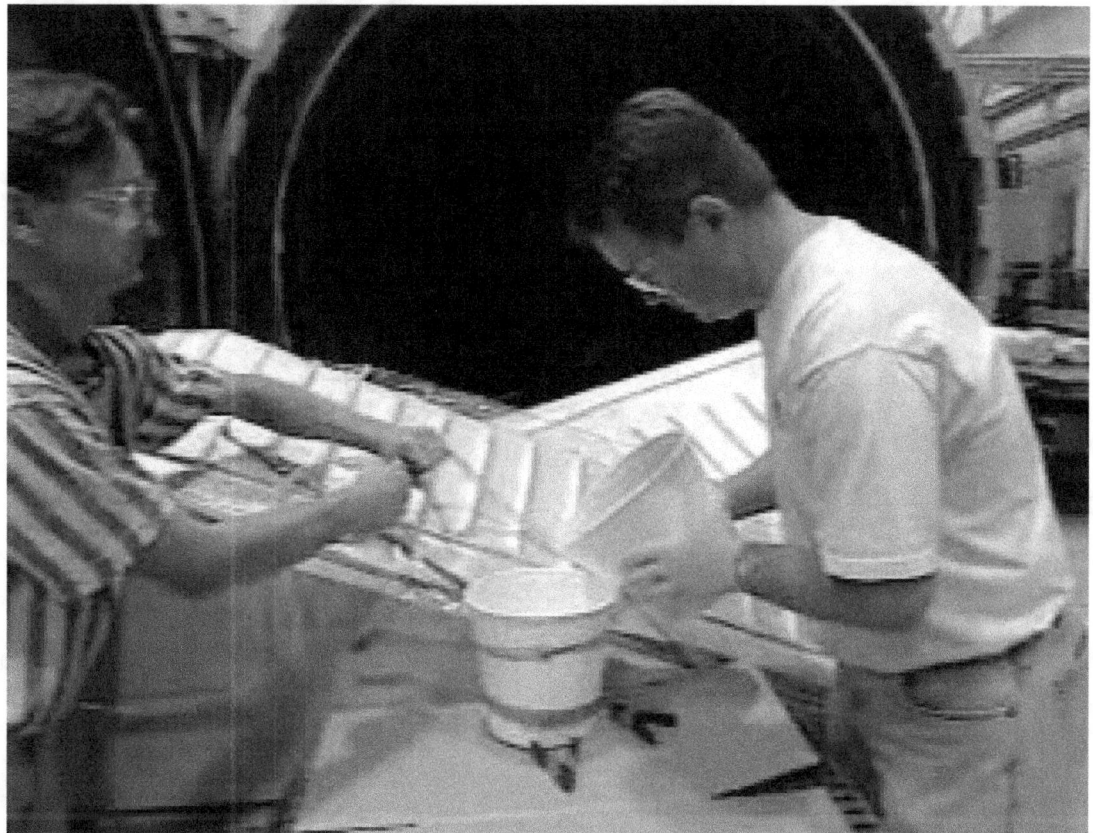

X-38: Low-Cost, High-Tech Space Rescue

31.5 Specifications

(For operational X-38 Crew Return Vehicle)*[7]

General characteristics

- **Crew:** seven astronauts (438 cu.ft / 12.4 cubic meters cabin)

- **Length:** 30 ft 0 in (36 ft including Deorbit Propulsion System rocket) (9.144 m (11 m including Deorbit Propulsion System rocket))

- **Wingspan:** 14 ft 6 in (15.5 ft with DPS) (4.42 m (4,729 m with DPS))

- **Height:** ()

- **Empty weight:** 23,500 lb (10,660 kg)

- **Loaded weight:** 25,000 lb (31,000 lb with deorbit propulsion module) (11,340 kg (14,061 kg with deorbit propulsion module))

- **Powerplant:** ×

Performance

31.6 See also

- Crew Return Vehicle

- International Space Station

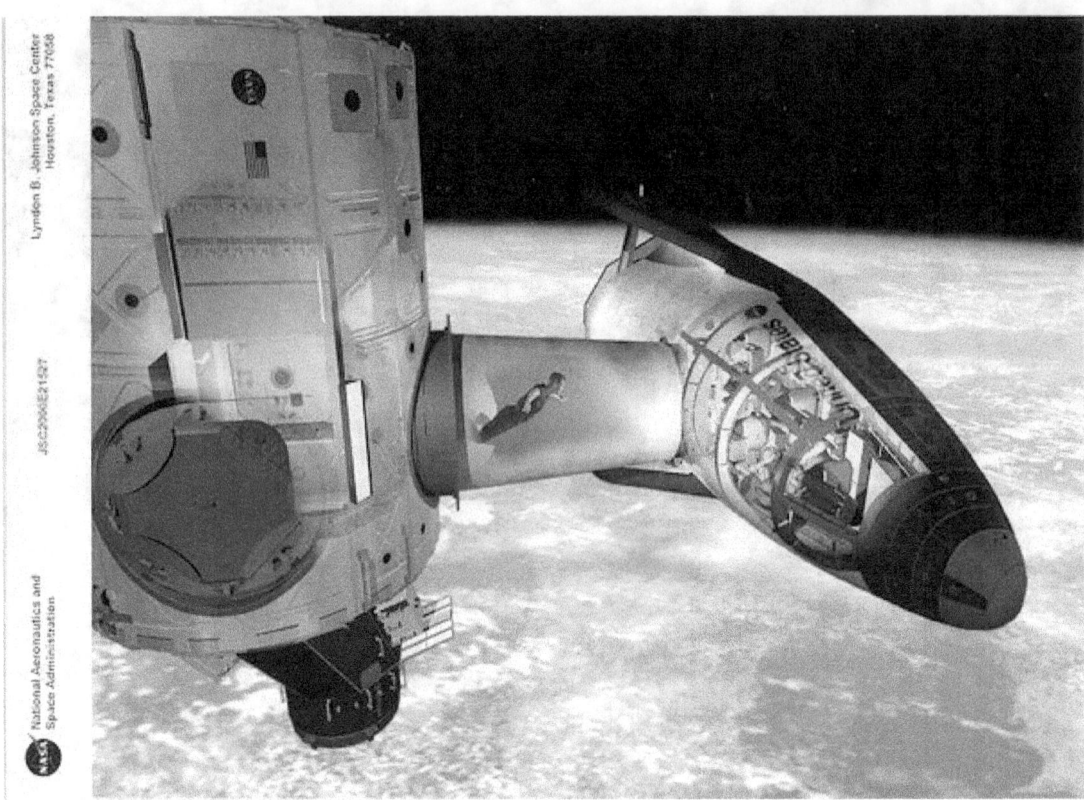

Artist's rendering of a docked X-38 being ingressed by a crew member through a docking mechanism.

Conceptual depiction of the deorbital propulsion system (DPS) attached to the rear of a crew return vehicle. The DPS would fire its eight thrusters to slow the spacecraft to below orbital velocity in order to re-enter Earth's atmosphere.

The X-38 Development Team with V131R, V132, and V201 on the east side of B220 at the Johnson Space Center at the close of the project (2003)

- Lifting body

- Pioneer Aerospace Corporation

Related development

- Martin-Marietta X-24

- HiMAT Remotely Piloted Vehicle

Aircraft of comparable role, configuration and era

- NASA M2-F1

- Northrop M2-F2

- Northrop M2-F3

- Northrop HL-10

- MiG-105

- SpaceDev Dream Chaser

31.7 Notes

[1] "NASA X-38 Project Description". NASA. Retrieved 2015-04-18.

X-38 V-201 orbital test vehicle previously located at Bldg. 220 at Johnson Space Center. Now held in the South end of Building 10, Houston, Texas

[2] "X-38". Federation of American Scientists. Retrieved 2006-09-20.

[3] Marcus Lindroos. "Nasa acrv". Encyclopedia Astronautica. Retrieved 2007-01-05.

[4] Carreau, Mark (June 9, 2002). "X-38 project's cancellation irks NASA, partners". *chron.com*. Houston Chronicle. Retrieved 2015-10-06. a serious illness or injury to a station astronaut; a serious fire or collision with space debris; or grounding of the space shuttle so that it could not deliver life-sustaining supplies.

[5] "NASA - Current Research Projects - X-38 CRV". NASA. Retrieved 2006-09-13.

[6] "X-38 Team Successfully Flies Largest Parafoil Parachute in History". NASA. Retrieved 2010-12-19.

[7] http://www.scribd.com/doc/46463919/The-X-38-Low-Cost-High-Tech-Space-Rescue#scribd

31.8 References

- Catchpole, John E. (2008). *The International Space Station: Building for the Future*. Praxis. p. 79. ISBN 978-0387781440.

- "NASA Dryden Fact Sheets". NASA. Retrieved 2006-09-13.

- "NASA - Current Research Projects - X-38 CRV". NASA. Retrieved 2006-09-13.

- "X38/CRV FDIR". NASA's Smart Systems Research Lab. Retrieved 2006-09-13.

- "Crew Return Vehicle (CRV)". ESA. Retrieved 2006-09-14.

X-38 V-201 orbital test vehicle as currently displayed atop its ground mobility carrier at NASA-Johnson Space Center behind Building 49.

31.9 External links

- NASA Dryden X-38 Photo Collection
- NASA Dryden X-38 Movie Collection

The fifth test drop flight of X-38. The aircraft is released from a B-52 mothership, free falls for a while, opens and fully deploys the parafoil and finally makes a gentle landing

Chapter 32

North American DC-3

The **DC-3** was a proposed space shuttle designed by Maxime Faget at the Manned Spacecraft Center (MSC) in Houston. The design was nominally developed by North American Aviation (NAA), although it was a purely NASA-internal design.

Unlike the eventual Space Shuttle design that emerged, the DC-3 was a fully reusable two-stage-to-orbit design with a smaller payload capacity of about 12,000 lbs and limited maneuverability. Its inherent strengths were good low-speed handling during landing, and a low-risk development that was relatively immune to changes in weight and balance.

Work on the DC-3 program ended when the US Air Force joined the Shuttle program; they demanded a much greater "cross-range" maneuverability than the DC-3 could deliver, and expressed serious concerns about its stability during re-entry. NAA eventually won the Shuttle Orbiter contract, although it was based on a very different design from another team at MSC.

32.1 History

32.1.1 Background

In the mid-1960s the US Air Force conducted a series of classified studies on next-generation space transportation systems. Among their many goals, the new launchers were intended to support a continued manned military presence in space, and so needed to dramatically lower the cost of launches and increase launch rates. Selecting from a series of proposals, the Air Force concluded that semi-reusable designs were the best choice from an overall cost basis, and the Lockheed Star Clipper design was one of the most-studied examples. They proposed a development program with an immediate start on a "Class I" vehicle based on expendable boosters, followed by a slower development of a "Class II" semi-reusable design, and perhaps a "Class III" fully reusable design in the further future. Although is it estimated that the Air Force spent up to $1 billion on the associated studies, only the Class I program that proceeded to development, as the X-20 Dyna-Soar, which was later cancelled.

Not long after the Air Force studies, NASA started studying the post-Project Apollo era. A wide variety of projects were examined, many based on re-using Apollo hardware (Apollo X, Apollo Applications Program, etc.) Flush with the success of the moon landings, a series of ever-more ambitious projects gained currency, a process that was considerably expanded under the new NASA director, Thomas O. Paine. By about 1970 these had settled on the near-term launching of a 12-man space station in 1975, expanding this to a 50-man "space base" by 1980, a smaller lunar-orbiting station, and then eventually a manned mission to Mars in the 1980s. NASA awarded $2.9-million study contracts for the space stations to North American and McDonnell Douglas in July 1969.

Almost as an afterthought the idea of a small and inexpensive "logistics vehicle" for supporting these missions developed in the late 1960s. George Mueller was handed the task of developing plans for such a system, and held a one-day symposium at NASA headquarters in December 1967 to study various options. Eighty people attended and presented a wide variety of potential designs, many from the earlier Air Force work, from small Dyna-Soar like vehicles primarily carrying crew and launched on existing expendable boosters, to much larger fully reusable designs.

32.1.2   ILRV

On 30 October 1968 NASA officially began work on what was then known as the "Integrated Launch and Re-entry Vehicle" (ILRV), a name they borrowed from the earlier Air Force studies. The development program was to take place in four phases; Phase A: Advanced Studies; Phase B: Project Definition; Phase C: Vehicle Design; and Phase D: Production and Operations. Four teams were to participate in Phase A; two in Phase B; and then a single prime contractor for Phases C and D. A separate Space Shuttle Main Engine (SSME) competition was to run in parallel.

NASA Houston and Huntsville jointly issued the Request for Proposal (RFP) for eight-month Phase A ILRV studies. The requirements were for 5,000 to 50,000 lb of payload to be delivered into a 500 km altitude orbit. The re-entry vehicle should have a cross range of at least 450 miles, meaning that it could fly to the left or right of its normal orbital path. General Dynamics, Lockheed, McDonnell-Douglas, Martin Marietta, and (the newly named) North American Rockwell were invited to bid. In February 1969, following study of the RFPs, Martin Marietta's entry was dropped, although they continued work on their own. The other entries were all given additional Phase A funding.

Supported by Paine's ambitious plans, in August 1969 the ILRV program was re-defined to be a "maximum effort" design, and only fully reusable designs would be accepted. This led to a second series of Phase A studies. The designs that were returned varied widely, meeting the huge payload range specified in the original RFP. Two basic fuselage designs seemed to be the most common; lifting body designs that offered high cross-range but limited maneuverability after re-entry, and delta-winged designs that reversed these criteria.

32.1.3   DC-3

Faget felt that all of the proposed designs incorporated an unacceptable amount of development risk. Unlike a conventional aircraft, with separate fuselage and wings, the ILRV designs had blended wing-body layouts. This meant that changes in weight and balance, which are almost unavoidable during development, would require changes to the entire orbiter structure to compensate. He also felt that the poor low-speed handling of any of these layouts presented a real danger during landing. Upset by what he felt was a project that seemed to guarantee failure, he started work on his own design, and presented it as the DC-3.

Unlike the other entries, DC-3 was much more conventional in layout, with an almost cylindrical fuselage and low-mounted slightly swept wings. The design looked more like a cargo aircraft than a spacecraft. Re-entry was accomplished in a 60 degree nose-high attitude that presented the lower surface of the spacecraft to the airflow, using a ballistic blunt-body approach that was similar to the one Faget had successfully pioneered on the Mercury capsule. During re-entry, the wings provided little or no aerodynamic lift. After re-entry, when the spacecraft entered the lower atmosphere, it would pitch over into a conventional flying attitude, ducts would open, and jet engines would start up for landing.

The upside of this design approach was that changes in the weight and balance could be addressed simply by moving the wing or re-shaping it, a common solution that had been used for decades in aircraft design—including the original Douglas DC-3 whose wings were swept rearward for just this reason. The downside was that the spacecraft would have little hypersonic lift, so its ability to maneuver while re-entering would be limited and its cross-range would be about 300 miles. It could make up for some of this with its improved low-speed flying ability, but would still not be able to match the mandated 450 miles.

Although the DC-3 had never been part of the original ILRV plans, Faget's name was so well respected that others at NASA MSC in Houston quickly rallied around him. Other NASA departments all selected their own favorite designs, including recoverable versions of Saturn boosters developed at the Marshall Space Flight Center in Huntsville, lifting-bodies based on the HL-10 that were favored by the Langley Research Center and Dryden Flight Research Center (Edwards), and even a single-stage-to-orbit Aerospaceplane were also proposed. From then on, the entire program was beset with in-fighting between the various teams. On 1 June 1969, a report was published that attacked the DC-3 design, followed by several others over the remainder of the year. In spite of this, North American quickly took up the DC-3 design, having learned over the years that the best way to win a NASA contract was to make whatever design Faget favored. They won contract NAS9-9205 to develop the DC-3 in December 1969.

In order to clear the logjam developing between the departments, on 23 January 1970 a meeting was held in Houston to study all of the in-house concepts. Over the next year a number of proposed designs would be dropped, including the entire series of lifting-body-derived vehicles as it proved too difficult to fit cylindrical tanks into the airframe. This left two basic approaches, delta wings and Faget's DC-3 series. Development of the DC-3 continued, with a drop test of a 1/10-scale model starting on 4 May.

32.1.4   Space Task Group

On 12 February 1969 Richard Nixon formed the Space Task Group under the direction of Vice President Spiro Agnew, giving them the task of selecting missions for a post-Apollo NASA. Agnew quickly became a proponent of NASA's ambitious plans that would culminate in a Mars attempt. The Task Group's final report, delivered on 11 September 1969, outlined three broad plans; the first required funding at $8 to $ 10 billion a year and would fulfill all of NASA's goals, the second would reduce this to $8 billion or less if the manned lunar orbiting station was dropped, and finally the third would require only $5 billion a year and would develop only the space stations and shuttle.

At first Nixon did not comment on the plans. Later he demanded that the program be greatly reduced even from the smallest of the Task Group's proposals, forcing them to select either the space base *or* the shuttle. Discussing the problem, NASA engineers concluded that the development of a shuttle would lower the cost of launching portions of the space station, so it seemed that proceeding with the shuttle might make the future development of the station more likely. However, NASA's estimates of the shuttle development costs were met with great skepticism by the Office of Management and Budget (OMB). Studies by RAND in 1970 showed that there was no benefit to developing a reusable spacecraft when development costs were taken into account. The report concluded that a manned station would be more cheaply supported with expendable boosters.

By this time Paine had left NASA to return to General Electric, and had been replaced by the more pragmatic James Fletcher. Fletcher ordered independent reviews of the shuttle concept; Lockheed was to prepare a report on how the shuttle could reduce payload costs, Aerospace Corporation was to make an independent report on development and operational costs, and Mathematica would later combine these two into a final definitive report. Mathematica's report was extremely positive; it showed that development of a fully reusable design would lower the per-launch cost, thereby reducing payload costs and driving up demand. However, the report was based on a greatly increased rate of launch; inherent in the math was the fact that lower launch rates would completely upset any advantage. Nevertheless, the report was extremely influential, and made the shuttle program an ongoing topic of discussion in Washington.

Looking to shore up support for the program, Fletcher directed NASA to develop the shuttle to be able to support the Air Force's requirements as well, as initially developed in their "Class III" fully reusable vehicles. If the shuttle became vital to the Air Force as well as NASA, it would be effectively unkillable. The Air Force's requirements were based about a projected series of large spy satellites then under development, which were 60 feet long and weighed 40,000 lbs. They needed to be launched into polar orbits, corresponding to a normal launch from Kennedy Space Center (KSC) of 65,000 lbs (launches to the east receive a free boost from the Earth's natural rotation).

The Air Force also demanded a cross-range capability of 1,500 miles, meaning that the spacecraft would have to be able to land at a point 1,500 miles (2,400 km) to either side of its orbital path when it started re-entry. This was due to the desire to be able to land again after one orbit, the so-called "orbit-once-around".

32.1.5   End of DC-3

The new Air Force cross-range requirements doomed the DC-3 design.

Spacecraft orbit around the center of the Earth, not the surface. If a spacecraft is launched due East from the equator into a 90 minute low-Earth orbit, it will circle the Earth and return to the spot where it was launched 90 minutes later. During this time, however, the launch site will have moved due to the Earth's rotation. Over the 90 minute period, the Earth would rotate about 1,500 kilometres (930 mi) to the west, towards the spacecraft. Given a spacecraft speed of about 17,000 miles per hour (27,000 km/h), simply starting the re-entry a few minutes earlier than the complete 90 minute orbit would make up this difference.

At KSC's ~30 degree north latitude the picture is similar. Over the same 90 minute orbit KSC will rotate about 1,350 miles (2,170 km). Unlike the equatorial orbit case, however, letting the spacecraft stay in the inclined orbit a little longer will start taking it south of the launch site, its closest point of approach being about 300 miles (480 km) to the southwest. A spacecraft wishing to return to its launch site will need about 300 miles of cross-range maneuverability during re-entry, and the NASA shuttle designs demanded about 450 miles in order to have some working room.

Polar orbits from the Air Force's Vandenberg Air Force Base are another matter entirely. Located slightly north of KSC, the distance it would move over a single orbit would be similar, but critically, the shuttle would be traveling south, not east. This meant that it was not flying toward the launch point as it traveled in its orbit, and when it completed one orbit it would have to make up the entire 1,350 miles during re-entry. These missions required a dramatically improved cross-range capability, set at 1,500 miles to give it a slight excess capability. The ballistic re-entry profile of the DC-3 series simply could not come close to matching this requirement.

On 1 May 1971 the OMB finally released a budget plan, limiting NASA to $3.2 billion per year for the next five years. Given existing project budgets, this limited any spending on the shuttle to about $1 billion a year, far less than required to develop any of the completely reusable designs. Based on these constraints, NASA returned to a Class II-like vehicle with external tankage, which led to the MSC-020 design. Later that year all straight-wing designs were officially abandoned, although Faget's team continued to work on them for some time in spite of this.

32.2   Description

The DC-3 was a two-stage vehicle with a large booster and smaller shuttle of overall similar design. Both were similar to "jumbo jets" in layout in general terms, with their large cylindrical fuselage containing fuel tanks instead of passengers or cargo. The bottom of the fuselage was flattened for re-entry aerodynamics, with a slight upward curve as you approached the nose in early models. The wings were low-mounted, in-line with the bottom of the fuselage, with a 14 degree rearward sweep on the front and no sweep on the back. The general layout of the wing planform was similar to the original DC-3. The empennage was a conventional three-surface unit, although in the original MSC-001 design the delta-shaped horizontal stabilizer was located at the bottom of the fuselage and served double-duty in protecting the rear-mounted engines during re-entry. Later versions did not generally include this feature, and used more conventional surfaces mid-mounted on the fuselage.

The orbiter carried a crew of two, and had accommodations for up to ten passengers. A cargo area was mounted in the middle of the craft between the liquid hydrogen (LH2) tank behind it, and a combined LH2/liquid oxygen tank in front of it. This arrangement was used in order to center the cargo over the wing, with the heavier oxygen and crew compartment balancing the weight of the engines. The lighter weight hydrogen then filled out the rest of the internal space. The booster had no cargo area, so it used a simpler arrangement of tankage with a single LH2 tank at the rear. The booster normally flew unmanned, but included a two-man cockpit area that was used during ferry flights.

The orbiter was powered by two modified XLR-129 engines with the thrust increased from 250,000 to 300,000 lbf, two 15,000 lbf RL-10 orbital manoeuvring engines, and six Rolls-Royce RB162 jet engines for landing. The booster used eleven of the same XLR-129 engines, and four Pratt & Whitney JT8D for landing. XLR-129s on both the shuttle and booster were fired for take-off. The orbiter was mounted relatively far forward for launch, its tail in-line with the booster's wings. The combined weight at launch would be about 2,030 tons.

The orbiter would re-enter nose-high at an angle of about 60 degrees above horizontal, decelerating at a peak of 2G until it reached low subsonic speeds at 40,000 ft. At this point the forward speed of the craft would be very low, so the nose was pitched down and the orbiter dove to pick up airspeed over the wings and transition to level flight. Expected re-entry heating rates on the orbiter were 1650 deg C on the leading edge, and 790 deg C over 80% of the lower surface.

In order to maximize overall performance, the booster released the orbiter at Mach 10 and 45 miles altitude. This required the booster to carry a complete thermal protection system in order to re-enter for landing. Both the orbiter and booster were to be protected with the LI-1500 silica tiles similar to those eventually used on the Space Shuttle, a design that had recently been introduced by Lockheed and quickly became a baseline design for all of the shuttle contenders. As a result, both airframes were able to be built out of aluminum, greatly reducing airframe cost.

Both craft carried just enough JP-4 for landing go-around. Both could also carry increased loads of JP-4 for test flights or ferrying. After dispatching the orbiter the booster would be too far down-range to easily turn around and return to Kennedy, so the normal mission profile had it coast across the ocean, land automatically, refuel and pick up a crew, and then be flown back to Kennedy on its JT8D engines.

Lockheed estimated that development and initial production would cost $5.912 billion over a period from 1970 to 1975. A fleet of six orbiters and four boosters would have supported a launch rate of 50 flights per year.

32.3   References

- Maxime Faget, "Space Shuttle: A New Configuration", *Astronautics & Aeronautics*, January 1970, p. 52

- Marcus Lindroos, "MSC/North America Concept-A, 'DC-3'", 21 January 2003

- "Shuttle", *astronautix.com*

Chapter 33

North American X-15

"X-15" redirects here. For other uses, see X-15 (disambiguation).

The **North American X-15** was a hypersonic rocket-powered aircraft operated by the United States Air Force and the National Aeronautics and Space Administration as part of the X-plane series of experimental aircraft. The X-15 set speed and altitude records in the 1960s, reaching the edge of outer space and returning with valuable data used in aircraft and spacecraft design. As of September 2015, the X-15 holds the official world record for the highest speed ever reached by a manned, powered aircraft. Its maximum speed was 4,520 miles per hour (7,274 km/h), or Mach 6.72.[1][2]

During the X-15 program, 13 flights by eight pilots met the Air Force spaceflight criterion by exceeding the altitude of 50 miles (80 km), thus qualifying the pilots for astronaut status. The Air Force pilots qualified for astronaut wings immediately, while the civilian pilots were awarded NASA astronaut wings in 2005, 35 years after the last X-15 flight. The sole Navy pilot in the X-15 program never took the aircraft above the requisite 50 mile altitude.[3][4]

Of the 199 X-15 missions, two flights (by the same pilot) qualified as space flights per the international (*Fédération Aéronautique Internationale*) definition of a spaceflight by exceeding 100 kilometers (62.1 mi) in altitude.

33.1 Design and development

The X-15 was based on a concept study from Walter Dornberger for the National Advisory Committee for Aeronautics (NACA) for a hypersonic research aircraft.[5] The requests for proposal were published on 30 December 1954 for the airframe and on 4 February 1955 for the rocket engine. The X-15 was built by two manufacturers: North American Aviation was contracted for the airframe in November 1955, and Reaction Motors was contracted for building the engines in 1956.

Like many X-series aircraft, the X-15 was designed to be carried aloft and drop launched from under the wing of a NASA B-52 mother ship. Air Force NB-52A, "The High and Mighty One" (serial 52-0003, AKA *Balls Three*), and NB-52B, "The Challenger" (serial 52-0008, AKA *Balls 8*) served as carrier planes for all X-15 flights. Release took place at an altitude of about 8.5 miles (13.7 km) and a speed of about 500 miles per hour (805 km/h).[6] The X-15 fuselage was long and cylindrical, with rear fairings that flattened its appearance, and thick, dorsal and ventral wedge-fin stabilizers. Parts of the fuselage were heat-resistant nickel alloy (Inconel-X 750).[5] The retractable landing gear comprised a nose-wheel carriage and two rear skids. The skids did not extend beyond the ventral fin, which required the pilot to jettison the lower fin (fitted with a parachute) just before landing.

33.1.1 Cockpit and pilot systems

The X-15 was a research program and changes were made to various systems over the course of the program and between the different models. The X-15 was operated under several different scenarios including attachment to a launch aircraft, drop, main engine start and acceleration, a ballistic flight into thin air/space, re-entry into thicker air, and an unpowered glide to landing. Alternatively, if the main engine was not started the pilot went directly to a landing. The main rocket engine operated only for a relatively short part of the flight, but was capable of boosting

X-15 after igniting rocket engine

X-15A2, with sealed ablative coating, external fuel tanks, and ramjet dummy test

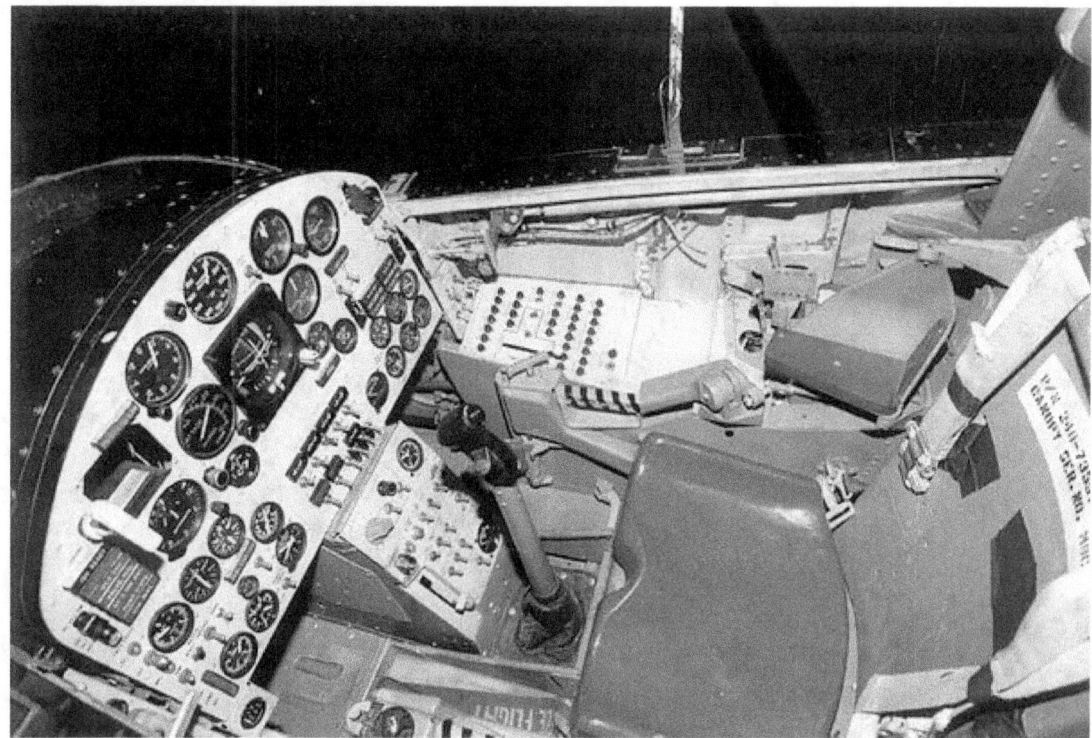

Cockpit of an X-15

the X-15 to its high speeds and altitudes. Without main engine thrust, the X-15's instruments and control surfaces remained functional, but the aircraft could not maintain altitude.

Because the X-15 also had to be controlled in an environment where there was too little air for aerodynamic flight control surfaces, it had a reaction control system (RCS) that used rocket thrusters.*[7] There were two different X-15 pilot control setups: one used three joysticks; the other, one joystick.*[8]

The X-15 type with multiple control sticks for the pilot included a traditional rudder and stick, and another joystick on the left which sent commands to the reaction control system.*[9] A third joystick on the right side was used during high-G maneuvers to augment the center stick.*[9] In addition to pilot input, the X-15 "Stability Augmentation System" (SAS) sent inputs to the aerodynamic controls to help the pilot maintain attitude control.*[9] The reaction control system could be operated in two modes, manual and automatic.*[8] The automatic mode used a feature called "Reaction Augmentation System" (RAS) that helped stabilize the vehicle at high altitude.*[8] The RAS was typically used for approximately three minutes of an X-15 flight before automatic power off.*[8]

The second setup used the MH-96 flight control system which allowed one joystick in place of three and simplified pilot input.*[10] The MH-96 could automatically blend aerodynamic and rocket controls depending on how effective each system was at controlling the aircraft.*[10]

Among the many controls, were the rocket engine throttle and a control for jettisoning the ventral tail fin.*[9] Other features of the cockpit were heated windows to prevent icing, and a forward headrest for periods of high deceleration.*[9]

The X-15 had an ejection seat that allowed ejection at speeds up to Mach 4 and/or 120,000 feet (37 km) altitude, although it was not used during the program.*[9] In the event of ejection, the seat had deployable fins which were used until it reached a safer speed/altitude, where it could deploy its main parachute.*[9] Pilots wore a pressure suit, which could be pressurized with nitrogen gas.*[9] Above 35,000 feet (11 km) altitude, the cockpit was pressurized to 3.5 psi (0.24 atm) with nitrogen gas, and oxygen for breathing was fed separately to the pilot.*[9]

33.1.2 Propulsion

Early flights used two Reaction Motors XLR11 liquid-propellant rocket engines. Later flights were undertaken with a single XLR99 rocket engine generating 57,000 pounds-force (250 kN) of thrust. The XLR11 used ethyl alcohol and

X-15 tail with XLR-99

liquid oxygen. The XLR99 engine used anhydrous ammonia and liquid oxygen as propellant, and hydrogen peroxide to drive the high-speed turbopump that delivered propellants to the engine.[7] It could burn 15,000 pounds (6,804 kg) of propellant in 80 seconds.[7] The XLR99s could be throttled, and were the first such controllable engines that were man-rated.

The X-15 reaction control system (RCS), for maneuvering in low-pressure/density environment, used high-test per-oxide (HTP), which decomposes into water and oxygen in the presence of a catalyst and could provide a specific impulse of 140 seconds.[8][11] The HTP also fueled a turbopump for the main engines and auxiliary power units (APUs).[7] Additional tanks for helium and liquid nitrogen performed other functions, for example the fuselage interior was purged with helium gas, and the liquid nitrogen was used as coolant for various systems.[7]

33.1.3   Wedge tail and hypersonic stability

The X-15 had a thick wedge tail for stability at hypersonic speeds.[12] This produced a significant amount of drag at lower speeds;[12] the blunt end at the rear of the X-15 could produce as much drag as an entire F-104 Starfighter.[12]

> A wedge shape was used because it is more effective than the conventional tail as a stabilizing surface at hypersonic speeds. A vertical-tail area equal to 60 percent of the wing area was required to give the X-15 adequate directional stability.
>
> —Wendell H. Stillwell, *X-15 Research Results (SP-60)*

Stability at hypersonic speeds was aided by side panels that could extend out from the tail to increase area, and the panels doubled as air-brakes.[12]

X-15 attached to its B-52 mother ship with a T-38 flying nearby

33.2 Operational history

Altitudes attained by X-15 aircraft do not match those of Alan Shepard's and Gus Grissom's Project Mercury space capsules in 1961, nor of any other manned spacecraft. The X-15 ranks supreme among manned rocket-powered aircraft, becoming the world's first operational spaceplane in the early 1960s.

Before 1958, United States Air Force (USAF) and NACA officials discussed an orbital X-15 spaceplane, the **X-15B** that would launch into outer space from atop an SM-64 Navaho missile. This was canceled when the NACA became NASA and adopted Project Mercury instead.

By 1959, the Boeing X-20 Dyna-Soar space-glider program was to become the USAF's preferred means for launching military manned spacecraft into orbit. This program was canceled in the early 1960s before an operational vehicle could be built.[*][3] Various configurations of the Navajo were considered, and another proposal involved a Titan I stage.[*][13]

Three X-15s were built, flying 199 test flights, the last on 24 October 1968.

The first X-15 flight was a captive-carry unpowered test by Albert Scott Crossfield, on 8 June 1959. Crossfield also piloted the first powered flight, on 17 September 1959, and his first flight with the XLR-99 rocket engine on 15 November 1960. Twelve test pilots flew the X-15. Among these were Neil Armstrong, later a NASA astronaut and first man to set foot on the Moon, and Joe Engle, later a commander of NASA Space Shuttle test flights.

In a 1962 proposal, NASA considered using the B-52/X-15 as a launch platform for a Blue Scout rocket to place satellites up to 150 pounds (68 kg) into orbit.[*][13][*][14]

In July and August 1963, pilot Joseph A. Walker exceeded 100 km in altitude, joining NASA astronauts and Soviet cosmonauts as the first human beings to cross that line on their way to outer space. The USAF awarded astronaut wings to anyone achieving an altitude of 50 miles (80 km), while the FAI set the limit of space at 100 kilometers (62.1 mi).

On 15 November 1967, U.S. Air Force test pilot Major Michael J. Adams was killed during X-15 Flight 191 when the

X-15-3, AF Ser. No. 56-6672, entered a hypersonic spin while descending, then oscillated violently as aerodynamic forces increased after re-entry. As his aircraft's flight control system operated the control surfaces to their limits, acceleration built to 15 g vertical and 8.0 g lateral. The airframe broke apart at 60,000 feet (18 km) altitude, scattering the X-15's wreckage for 50 square miles (130 km^2). On 8 May 2004, a monument was erected at the cockpit's locale, near Randsburg, California.[15] Major Adams was posthumously awarded Air Force astronaut wings for his final flight in X-15-3, which had reached an altitude of 50.4 miles (81.1 km). In 1991, his name was added to the Astronaut Memorial.[15]

The second X-15A was rebuilt after a landing accident. It was lengthened 2.4 feet (0.73 m), a pair of auxiliary fuel tanks attached beneath its fuselage and wings, and a complete heat-resistant ablative coating was added. Renamed the **X-15A-2**, this plane first flew on 28 June 1964. It reached a maximum speed of 4,520 miles per hour (7,274 km/h) in October 1967 while being flown by William "Pete" Knight of the U.S. Air Force.

Five aircraft were used during the X-15 program: three X-15s planes and two B-52 bombers:

- **X-15A** – *56-6670*, 82 powered flights

- **X-15A (later X-15A-2)** – *56-6671*, 31 powered flights as X-15A, 22 powered flights as X-15A-2, and 53 in total

- **X-15A** – *56-6672*, 64 powered flights

- **NB-52A** – *52-003* (retired in October 1969)

- **NB-52B** – *52-008* (retired in November 2004)

A 200th flight over Nevada was first scheduled for 21 November 1968, to be flown by William "Pete" Knight. Numerous technical problems and outbreaks of bad weather delayed this proposed flight six times, and it was permanently canceled on 20 December 1968. This X-15 was detached from the B-52 and then put into indefinite storage. The aircraft was later donated to the Air Force Museum at Wright-Patterson Air Force Base for display.

- NB-52A (s/n 52-003), permanent test variant, carrying an X-15, with mission markings; horizontal X-15 silhouettes denote glide flights, diagonal silhouettes denote powered flights.

- X-15 just after release.

- X-15 touching down on its skids, with the lower ventral fin jettisoned.

- X-15A2 (56-6671) with external fuel tanks

33.3   Current static displays

- X-15A-1 (AF Ser. No. 56-6670) is on display in the National Air and Space Museum "Milestones of Flight" gallery, Washington, D.C.

- X-15A-2 (AF Ser. No. 56-6671) is at the National Museum of the United States Air Force, at Wright-Patterson Air Force Base, near Dayton, Ohio. It was retired to the Museum in October 1969.[16] The aircraft is displayed in the Museum's Research & Development Hangar alongside other "X-planes", including the Bell X-1B and Douglas X-3 Stiletto.

33.3.1   Mockups

- Dryden Flight Research Center, Edwards AFB, California, USA (painted with AF Ser. No. 56-6672)

- Pima Air & Space Museum, adjacent to Davis-Monthan AFB, Tucson, Arizona (painted with AF Ser. No. 56-6671)

- Evergreen Aviation & Space Museum, McMinnville, Oregon (painted with AF Ser. No. 56-6672). A full-scale wooden mockup of the X-15, it is displayed along with one of the rocket engines.

33.3.2 Stratofortress mother ships

- NB-52A (AF Ser. No. 52-0003) is displayed at the Pima Air & Space Museum adjacent to Davis–Monthan AFB in Tucson, Arizona. It launched the X-15-1 30 times, the X-15-2, 11 times, and the X-15-3 31 times (as well as the M2-F2 four times, the HL-10 11 times and the X-24A twice).

- NB-52B (AF Ser. No. 52-0008) is on permanent display outside the north gate of Edwards AFB, California. It launched the majority of X-15 flights.

33.4 Record flights

33.4.1 Highest flights

The FAI set the limit of space at 100 kilometers (62.1 mi). But in the 1960s, the USAF considered an altitude of 50 miles (80 km) as the limit of space; USAF and NASA pilots and crew exceeding that altitude at that time could be awarded the Astronaut Badge. Thirteen X-15 flights went higher than 50 miles and two of these exceeded 100 kilometers.

*† fatal

33.4.2 Fastest flights

33.5 X-15 pilots

33.6 Specifications (X-15)

Other configurations include the Reaction Motors XLR11 equipped X-15, and the long version.

General characteristics

- **Crew:** one
- **Length:** 50 ft 9 in (15.45 m)
- **Wingspan:** 22 ft 4 in (6.8 m)
- **Height:** 13 ft 6 in (4.12 m)
- **Wing area:** 200 ft^2 (18.6 m^2)
- **Empty weight:** 14,600 lb (6,620 kg)
- **Loaded weight:** 34,000 lb (15,420 kg)
- **Max. takeoff weight:** 34,000 lb (15,420 kg)
- **Powerplant:** 1 × Reaction Motors XLR99-RM-2 liquid propellant rocket engine, 70,400 lb$_f$ at 30 km (313 kN)

Performance

- **Maximum speed:** 4,520 mph (7,274 km/h)
- **Range:** 280 mi (450 km)
- **Service ceiling:** 67 mi (108 km, 354,330 ft)
- **Rate of climb:** 60,000 ft/min (18,288 m/min)
- **Wing loading:** 170 lb/ft^2 (829 kg/m^2)
- **Thrust/weight:** 2.07

33.7   See also

Aircraft of comparable role, configuration and era

- Bell X-2

- Douglas D-558-2 Skyrocket

Related lists

- List of rocket aircraft

- List of X-15 flights

- List of spaceflight-related accidents and incidents

33.8   References

Notes

[1]   "North American X-15 High-Speed Research Aircraft". *Aerospaceweb.org*. 2010. Retrieved 24 November 2008.

[2]   Gibbs, Yvonne, ed. (28 February 2014). "NASA Armstrong Fact Sheet: X-15 Hypersonic Research Program". NASA. Retrieved 4 October 2015.

[3]   Jenkins 2001, p. 10.

[4]   Thompson, Elvia H.; Johnsen, Frederick A. (23 August 2005). "NASA Honors High Flying Space Pioneers" (Press release). NASA. Release 05-233.

[5]   Käsmann 1999, p. 105.

[6]   "X-15 launch from B-52 mothership". Armstrong Flight Research Center. 6 February 2002. Photo E-4942.

[7]   Raveling, Paul. "X-15 Pilot Report, Part 1: X-15 General Description & Walkaround". *SierraFoot.org*. Retrieved 30 September 2011.

[8]   Jarvis, Calvin R.; Lock, Wilton P. (1965). *Operational Experience With the X-15 Reaction Control and Reaction Augmentation Systems* (PDF). NASA. OCLC 703664750. TN D-2864.

[9]   Raveling, Paul. "X-15 Pilot Report, Part 2: X-15 Cockpit Check". *SierraFoot.org*. Retrieved 1 October 2011.

[10]   "Forty Years ago in the X-15 Flight Test Program, November 1961–March 1962". Goleta Air & Space Museum. Retrieved 3 October 2011.

[11]   Davies 2003, p. 8.28.

[12]   Stillwell, Wendell H. (1965). *X-15 Research Results: With a Selected Bibliography*. NASA. OCLC 44275779. NASA SP-60.

[13]   Wade, Mark. "X-15/Blue Scout". *Encyclopedia Astronautica*. Retrieved 30 September 2011.

[14]   "Historical note: Blue Scout / X-15". *Citizensinspace.org*. 21 March 2012.

[15]   Merlin, Peter W. (30 July 2004). "Michael Adams: Remembering a Fallen Hero". *The X-Press* **46** (6).

[16]   USAF Museum Guidebook 1975, p. 73.

Bibliography

- Davies, Mark, ed. (2003). *The Standard Handbook for Aeronautical and Astronautical Engineers*. New York: McGraw-Hill. pp. 8–28. ISBN 978-0-07-136229-0.

- Godwin, Robert, ed. (2001). *X-15: The NASA Mission Reports*. Burlington, Ontario: Apogee Books. ISBN 1-896522-65-3.

- Hallion, Richard P. (March–June 1978). "Saga of the Rocket Ships". In Green, William; Swanborough, Gordon. *Air Enthusiast Six*. Bromley, Kent, UK: Pilot Press.

- Jenkins, Dennis R. (2001). *Space Shuttle: The History of the National Space Transportation System: The First 100 Missions* (3rd ed.). Stillwater, Minnesota: Voyageur Press. ISBN 0-9633974-5-1.

- Jenkins, Dennis R.; Landis, Tony; Miller, Jay (June 2003). *American X-Vehicles: An Inventory—X-1 to X-50* (PDF). Monographs in Aerospace History No. 31. NASA. OCLC 68623213. SP-2003-4531.

- Jenkins, Dennis R. (2007). *X-15: Extending the Frontiers of Flight* (PDF). NASA. ISBN 9780160792854.

- Käsmann, Ferdinand C. W. (1999). *Die schnellsten Jets der Welt: Weltrekord-Flugzeuge [The Fastest Jets in the World: World Record Aircraft]* (in German). Kolpingring, Germany: Aviatic Verlag. ISBN 3-925505-26-1.

- Price, A. B. (12 January 1968). *Design Report - Thermal Protection System, X-15A-2*. Denver, Colorado: Martin Marietta Corporation. NASA CR-82003.

- Thompson, Milton O. (1992). *At the Edge of Space: The X-15 Flight Program*. Washington, D.C.: Smithsonian Institution Press. ISBN 1-56098-107-5.

- Tregaskis, Richard (2000). *X-15 Diary: The Story of America's First Space Ship*. Lincoln, Nebraska: iUniverse. ISBN 0-595-00250-1.

- *United States Air Force Museum Guidebook*. Wright-Patterson AFB, Ohio: Air Force Museum Foundation. 1975.

- Watts, Joe D. (October 1968). *Flight Experience With Shock Impingement and Interference Heating on the X-15-2 Research Airplane* (PDF). NASA. NASA-TM-X-1669.

33.9 External links

NASA

- NASA's X-15 website

- "Transiting from Air to Space: The North American X-15" (1998)

- "Proceedings of the X-15 First Flight 30th Anniversary Celebration, 8 June 1989"

- *Hypersonics Before the Shuttle: A Concise History of the X-15 Research Airplane* (NASA SP-2000-4518, 2000)

- Interview with Neil Armstrong about his experience in the X-15

- NASA Armstrong Fact Sheet: X-15 Hypersonic Research Program

- X-15 photos at NASA Armstrong

- X-15 movies at NASA Armstrong

- Interactive X-15 portal by NASA (uses Macromedia Flash)

Non-NASA

- X-15 Advanced Research Airplane Design Summary (NA-55-221, 1955)

- "Recovery of the X-15 Supersonic Experimental Aircraft" at History in Pieces

- A film clip "X-15 Aloft: First Free Flight Of Manned Space Plane" (1959) is available for free download at the Internet Archive

- A film clip "X-15 Space Record: Plane Flown To 59-mile Mark" (1962) is available for free download at the Internet Archive

- A film clip "Space Triumph: Discoverer Capsule Recovered From Orbit" (1960) is available for free download at the Internet Archive

- A film clip "Nuclear Navy: First Polaris A-Sub Sails On Ocean Patrol" (1960) is available for free download at the Internet Archive

- The short film *History of the Air Force, 1954-1964: From Missile Development to Space (1977)* is available for free download at the Internet Archive

- X-15 at Encyclopedia Astronautica

- *X-15* at the Internet Movie Database

Neil Armstrong with X-15 number 1

Members of the X-15 flight crew, left to right: Engle, Rushworth, McKay, Knight, Thompson, and Dana.

X-15 at the National Air and Space Museum in Washington, D.C.

NB-52B takes off with an X-15

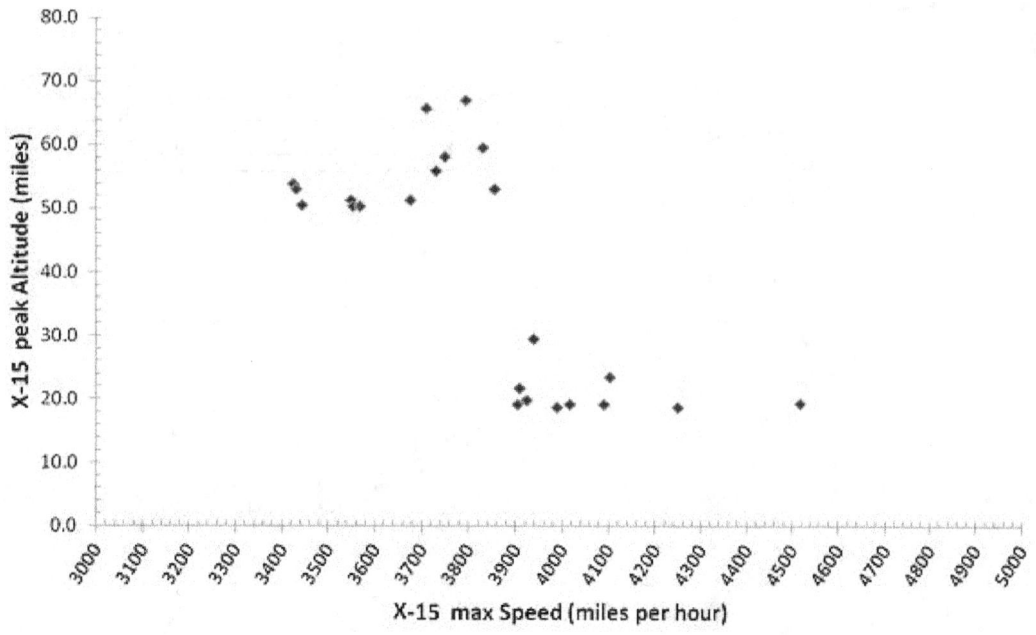

Key speed and altitude benchmarks of the X-15.

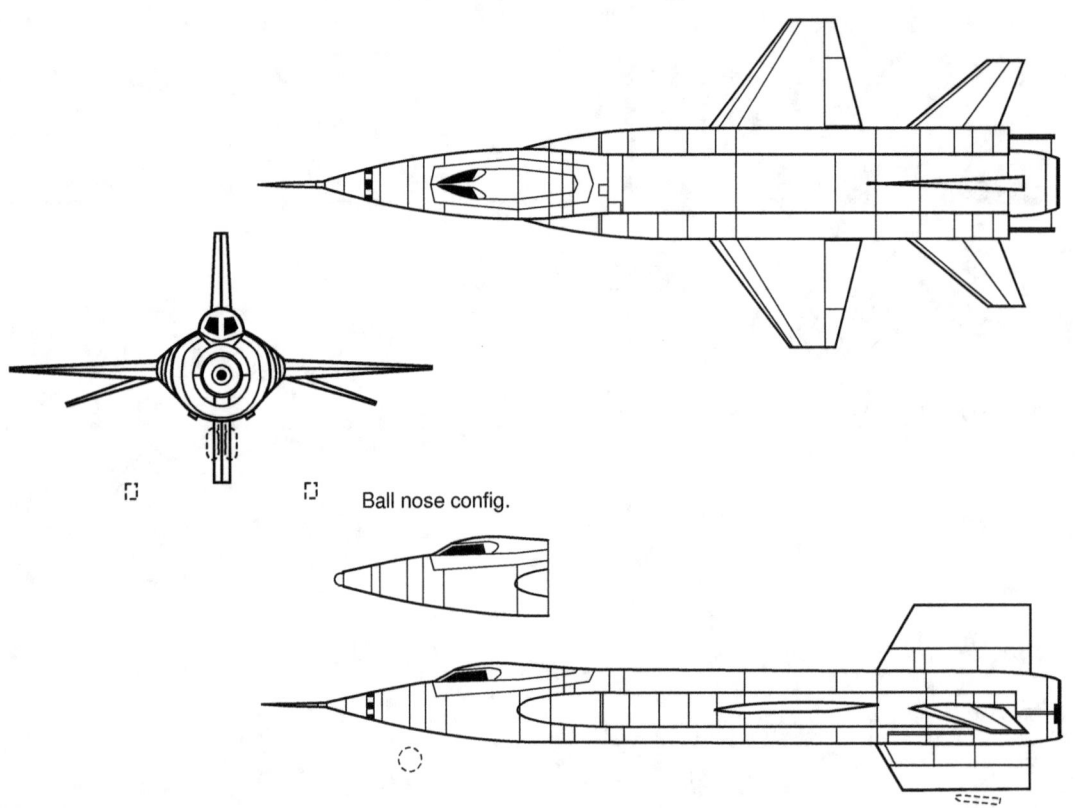

Ball nose config.

X-15 3-view

Chapter 34

Orbital Sciences X-34

The **Orbital Sciences X-34** was intended to be a low-cost testbed for demonstrating "key technologies" which could be integratad into the Reusable Launch Vehicle program. It was intended to be an autonomous pilotless craft powered by a 'Fastrac' liquid-propellant rocket engine, capable of reaching Mach 8, and performing 25 test flights per year.

The X-34 began as a program for a suborbital reusable-rocket technology demonstrator. In early 2001, the first flight vehicle was near completion, but the program was ended after NASA demanded sizable design changes without further funding. The contractor, Orbital Sciences, refused to make the changes. Up to this point, the project had encompassed spending of just under $112 million: $85.7M from the original contract with designer Orbital Sciences, $16M from NASA and various government agencies for testing, and an additional $10M for Orbital Sciences to adapt its L-1011 carrier to accommodate the X-34. The program was officially canceled by NASA on March 1, 2001. The unpowered prototype had been used only for towing and captive flight tests when the project was canceled.

The two demonstrators remained in storage at Edwards Air Force Base*[2] until November 16, 2010, when both X-34s were moved with their vertical tails removed from Dryden to a hangar owned by the National Test Pilot school in Mojave, California. They are to be inspected, and NASA is investigating the possibility of restoring them to flight status.*[3]

34.1 See also

- List of experimental aircraft
- Cygnus (spacecraft)

34.2 References

[1] Wikisource:X-34: Demonstrating Reusable Launch Vehicle Technologies

[2] Orbital Sciences Corporation X-34 - 2007 photo. Airliners.net

[3] http://www.flightglobal.com/articles/2010/11/19/349997/photos-nasa-moves-x-34s-out-of-storage-considers-return-to-flight.html

34.3 External links

- NASA Dryden X-34 Technology Testbed Demonstrator Photo Collection
- Federation of American Scientists

Chapter 35

Orbital Space Plane Program

Conceptual drawings of the Orbital Space Plane

The **Orbital Space Plane (OSP) program** was a NASA concept in the early 2000s designed to support the International Space Station requirements for crew rescue, crew transport and contingency cargo such as supplies, food and other needed equipment.

With the initiation of the Constellation program in 2004, NASA transferred the knowledge gained on the OSP to the development of Crew Exploration Vehicle,[*][1] an Apollo-style capsule with separate crew and service modules.

35.1 Origin

The initial plans for the International Space Station envisaged a small, low-cost 'Assured Crew Return Vehicle' (ACRV) which would provide emergency evacuation capability; the X-38 was the prototype of this.*[2] Following cancellation of the ACRV in 2002, the program led to the more capable Orbital Space Plane concept.

The first variant of the Orbital Space Plane was designed to serve as a crew rescue vehicle for the ISS; this replaced the previous plans for a dedicated station Crew Return Vehicle, which had been sidelined by budget cuts.

This early version of the plane had been expected to enter service by 2010.*[3]

35.2 Function

Future versions of the Orbital Space Plane would have been launched on an existing EELV rocket to carry crews to the International Space Station. It was envisaged that the OSP would operate alongside the Space Shuttle with the OSP responsible for crew flights and the shuttle handling construction and cargo flights. At the time, the shuttle program was not yet set for retirement and was thought to be technically viable up until the 2030s. Thus it was expected that the two spacecraft would complement each other throughout the lifespan of the ISS. One advantage of this approach would have been assured manned access to space; the lack of this capability was to be highlighted starkly with the loss of Space Shuttle Columbia.

Top level requirements for the Orbital Space Plane and its related systems were approved in February 2003.*[4] In March 2003, the program began evaluating system operations to ensure the alignment of systems design between the NASA mission and the contractor design.

35.3 Transfer to the CEV program

The Crew Exploration Vehicle program was based on four groups of concepts considered for the physical design of the space plane itself —or the vehicle architecture: a capsule, a lifting body, a sharp body with wings and a blunt body with wings (see image).

After the Columbia accident investigation, the capsule design with a separate escape system was considered the optimal design for crew safety.

35.4 Other program components

Other components of the OSP program were the X-37 and the DART.*[5]

The X-37 vehicle was designed to flight test advancing technologies to reduce the risk of future reusable launch vehicle systems, including the Orbital Space Plane.

The Demonstration for Autonomous Rendezvous Technology or *DART*, was another flight demonstrator vehicle designed to test technologies required to locate and rendezvous with other spacecraft. Although Russia has mastered this technology for years, this is the first for NASA. Using onboard guidance sensors, DART would have performed a series of maneuvers around a retired satellite. However, after a successful launch, unknown problems with the guidance system caused the vehicle to run out of thruster fuel prematurely, ending the mission before all objectives could be carried out.

In 2010, Orbital Sciences Corporation reused some work done under OSP program contracts for its Prometheus spacecraft proposal to NASA under phase 2 of the Commercial Crew Development program.*[6]

35.5 References

[1] "VA-HUD and Independent Agencies Subcommittee Hearing of FY05 Budget Request for NASA: Testimony of Sean O'Keefe, Administrator, NASA". *Hearings & Testimony*. U.S. Senate Committee on Appropriations. 2004-03-11. Archived from the original on 2006-12-27. Retrieved 2007-01-29.

[2] Lindroos, Marcus. "NASA ACRV". *Encyclopedia Astronautica*. Retrieved 2011-03-18.

[3] Bray, Becky; Meyer, Patrick (2003-06-17). "Orbital Space Plane--Commuting To Space". *Liftoff to Space Exploration*. Marshall Space Flight Center. Archived from the original on 2006-12-11. Retrieved 2007-01-29.

[4] "Beginning a New Era of Space Flight: The Orbital Space Plane". Marshall Space Flight Center. May 2003. Retrieved 2007-01-29.

[5] Orbital Space Plane (OSP) Program at Lockheed Martin (Sept. 2003)

[6] "Orbital Submits Proposal for NASA's Commercial Crew Development Program". *Press Release*. Orbital Sciences Corporation. 2010-12-14. Retrieved 2011-03-08.

Chapter 36

Programme for Reusable In-orbit Demonstrator in Europe

The **Programme for Reusable In-orbit Demonstrator in Europe** (**PRIDE**) is a European Space Agency (ESA) programme that aims to develop a reusable robotic spacecraft. PRIDE was approved at the ESA Ministerial Council in Naples, Italy on November 21, 2012. PRIDE spaceplane will be similar to, but smaller and cheaper than, the Boeing X-37. It will be launched by the Vega light rocket, operate robotically in orbit, and land automatically on a runway.[2]

36.1 History

PRIDE was initially funded by the ESA on November 21, 2012 at the ESA Ministerial Council in Naples, Italy.[3] The project was created with the objective of creating an small unmanned spaceplane that was also affordable and reusable. During the initial design stage the vehicle was referred to as PRIDE-ISV. The suffix ISV stands for *Innovative Space Vehicle*.[4] It is projected that from September 2015, the PRIDE development team will begin industrial activities. In December 2015 a ministerial-level meeting will make a decision regarding the funding for the project as around €200 million is required to finalize the project, excluding launch costs. If funding is successful, the first launch is expected around 2020.[5]

The European Space Agency has developed two test vehicles: the Atmospheric Reentry Demonstrator (launched in 1998), and the Intermediate eXperimental Vehicle (IXV, launched on Feb 11th 2015 [6] and with a second launch planned for 2019 or 2020).[1][4][7]

36.2 Design

With affordability in mind, the PRIDE spaceplane will be based on technologies developed and tested on the IXV. Final specifications of the spaceplane have not yet been determined; both winged and lifting body variants are under consideration.

The PRIDE spaceplane will be capable of operating up to 300 kilograms (660 lb) of payload, and it will be equipped with solar panels, allowing for extended in-orbit operations. Vega will be used as a launch vehicle.[1][8]

The PRIDE spaceplane will be used as an orbital test platform for re-usable launcher stages, Earth observation, robotic exploration, servicing of orbital infrastructures, and microgravity experiments.[8]

36.3 See also

- Boeing X-37 - A comparable United States Air Force spaceplane

- Atmospheric Reentry Demonstrator (ARD) - ESA reentry testbed flown in 1998

- Avatar space shuttle (India)

- European eXPErimental Re-entry Testbed (EXPERT) - Research programme developing materials used in IXV

- Future Launchers Preparatory Programme - parent programme for IXV

- Hopper - an earlier ESA project on developing manned spaceplane, cancelled

36.4   References

[1] Howell, Elizabeth (23 February 2015). "Europe's Newly-Tested Space Plane Aims for Next Launch in 2019". *Space.com*. Retrieved 16 June 2015.

[2] "Frequently Asked Questions on IXV". ESA. 9 February 2015. Retrieved 4 July 2015.

[3] "N° 37–2012: European Ministers decided to invest in space to boost Europe's competitiveness and growth" (Press release). ESA. 21 November 2012. Retrieved 16 June 2015.

[4] Rob Coppinger (25 October 2012). "IXV's Pride: Europe's spaceplane homecoming prelude to future goals". Space.com. Retrieved 16 June 2015.

[5] "Replay of IXV conference". ESA. 16 June 2015. Retrieved 17 July 2015.

[6] http://www.space.com/28520-europe-launches-mini-shuttle-ixv.html

[7] "Europe's mini-space shuttle returns". *BBC News*. 11 February 2015. Retrieved 16 June 2015.

[8] "ESA spaceplane on display" (Press release). ESA. 16 June 2015. Retrieved 17 June 2015.

36.5   External links

- PRIDE mission image

Chapter 37

Project 921-3

Project 921-3 is Manned Spacecraft sub-system of Project 921. The term 921-3 is often used for the Chinese space shuttle program.

37.1 History

The Chinese National Manned Space Program was given the designation of Project 921 in 1992. This broad project was divided into three phases: 921/1 to launch a manned mission by 2002 in a craft that became the Shenzhou, the Project 921/2 temporary space station by 2010, and the 921/3 permanent space station by 2020. Care must be taken not to confuse the 3 phases of Project 921 with its 7 sub-systems (921-1, 921-2 ... 921-7).

Early planning of Project 921 included 6 different proposals for a manned space transportation system. Five of these proposals were of a space-earth transportation system using a delta winged orbiter. By 1990, the proposal for the Soyuz-like capsule Shenzhou had won out.

Some small models for a space plane were made public, but the concept was rejected in favor of a Soyuz-like capsule which became Shenzhou. Concepts for a space shuttle now are only academic. There is no known Chinese government support beyond very basic research for a space plane.

Photographs of a two-seat spaceplane simulator were published after 1980, probably belonging to a Chinese Dynasoar-like vehicle. Reports of the existence of a wind tunnel model have continued since then.

37.2 869 Project

After 1986 the Air Ministry starts its 869 Project regarding space plane concepts. Up to 1990, the several space-shuttle proposals studied were:[*][1]

- Tianjiao-1 space shuttle, proposed by China Academy of Launch Vehicle Technology. Totally dependent on the parent rocket booster to reach orbit. 25 ton orbiter, 2 ton payload.

- Chang Cheng-1 (Great Wall-1) space shuttle, proposed by the Shanghai Academy of Spaceflight Technology and 640 Institute of the Air Ministry. 94 ton orbiter launched atop three-parallel HT-1 SLVs to a 200–500 km orbit. 5-ton payload and maximum of 5 crew members.

- V-2 rocket plane, proposed by the 11th Aeronautics Institute;

- H-2 aerospace plane, proposed by Shenyang Aircraft Design Institute. Reusable launch stage weighs 198 tons using dual rocket and ramjet engine for Mach 5+ launch speed. 132 tons orbiter.

- Mini space shuttle, proposed by 611 Aircraft Design Institute;

37.3   Shenlong (Divine Dragon) Test Platform

The latest academic models shown in 2000, reveal a delta winged spaceplane with a single vertical stabilizer, equipped with three high-expansion engines. Presuming a seating arrangement of two crew members siting side-by-side in the cockpit, dimensions could be very roughly estimated as a wingspan of 8 m, a length of 12 m and a total mass of 12 tonnes. This is within the payload capability of the Chinese CZ-2E(A) or Type A launch vehicles.

37.4   HTS Maglev Launch Assist Technology

During the 2006 Zhuhai Airshow, pictures of a totally new space vehicle developed by the Beijing University of Aeronautics and Astronautics(北京航空航天大学) were published.[2]

This new Chinese space shuttle was based on the HTS (High Temperature Superconductor) Maglev Launch Assist Technology for Space Flight Vehicle (航天运载器高温超导磁悬浮助推发射技术), with an initial take off speed of 1000 km/h.[3]

37.5   Reusable launch vehicle Space Plane

Concept proposed by China Academy of Launch Vehicle Technology. A 140 tons, 32m length orbiter launched atop a Long March 5 rocket with a payload of 7 tons.[1]

37.6   Shenlong Space Plane

Main article: Shenlong (spacecraft)

Images of an aerodynamic scaled model, ready to be launched from under the fuselage of a H-6K bomber, were first published in the Chinese media on 11 December 2007. Code named Project 863-706, the Chinese name of this spacecraft was revealed as "神龙" 空天飞机 or "Shenlong Space Plane", meaning Divine Dragon in Mandarin. These images, possibly taken in late 2005, show the vehicle's black reentry heat shielding, indicating a reusable design, and its engine assembly.[4] First sub-orbital flight of the Shenlong reportedly took place on 8 January 2011.[5]

It has been proposed that the vehicle is fitted with a Russian-designed D-30K turbofan engine, which would likely not provide enough power to reach Low Earth orbit. A larger Shenlong model, however, would be capable of carrying a payload to orbit. Analysts had previously reported on a late 2006 Chinese test flight of what is believed to be a scramjet demonstrator, possibly related to the Shenlong vehicle.[4]

Earlier, images of the High-enthalpy Shock Waves Laboratory wind tunnel of the CAS Key Laboratory of high-temperature gas dynamics (LHD) were published in the Chinese media. Test with speed up to Mach 20 where reached around 2001.[6]

As of 2007, the CAS academician Zhuang Fenggan (莊逢甘) said that a first test flight of the spaceplane would be conducted during the "Eleventh Five-Year Plan", meaning from 2006 to 2010.[7]

37.7   Hypersonic Vehicle

According to 'informal sources', another hypersonic vehicle has been tested, which is equivalent to the X-43.[8]

37.8   References

[1] http://www.strategycenter.net/docLib/20111024_ChinasSpacePlaneProgram.pdf

[2] " 航天运载器高温超导磁悬浮助推发射技术". 虚幻军事天空. 2006-11-11. Retrieved 2008-04-20.

[3] "航天发射用磁悬浮助推发射系统概念研究". 维普资讯网. 2005-01-31. Retrieved 2008-04-20.

[4] "中国" 神龙" 飞行器首度曝光身世扑朔迷离". SOHU.com. 2008-01-11. Retrieved 2008-04-13.

[5] "Shenlong 'Divine Dragon' Takes Flight: Is China developing its first spaceplane?". China Signpost. 2012-05-04. Retrieved 2012-06-19.

[6] "氢氧爆轰驱动激波高焓风洞". 中国科学院高温气体动力学重点实验室. 2005-03-17. Retrieved 2008-04-16.

[7] "國産空天飛機 3 年内試飛". 香港文匯報. 2007-12-11. Retrieved 2008-04-16.

[8] "International Assessment and Strategy Center > Research > PLA and U.S. Arms Racing in the Western Pacific". Strategycenter.net. 2011-06-29. Retrieved 2013-11-16.

37.9 External links

- http://www.friends-partners.org/partners/mwade/craft/prot9213.htm

- http://web.archive.org/web/20091027100240/http://geocities.com/CapeCanaveral/Launchpad/1921/

- http://www.sinodefence.com/strategic/mannedspace/project921.asp

- http://www.worldspaceflight.com/china/921.htm

- http://www.strategycenter.net/docLib/20111024_ChinasSpacePlaneProgram.pdf

Chapter 38

Prometheus (spacecraft)

This article is about a proposed spaceplane. For other uses, see Prometheus (disambiguation).
For the Alien franchise ship, see Prometheus (2012 film). For the Star Trek franchise starships, see USS Prometheus.
For the Star Trek franchise ship class, see Prometheus class starship. For the Stargate franchise spaceship, see
Prometheus (Stargate).

Prometheus was a proposed manned vertical-takeoff, horizontal-landing (VTHL) spaceplane concept put forward

Rendering of Prometheus in orbit

by Orbital Sciences Corporation in late 2010 as part of the second phase of NASA's Commercial Crew Development (CCDev) program.*[1]*[2]

38.1 Design

The Prometheus design was based on an earlier NASA design, the HL-20 Personnel Launch System. Prometheus also included other NASA-funded design improvements to HL-20 by Orbital Sciences that were done some years ago as part of NASA's Orbital Space Plane program.*[1] Whereas the HL-20 was a pure lifting body, the Prometheus design was for a Blended Lifting Body (BLB).*[1] This design combines volumetric efficiency with superior aerodynamic qualities.*[3] Prometheus could have initially carried four astronauts to the International Space Station or future commercial space stations but further development could have increased the seating capacity to six.*[4] The baselined launch vehicle was the Atlas V, but the design could have accommodated other launch vehicles. The cost of the development of the Prometheus spacecraft and of upgrading the Atlas V would be between $3.5 and $4 billion.*[5]

38.2 Commercial Crew Development program

Failing to be selected in NASA's CCDev phase 2 program, Orbital Sciences announced in April 2011 that they will likely wind down their efforts to develop a commercial crew vehicle.*[6]

38.3 See also

- Dream Chaser (spacecraft) - another HL-20 derived proposed spacecraft put forward in 2010 by Sierra Nevada Corporation

38.4 References

[1] "Orbital Submits Proposal for NASA's Commercial Crew Development Program". *Press Release*. Orbital Sciences Corporation. 2010-12-14. Retrieved 2011-02-06.

[2] "The Shape of Things to Come - Orbital's Prometheus Space Plane Ready for NASA's Commercial Crew Development Initiative" (PDF).

[3] Leo, Ryan D. (2004-05-15). "Evolution of a Blended Lifting Body for the Orbital Space Plane". *no. 3326*. Society of Allied Weight Engineers. Retrieved 2011-02-28. *issues of volumetric efficiency, high L/D for cross range, low wing loading for reduced landing speed, and passive stability for all abort conditions were addressed. ... As the optimization process continued, the HL-20 initial reference shape eventually evolved into the Blended Lifting Body (BLB). The BLB combines volumetric efficiency with superior aerodynamic qualities and was designed to launch vertically and land horizontally.*

[4] Wall, Mike (2011-02-09). "Proposed Private Space Plane Gets Fiery Name: Prometheus". Space.com. Retrieved 16 August 2011.

[5] Businesses Take Flight, With Help From NASA, New York Times, 2011-1-31, accessed 2011-02-28

[6] Orbital may wind down its commercial crew effort "Orbital may wind down its commercial crew effort" Check |url= scheme (help). *NewSpace Journal*. 2011-04-22. Retrieved 2011-04-25. *CEO Dave Thompson said ... "I don' t, at this time, anticipate that we' ll continue to pursue our own project in that race. We' ll watch it and if an opportunity develops we may reconsider. But at this point, I would not anticipate a lot of activity on our part in the commercial crew market."*

38.5 External links

- A Blended Lifting Body Aerodynamic Design for the Orbital Space Plane, Henri D. Fuhrmann, AIAA-2003-3807, June 2003.

Chapter 39

Rocketplane XP

The **Rocketplane XP** was a suborbital spaceplane design that was under development circa 2005 by Rocketplane Kistler. The vehicle was to be powered by two jet engines and a rocket engine, intended to enable it to reach suborbital space. The XP would have operated from existing spaceports in a manner consistent with established commercial aviation practices. Commercial flights were projected to begin in 2009.[*][1] Rocketplane Global declared bankruptcy in mid-June 2010.[*][2] Their assets were auctioned off in 2011.[*][3]

39.1 Design and development

As envisioned, the Rocketplane XP would carry a pilot and five passengers on a flight profile from a runway using jet engines like a conventional aircraft. It would then climb to about 12 km (40,000 feet). At this point, a reusable rocket engine would power the XP on a suborbital trajectory reaching altitudes of over 100 km (62 mi) after burnout. The XP was to then reenter Earth's atmosphere and land at the same spaceport under conventional jet power. The relatively low speeds involved meant that heat shielding was not a major concern. The XP was expected to operate from the Clinton-Sherman Industrial Airpark near Burns Flat, Oklahoma.

On January 24, 2006 Rocketplane Limited announced a Space Act agreement with NASA Johnson Space Center for the loan of a Rocketdyne RS-88 rocket engine for three years, for use in flight tests of the XP vehicle.[*][4]

39.2 See also

- EADS Astrium Space Tourism Project

- Lynx (spacecraft)

- Dream Chaser

- SpaceShipTwo

- Blue Origin New Shepard

39.3 References

Notes

[1] Popsci article (Up and Away) - April 2007

[2] Foust, Jeff. "The gap in NewSpace business plans." *The Space Review*, July 12, 2010. Retrieved: July 17, 2010.

[3] http://nasawatch.com/archives/2011/10/rocketplane-kis.html

[4] "Small steps forward for NewSpace". The Space Review.

Bibliography

- Belfiore, Michael. "It's a Rocket! It's a Plane! It's...Rocket Plane!" *Popular Science*, January 8, 2006.

- "Model XP Specifications." *Rocketplane XP*, February 21, 2011.

39.4 External links

- Rocketplane official website

- astronautix.com

- Video animation - Rocketplane XP concept

Chapter 40

Rockwell X-30

The **Rockwell X-30** was an advanced technology demonstrator project for the **National Aero-Space Plane** (NASP), part of a United States project to create a single-stage-to-orbit (SSTO) spacecraft and passenger spaceliner.[*][1] It was cancelled in the early 1990s, before a prototype was completed, although much development work in advanced materials and aerospace design was completed. While a goal of a future NASP was a passenger liner capable of two hour flights from Washington to Tokyo, the X-30 was planned for a crew of two and oriented towards testing.

40.1 Development

The NASP concept is thought to have been derived from the "Copper Canyon" project of the Defense Advanced Research Projects Agency (DARPA), from 1982 to 1985. In his 1986 State of the Union address, President Ronald Reagan called for "a new Orient Express that could, by the end of the next decade, take off from Dulles Airport, accelerate up to 25 times the speed of sound, attaining low earth orbit or flying to Tokyo within two hours."

Research suggested a maximum speed of Mach 8 for scramjet based aircraft, as the vehicle would generate heat due to atmospheric friction, which would expend considerable energy. The project showed that much of this energy could be recovered by passing hydrogen over the skin and carrying the heat into the combustion chamber: Mach 20 then seemed possible. The result was a program funded by NASA, and the United States Department of Defense (funding was approximately equally divided among NASA, DARPA, the US Air Force, the Strategic Defense Initiative Office (SDIO) and the US Navy).[*][2]

McDonnell Douglas, Rockwell International, and General Dynamics competed to develop technology for a hypersonic air-breathing SSTO vehicle. Rocketdyne and Pratt & Whitney competed to develop engines.

In 1990, the companies joined under the direction of Rockwell International to develop the craft to deal with the technical and budgetary obstacles. Development on the X-30, as it was then designated.

Despite progress in developing the necessary structural and propulsion technology, NASA still had substantial problems to solve. The Department of Defense wanted it to carry a crew of two and even a small payload. The demands of being a human-rated vehicle, with the instrumentation, environmental control system, and safety equipment, made X-30 larger, heavier, and more expensive than required for a technology demonstrator. The X-30 program was terminated amid budget cuts and technical concerns in 1993.

A more modest hypersonic program that culminated in the unmanned X-43 "Hyper-X".

A detailed 1/3rd scale (50 foot long) mock-up of the X-30 was built by engineering students at Mississippi State University's Raspet Flight Research Lab in Starkville, Mississippi.[*][3][*][4][*][5] The mock-up is on display at the Aviation Challenge campus of the U.S. Space Camp facility in Huntsville, Alabama.

40.2 Design

The X-30 configuration was a highly integrated engine. The shovel-shaped forward fuselage generated a shock wave to compress air before it entered the engine. The aft fuselage formed an integrated nozzle to expand the exhaust. The engine between was a scramjet engine. At the time, however, no scramjet engine of the kind was close to operational.

1986 artist's concept of the X-30 on liftoff

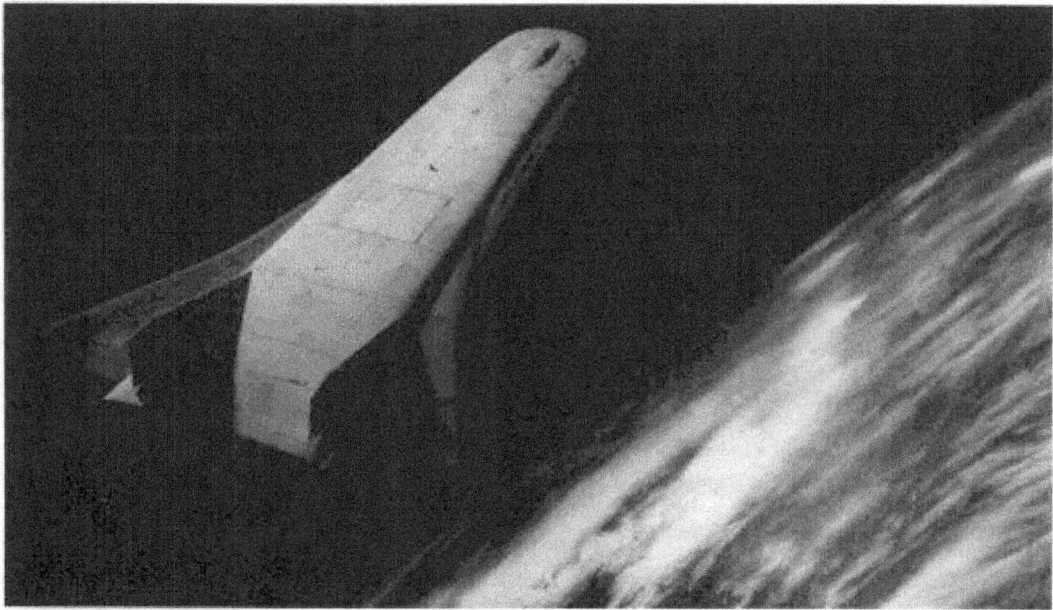

Artist's Concept of the X-30 in orbit

The aerodynamic configuration was an example of a waverider. Most of the lift was generated by the fuselage by compression lift. The "wings" were small fins providing trim and control. This configuration was efficient for high-speed flight, but would have made take-off and slow-speed flight difficult.

Temperatures on the airframe were expected to be 980 °C (1800 °F) over a large part of the surface, with maxima of more than 1650 °C (3000 °F) on the leading edges and portions of the engine. This required the development

Artist's Concept of the X-30 on reentry

of high temperature lightweight materials, including alloys of titanium and aluminum known as gamma and alpha titanium aluminide, advanced carbon/carbon composites, and titanium metal matrix composite (TMC) with silicon carbide fibers. Titanium matrix composites were used by McDonnell Douglas to create a representative fuselage section called "Task D". The Task D test article was four feet high by eight feet wide by eight feet long. A carbon/epoxy cryogenic hydrogen tank was integrated with the fuselage section and the whole assembly, including volatile and combustible hydrogen, was successfully tested with mechanical loads and a temperature of 820 °C (1500 °F) in 1992, just before program cancellation.

40.3   Specifications (X-30 as designed)

General characteristics

- **Length:** 160.0 ft (48.8 m)

- **Wingspan:** 74.0 ft (22.6 m)

- **Gross weight:** 300,000 lb (136,078 kg)

- **Powerplant:** 1 × Scramjet

Performance

- **Propellants:** Air/Slush LH2

40.4   See also

- Scramjet

Aircraft of comparable role, configuration and era

X-30 model in a wind tunnel

- NASA X-43 (essentially a down-scaled model)

- Tupolev Tu-2000

- HOTOL

40.5 References

[1] Chang, Kenneth (20 October 2014). "25 Years Ago, NASA Envisioned Its Own 'Orient Express' ". *New York Times*. Retrieved 21 October 2014.

[2] National Aero-space Plane Program Fact Sheet

[3] Mississippi State Wins Aero-Space Plane Mockup Competition

[4] NASP X-30

[5] X-30 National Aero-Space Plane Mockup Rolls Out

- American X-Vehicles

- Air Force History of the NASP(pdf)

40.6   External links

- Photo archive of the X-30 mockup

- Video Aero-Space Plane: Flexible Access to Space (NASA) 4 min

Chapter 41

Rotary Rocket

Rotary Rocket Company was a rocketry company that developed the **Roton** concept in the late 1990s as a fully reusable single-stage-to-orbit (SSTO) manned spacecraft. The design was initially conceived by Gary Hudson, who formed the company to commercialize the concept. The Roton was intended to reduce costs of launching payloads into low earth orbit by a factor of ten.

The company gathered considerable venture capital from angel investors and opened a factory headquartered in a 45,000-square-foot (4,200 m^2) facility at Mojave Air and Space Port in Mojave, California. The fuselage for their vehicles was made by Scaled Composites, at the same airport, while the company developed the novel engine design and helicopter-like landing system. A full-scale test vehicle made three hover flights in 1999, but the company exhausted its funds and closed its doors in early 2001.

41.1 Rotary Rocket design evolution

41.1.1 Helicopter to orbit

Gary Hudson's and Bevin McKinney's initial concept was to merge a launch vehicle with a helicopter: spinning rotor blades, powered by tip jets, would lift the vehicle in the earliest stage of launch. Once the air density thinned to the point that helicopter flight was impractical, the vehicle would continue its ascent on pure rocket power, with the rotor acting as a giant turbopump.[*][1]

Calculations showed that the helicopter blades modestly increased the effective specific impulse (I_{sp}) by approximately 20–30 seconds, essentially only carrying the blades into orbit "for free". Thus, there was no overall gain from this method during ascent. However, the blades could be used to soft land the vehicle, so its landing system carried no additional cost.

One problem found during research at Rotary was that once the vehicle left the atmosphere additional thrust would be necessary. Thus multiple engines would be needed at the base as well as at the rotor tips. Another issue would be noise, since the rocket tipped rotor is likely to be extraordinarily noisy, as was the Fairey Rotodyne.

This initial version of the Roton had been designed with the small communications satellite market in mind. However, this market crashed, signaled by the failure of Iridium Communications. Consequently, the Roton concept needed to be redesigned for heavier payloads.

41.1.2 Helicopter from orbit

The revised and redesigned **Roton** concept was a cone-shaped launch vehicle, with a helicopter rotor on top for use only during landing. An internal cargo bay could be used both for carrying payloads to orbit and bringing others back to Earth. The projected price to orbit of this design was given as $1,000 per kg of payload, less than one-tenth of the then-current launch price. Payload capacity was limited to a relatively modest 6,000 pounds (2,700 kg).

The revised version would have used a unique rotating annular aerospike engine: the engine and base of the launch vehicle would spin at high speed (720 rpm) to pump fuel and oxidizer to the rim by the rotation. Unlike the landing rotor, due to the shallow angle of the nozzles in the base rotor, the rotation speed self limited and required no control

The Rotary Rocket Roton ATV on permanent display at the Mojave Spaceport.

system. Since the density of the LOX (liquid oxygen) was higher than that of the Kerosene, extra pressure was available with the LOX, so it would have been used to cool the engine's throat and other components, rather than using the kerosene as the coolant as in a conventional LOX/kerosene rocket. However, at the high G levels at the outer edge of the rotating engine block, clarity on how LOX would work as a coolant was both unknown and difficult to validate. That added one layer of risk.

In addition, the rotating exhaust acted as an effective wall at the outer edge of the engine base, and the entire base area effectively is pumped down below ambient due to ejector pump effect, creating an effective suction cup at the bottom in atmosphere. This could be alleviated using makeup gas to develop base pressure, requiring effectively an additional rocket engine to fill up the base of the main rocket engine. (Similar problems would have occurred in a conventional aerospike engine, but there, natural recirculation plus use of the turbopump gas-generator's exhaust as the makeup gas would have largely alleviated the problem "for free.")

At the rim, 96 miniature jets would exhaust the burning propellants (LOX and kerosene) around the rim of the base of the vehicle, which gained the vehicle extra thrust at high altitude –effectively acting as a zero-length truncated aerospike nozzle.[*][2] A similar system with non-rotating engines was studied for the N1 rocket. That application had a much smaller base area, and did not create the suction effect a larger peripheral engine induces. The Roton engine had a projected vacuum I_{SP} (specific impulse) of ~355 seconds (3.5 kN·s/kg), which is very high for a LOX/kerosene engine –and a thrust to weight ratio of 150, which is extremely light.[*][3]

During reentry, the base also served as a water-cooled heatshield. This was theoretically a good way to survive reentry, particularly for a lightweight reusable vehicle. However, using water as a coolant would require converting it into superheated steam, at high temperatures and pressures, and there were concerns about micrometeorite damage on orbit puncturing the pressure vessel, causing the reentry shield to fail. These concerns were resolved using a failure resistant massively redundant flow system, created using thin metal sheets chemically etched with a pattern of micropores forming a channel system that was robust against failure and damage.

In addition, cooling was achieved two different ways; one way was the vaporization of the water, but the second was even more significant, and was due to the creation of a layer of "cool" steam surrounding the base surface, reducing the ability to heat. Further, the water metering system would have to be extremely reliable, giving one drop per second per square inch, and was achieved via a trial/error design approach on real hardware. By the end of the ROTON program, some hardware had been built and tested. The reentry trajectory was to be trimmed, similar to the Soyuz, to minimize the G loads on the passengers. And the ballistic coefficient was better for the Roton and could be better tailored. When the Soyuz trim system failed and it went full ballistic, the G levels did rise significantly but without incident to the passengers.

The vehicle was also unique in planning to use its helicopter-style rotors for landing, rather than wings or parachutes. This concept allowed controlled landings (unlike parachutes), and it was 1/5 the weight of fixed wings. Another advantage was that a helicopter could land almost anywhere, whereas winged spaceplanes such as the Shuttle had to make it back to the runway. The rotor blades were to be powered by peroxide tip rockets. The rotor blades were to be deployed before reentry; some questions were raised about whether the blades would survive until landing.

The initial plan was to have them almost vertical, but that was found to be unstable as they needed to drop lower and lower and spin faster for stability, the heating rates went up dramatically and the air flow became more head on. The implication of that was that the blades went from a lightly heated piece of hardware to one that either had to be actively cooled or made of SiC or other refractory material. The idea of popping out the blades became much more attractive at this point, and initial studies were made for that option. This rotor design concept was not without precedent. In 1955, one of five Soviet designs for planned suborbital piloted missions was to include rocket-tipped rotors as its landing system. On May 1, 1958 these plans were dropped as a decision was made to proceed directly to orbital flights.

Rotary Rocket designed and pressure-tested an exceptionally lightweight but strong composite LOX tank. It survived a test program which involved it being pressure cycled and ultimately deliberately shot to test its ignition sensitivity. This composite construction was a world first.

41.1.3 A new engine

In June 1999, Rotary Rocket announced that it would use a derivative of the Fastrac engine under development at NASA's Marshall Space Flight Center, instead of the company's own unconventional spinning engine design. Reportedly, the company had been unable to convince investors that its engine design was viable; the composite structure and gyrocopter reentry was an easier sell.

At the same time as this change, the company laid off about a third of its employees, lowering approximate headcount from 60 to 40. At this point, the company planned to begin its commercial launch service sometime in 2001.[*][4] Although the company had raised $30 million, it still needed to raise an additional $120 million before entering service.

41.2   The Atmospheric Test Vehicle (ATV)

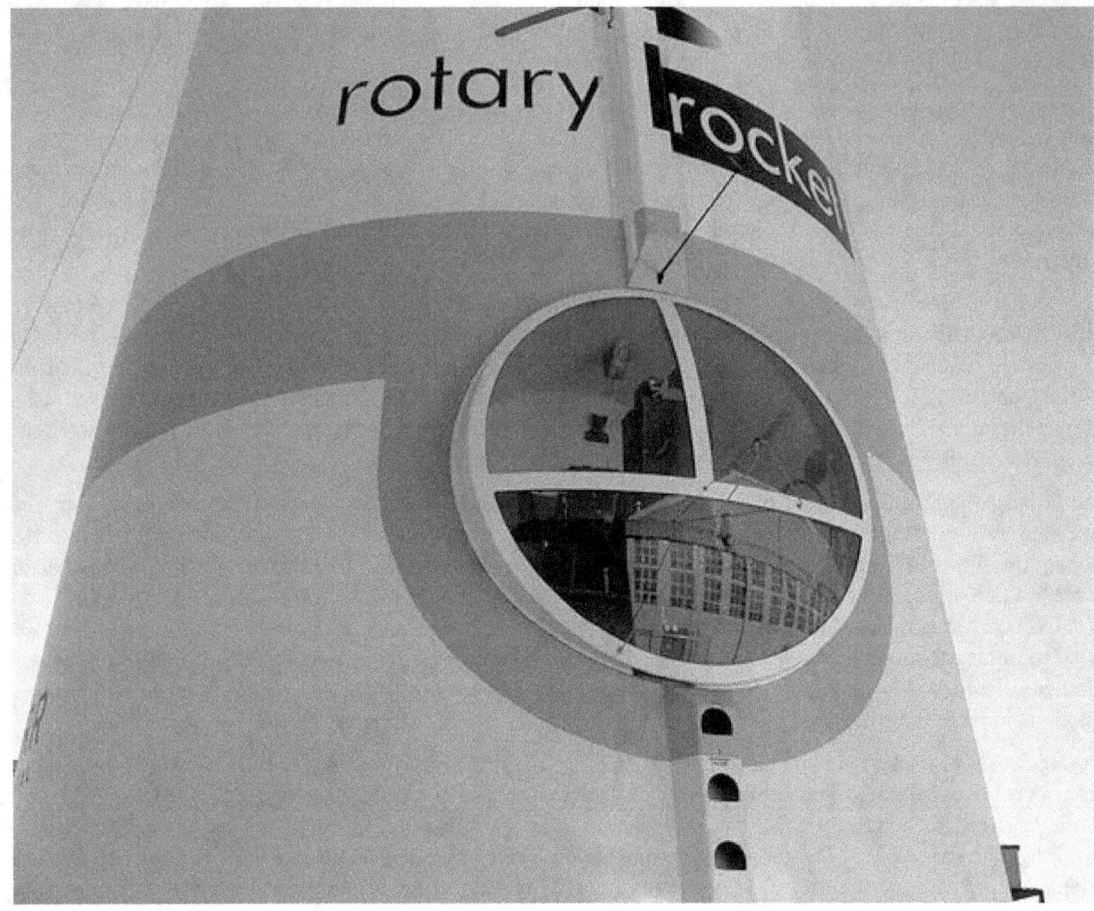

The ATV's cockpit was nicknamed the 'Batcave' by its pilots because of its restricted field of view.

A full size, 63 ft (19 m) tall, Atmospheric Test Vehicle (ATV) was built under contract by Scaled Composites for use in hover test flights. The $2.8 million ATV was not intended as an all-up test article, since it had no rocket engine and no heat shielding. The ATV was rolled out of its Mojave hangar on March 1, 1999, bearing an FAA registry of N990RR.

The rotor head was salvaged from a crashed Sikorsky S-58, at a price of $50,000 –as opposed to as much as $1 million for a new head. Each rotor was powered by a 350-lbf (1,560 N) hydrogen peroxide jet, as intended for the orbital vehicle.[*][5] The rotor assemblage was tested in a rock quarry before installation on the ATV.

The ATV flew three successful test flights in 1999. The pilot for these three flights was Marti Sarigul-Klijn and the copilot was Brian Binnie (who later gained fame as pilot of Scaled Composites' SpaceShipOne on its second X-Prize flight).

The ATV made its first flight on July 28. This flight consisted of three vertical hops totaling 4 min 40 sec in duration and reaching a maximum altitude of 8 ft (2.4 m). The pilots found the flying extremely challenging for a number of reasons. Visibility in the cockpit was so restricted that the pilots nicknamed it the Batcave. The view of the ground was entirely obstructed, so the pilots had to rely on a sonar altimeter to judge ground proximity. The entire craft had a low rotational inertia, and torque from the spinning rotor blades made the body spin, unless counteracted by yaw thrust in the opposite direction.[*][6]

The second flight, on September 16, was a continuous hover flight lasting 2 min 30 sec, reaching a maximum altitude of 20 ft (6.1 m). The sustained flight was made possible by the installation of more powerful rotor tip thrusters and an autothrottle.[*][7]

The third and last flight was made on October 12. The ATV flew down the flightline at Mojave Air and Space Port, covering 4,300 ft (1,310 m) in its flight and rising to a maximum altitude of 75 ft (23 m). The speed was as high as 53 mph (85 km/h). This test revealed some instability in translational flight.

A fourth test was planned to simulate a full autorotative descent. The ATV would climb to an altitude 10,000 ft (3,050 m) under its own power, before throttling back and returning for a soft landing.[8] At this point, given that further funding was then unlikely, safety considerations prevented the test being attempted.

41.3 Criticism of the design

Rotary Rocket failed due to lack of funding, but some have suggested that the design itself was inherently flawed.

On one hand, Rotary Rocket demonstrated its technical ability by flight testing of actual hardware. The ATV flew three test flights and a composite propellant tank survived a full test program. As Jim Ransom, Rotary Rocket consultant, pointed out at the demise of the company, this was more than could be said for Lockheed Martin's X-33, which had a budget 30 times larger.[9]

On the other hand, these tests revealed problems. For instance, the ATV demonstrated that a landing of the Rotary Rocket would be tricky, even dangerous. Test pilots have a rating system, the Cooper-Harper rating scale, for vehicles between 1 and 10 that relates to difficulty to pilot. The Roton ATV scored a 10 —the vehicle simulator was found to be practically unflyable by anyone except the Rotary test pilots, and even then there were expected to be short periods where the vehicle was out of control.

Other aspects of the flight plan remained unproven. It is not known whether Roton could in practice have developed enough overall performance to reach orbit with a single stage, and return – although on paper this might have been possible. These doubts led some of the aerospace community to dismiss the Rotary Rocket concept as an impossible pipe dream.

Whether a flight-article vehicle would have worked successfully remains open to speculation.

41.4 The venture's last days

The Rotary Rocket Hangars at Mojave Air and Space Port, as seen in 2005. The taller hangar on the left was the Rotary Rocket Assembly Building.

Engine development ceased in 2000, reportedly two weeks before a full-scale test was due. The vehicle failed to secure launch contracts and Rotary Rocket closed its doors in 2001.

The timing of the venture was unfortunate: the Iridium Communications venture was nearing bankruptcy, and the space industry in general was experiencing financial stress. Ultimately, the company did not attract sufficient funding

– even though numerous individuals provided a total of $33 million of support, including writer Tom Clancy.*[10]

Some of the engineers that worked there have since set up other rocketry ventures, notably XCOR Aerospace, t/Space and Space Launch.

The Atmospheric Test Vehicle was to be displayed at Classic Rotors Museum, a helicopter museum near San Diego, California, but an attempt to move it there on May 9, 2003 via a short-line sling-load under an Army Reserve CH-47 Chinook failed when the Roton began to oscillate at airspeeds above 35 knots (65 km/h). Instead, the Mojave Airport administration worked to keep this historic vehicle at Mojave, and on November 10, 2006, the Roton was moved to its permanent display location at the intersection of Airport Blvd and Sabovich Road. To many, the Roton represents the program that launched Mojave into the Space Age, and this theme was echoed during the dedication ceremony that took place during the Veterans' Day celebration on November 11, at which Brian Binnie was the keynote speaker.

The Rotary Rocket hangars are now occupied by the National Test Pilot School.

41.4.1   Roton C-9 specifications

- **Overall dimensions:**
    - **Height:** 63 ft (19.2 m)
    - **Maximum diameter:** 22 ft (6.7 m)
- **Cargo bay dimensions:**
    - **Length:** 16.7 ft (5.1 m)
    - **Diameter:** 12 ft (3.7 m)
- **Total mass (estimated):** 400,000 lb (180,000 kg)
- **Low Earth Orbit payload (projected):** 7,000 lb (3,180 kg)
- **Orbit apogee (projected):** 160 mi (260 km)
- **Engine thrust:** 6,950 lb (30,860 N)
- **Engine specific impulse (vac):** 340 sec
- **Number of engines:** 72

(Projections are based on estimates reported in *Aviation Week & Space Technology* on October 5, 1998.)   Also see:*[11]

41.5   References

Citations

[1] Wired - Insanely Great? or Just Plain Insane?

[2] United States Patent 5842665

[3] Anselmo, Joseph C., "Rotarians." *Aviation Week & Space Technology*, October 5, 1998, p. 17.

[4] Dornheim, Michael A., "Rotary Cuts Staff, Changes Engine." *Aviation Week & Space Technology*, June 28, 1999, p. 44.

[5] Dornheim, Michael A., "Roton Test Craft Rolled Out." *Aviation Week & Space Technology*, March 8, 1999, p. 40.

[6] Dornheim, Michael A., "Roton Hops Off Ground." *Aviation Week & Space Technology*, August 12, 1999, p. 36.

[7] Smith, Bruce A., "Roton Test." *Aviation Week & Space Technology*, October 11, 1999, p. 21.

[8] "Roton Achieves Forward Flight." *Aviation Week & Space Technology*, October 25, 1999, p. 40.

[9] Agle, D. C., "Sounding: Elegy in the High Desert." *Air&Space/Smithsonian Magazine*, May, 2001. A short article on the death of Rotary Rocket.

[10] Rotary Rocket Unveils New RLV Design Major Investor Tom Clancy Becomes a Director

[11] Rotary Rocket - Specifications

Bibliography

- Petit, Charles, "Rockets for the Rest of Us." *Air&Space/Smithsonian Magazine*, March, 1998. A look at the early design of Rotary Rocket.

- Sarigul-Klijn, Marti, "I Survived the Rotary Rocket." *Air&Space/Smithsonian Magazine*, March, 2002. The test pilot of the ATV describes the three test flights.

- Weil, Elizabeth, *They All Laughed At Christopher Columbus: An Incurable Dreamer Builds the First Civilian Spaceship*. Bantam, 2003. An insider's view of the development of Rotary Rocket. ISBN 978-0-553-38236-5

41.6 External links

- Roton article at Encyclopedia Astronautica

- Gary C. Hudson, Insanely Great? or Just Plain Insane? - Wired magazine article on Roton, May 1996

- Space.com on test flights; archived copy

- Space.com on helicopter museum trip

- QuickTime footage of final test flight of Roton ATV, from the Air&Space Magazine website

- Archives of original rotaryrocket.com website, from Internet Archive Wayback Machine

- Tom Brosz's personal account of Rotary Rocket and fallout

- Roton C-9 specs

- Photos of the project from the Mojave Virtual Museum

- Photos from the unveiling of the Roton

Coordinates: 35°03′19″N 118°09′30″W / 35.055321°N 118.158375°W

Chapter 42

Saenger (spacecraft)

Saenger or **Sänger** was a concept design of a spaceplane. Its first design, Saenger I, was developed between 1961 and 1974 from plans originating from Eugen Sänger at the West German company Junkers.[1]

42.1 Saenger I

The Saenger vehicle utilised a two stage concept similar to that of the space shuttle. The first stage with the second stage attached on top would takeoff horizontally using a runway and climb to an altitude of 30 km using airbreathing ramjet engines. The second stage would then detach and accelerate to orbital speeds and altitudes using its LOX/LH2 rocket engine. The advantage of this approach is that the first stage utilises the advantages of air breathing engines (such as higher specific impulse) until they are no longer viable due to low air pressure and high velocities. The second stage had dimensions of 31 m × 12 m and would have been capable of carrying two astronauts.[1]

42.2 Saenger II

The saenger-II project grew out of Saenger I in the late 1980s and was planned to be a European launch vehicle emulating the successes of the space shuttle in the USA. The development was undertaken at Messerschmitt-Bölkow-Blohm.[2] The vehicle would take off from a runway using ramjet engines and climb to 30 km altitude and reach Mach 7. The second stage would then detach, allowing the first stage to return to the original runway, and accelereate to orbital velocities and altitudes using its rocket engine. The spacecraft would have been able to deliver a payload of 10,000 kg or a crew module to low earth orbit.[2]

The project was discontinued in 1995 primarily due to concerns of development costs and limited gains in price and performance compared to the existing expendable launcher programs such as Ariane V.[2]

42.3 References

[1] Handbuch der Raumfahrt Technik by Hallmann willi, ISBN 3-446-15130-3 Dritte im Weltraum by Herbert Erdmann. schwann verlag, 1969

[2] "Saenger II" . Encyclopedia Astronautica. Retrieved 11 September 2014.

Chapter 43

Sharp Edge Flight Experiment

SHEFEX (Sharp Edge Flight Experiment), is an experiment conducted by the German Aerospace Center (DLR), for the development of some new, cheaper and safer design principles for space capsules, hypersonic vehicles and spaceplanes with re-entry capability in the atmosphere and their integration into a complete system.

DLR explained the objectives of SHEFEX: The aim of the research is a spaceplane that is usable for experiments under microgravity from 2020 on.[*][1] It is set to finish with a space plane project named REX Freeflyer(REX for Returnable experiment, dt. Rückkehrexperiment).[*][2]

During re-entry of spacecraft into the earth's atmosphere, the high velocity of the spacecraft together with friction and displacement of air molecules leads to temperatures of over 2000 °C.[*][1] In order to not burn up, spaceships need very expensive and sometimes failing heat shields.

43.1 First spacecraft with sharp corners and edges

The namesake idea for the sharp-edged flight experiment of Hendrik Weihs, coordinator for returning technologies DLR, is an entirely new form for a spacecraft, namely with sharp corners and edges instead of the rounded shapes ubiquitously used in space flight today. Few flat tile shapes can be produced at lower cost than highly individual rounded shapes.[*][1]

Dr. Klaus Hannemann, Head of the spacecraft department at the DLR Institute of Aerodynamics and Flow Technology in Göttingen explains the fundamental advantage of the concept:

"A space shuttle has more than 25,000 differently shaped tiles. The simple shape of Shefex tiles should lower the maintenance costs of the thermal protection system and a simple replacement of tiles in space would be possible." , Missing or empty |title= (help)

Additionally, the project is aimed at improving aerodynamics. General Project Manager Hendrik Weihs said:

"The capsule almost achieves the aerodynamic characteristics of a space shuttle, but is smaller and does not need wings" , Missing or empty |title= (help)[*][3]

Programmatically, the DLR, said:

"Judging from experience in the development of thermal protection systems, curved outer contours with high accuracy requirements were identified as a major cost driver. Large, curved fiber-ceramic structures require sophisticated production tools and require auxiliary molds and optimized manufacturing for each individual component. It is therefore possible to reduce costs through simplification by tessellating the outer contour with flat tiles with only few distinct shapes. It is possible in principle to produce different flat tiles from a basic tile by cropping. This also leads to significant savings in maintenance and replacement of damaged tiles. Problems arise, however, from the fluid dynamics around the sharp edges and corners, which give rise to very high temperatures that must be controlled by new technologies, such as actively cooled elements. Sharp edges have aerodynamic advantages as well, causing lower drag in hypersonic flight conditions." , Missing or empty |title= (help).

The assembled SHEFEX II body.

43.2 SHEFEX I

SHEFEX I was the first experimental vehicle of the SHEFEX project[*][4] and launched on Thursday, 27 October 2005 from the Andøya Rocket Range in Norway. Shefex I reached a height of about 200 km over the North Sea. Within 20 seconds, the vehicle re-entered Earth's atmosphere at almost seven times the speed of sound. The measured data and live images of the on-board camera were transferred directly to the ground station. However, during the activation of the parachute system an error occurred that led to the loss of the parachute system and consequently to the loss of the flight unit. According to the DLR, the evaluation of the data provided important insights so that SHEFEX I could be seen as a great success from the perspective of the DLR. For the flight, a missile system was used, which consisted of a combined Brazilian VS-30 lower level and a HAWK rocket as the second stage. The cost of the three-year project was approximately 4 million euros. It was part of the space program of the Helmholtz Association of German Research Centers (HGF) and the DLR.

43.3 SHEFEX II

With SHEFEX II, nine different thermal protection systems were to be evaluated on the facetted skin, mainly new fiber ceramics. Additionally, the aerospace companies EADS Astrium and MT Aerospace as well as Boeing used some of the surface of SHEFEX II for their own experiments. The vehicle was equipped with sensors to measure pressure, heat flux, and temperature in the vehicle tip.

On 22 June 2012, SHEFEX II was launched from the same launch station, the Andøya Rocket Range in Norway. It reached a height of about 180 kilometers and a speed of about 11,000 kilometers per hour (eleven times the speed of sound). The rocket used was the Brazilian VS-40. During its re-entry, SHEFEX II survived temperatures above 2500 °C, while sending data from the 300 different sensors to the ground station.

43.4 SHEFEX III

For 2016 the DLR plans SHEFEX III, a spaceplane-like vehicle. It should fly even faster and stay in the air for 15 minutes, far longer than the previous two experiments.[*][5]

43.5 REX Free Flyer (SHEFEX IV)

The REX-Free Flyer is planned as a first application of the experience collected from SHEFEX. This system should serve as a free-flying platform for high quality microgravity experiments over several days. The possibility of controlled return and a modular design of the experiment trays, that closely resemble those found on sounding rockets, should give experimenters quick and inexpensive access to their experiments.[*][6]

43.6 References

[1] "Raumfahrzeug SHEFEX II startet im September 2011 in Norwegen" (in German). DLR. 2011-04-07. Retrieved 2011-07-09.

[2] Seite des Institut für Bauweisen- und Konstruktionsforschungder DLR über das Projekt REX-Free Flyer, Retrieved 2012-06-28

[3] "Test für neues Raumfahrzeug" (in German). Astronews. 2010-05-10. Retrieved 2012-06-29.

[4] "DLR Portal - SHEFEX flight: Webcast replay". Dlr.de. 2005-10-27. Retrieved 2012-09-30.

[5] DLR/Redaktion Astronews.com: *SHEFEX II. Erfolgreicher Testflug lieferte wichtige Daten*. Retrieved 2012-06-29

[6] DLR-Handout: SHEFEX II Ein weiterer Schritt im Flugtestprogramm für Wiedereintrittstechnologie PDF file

43.7   External links

- SHEFEX at DLR website.

Chapter 44

Silbervogel

Silbervogel, German for **silver bird**, was a design for a liquid-propellant rocket-powered sub-orbital bomber produced by Eugen Sänger and Irene Bredt in the late 1930s for The Third Reich/Nazi Germany. It is also known as the RaBo (*Raketenbomber* or "rocket bomber"). It was one of a number of designs considered for the *Amerika Bomber* mission, which started out in the spring of 1942 being focused solely on trans-Atlantic range piston-engined strategic bombers, like the Messerschmitt Me 264 and Junkers Ju 390, the only two airframe types actually built and flown for the competition. When Walter Dornberger attempted to create interest in military spaceplanes in the United States after World War II, he chose the more diplomatic term *antipodal bomber*.

44.1 Concept

The design was a significant one, as it incorporated new rocket technology, and the principle of the lifting body, foreshadowing future development of winged spacecraft such as the X-20 Dyna-Soar of the 1960s and the Space Shuttle of the 1970s. In the end, it was considered too complex and expensive to produce. The design never went beyond mock up test.

The Silbervogel was intended to fly long distances in a series of short hops. The aircraft was to have begun its mission propelled along a 3 km (2 mi) long rail track by a large rocket-powered sled to about 800 km/h (500 mph). Once airborne, it was to fire its own rocket engine and continue to climb to an altitude of 145 km (90 mi), at which point it would be travelling at some 5,000 km/h (3,100 mph). It would then gradually descend into the stratosphere, where the increasing air density would generate lift against the flat underside of the aircraft, eventually causing it to "bounce" and gain altitude again, where this pattern would be repeated. Because of aerodynamic drag, each bounce would be shallower than the preceding one, but it was still calculated that the Silbervogel would be able to cross the Atlantic, deliver a 4,000 kg (8,800 lb) bomb to the continental United States, and then continue its flight to a landing site somewhere in the Empire of Japan–held Pacific, a total journey of 19,000 to 24,000 km (12,000 to 15,000 mi).

Postwar analysis of the Silbervogel design involving a mathematical control analysis unearthed a computational error and it turned out that the heat flow during the initial atmospheric re-entry would have been far higher than originally calculated by Sänger and Bredt; if the Silbervogel had been constructed according to their flawed calculations the craft would have been destroyed during re-entry. The problem could have been solved by augmenting the heat shield, but this would have reduced the craft's already small payload capacity.[*][1]

44.2 History

On 3 December 1941 Sänger sent his initial proposal for a suborbital glider to the Reich Air Ministry (RLM) as *Geheime Kommandosache Nr. 4268/LXXX5*. The 900-page proposal was regarded with disfavor at the RLM due to its size and complexity and was filed away. Then Sänger went to work on more modest projects such as the Skoda-Kauba Sk P.14 ramjet fighter.[*][2]

Professor Walter Gregorii had Sänger rework his report and a greatly reduced version was submitted to the RLM in September 1944, as UM 3538. It was the first serious proposal for a vehicle which could carry a pilot and payload to the lower edge of space.

Two manned and one unmanned version were proposed: the *Antipodenferngleiter* (antipodal long-range glider) and the *Interglobalferngleiter* (intercontinental long-range glider). Both were to be launched from a rocket-powered sled. The two manned versions were identical except in payload. The *Antipodenferngleiter* was to be launched at a very steep angle (which would shorten the range) and after dropping its bomb load on New York City was to land at a Japanese base in the Pacific.[*][3]

44.3   Postwar

After the war ended, Sänger and Bredt worked for the French government[*][4] and in 1949 founded the Fédération Astronautique. Whilst in France, Sänger was the subject of a botched attempt by Soviet agents to win him over. Joseph Stalin had become intrigued by reports of the Silbervogel design and sent his son, Vasily, and scientist Grigori Tokaty to kidnap Sänger and Bredt and bring them to the USSR.[*][5][*][6] When this plan failed, a new design bureau was set up by Mstislav Vsevolodovich Keldysh in 1946 to research the idea. A new version powered by ramjets instead of a rocket engine was developed, usually known as the Keldysh bomber, but not produced.[*][1] The design, however, formed the basis for a number of additional cruise missile designs right into the early 1960s, none of which were ever produced.

In the US, a similar project, the X-20 Dyna-Soar, was to be launched on a Titan II booster. As the manned space role moved to NASA and unmanned reconnaissance satellites were thought to be capable of all required missions, the United States Air Force gradually withdrew from manned space flight and Dyna-Soar was cancelled.

One lasting legacy of the Silverbird design is the "regenerative cooling—regenerative engine" design, in which fuel or oxidizer is run in tubes around the engine bell in order to both cool the bell and pressurize the fluid. Almost all modern rocket engines use this design today and some sources still refer to it as the **Sänger-Bredt** design.

44.4   Sänger II Space Plane

On 18 October 1985 Messerschmitt-Bölkow-Blohm (MBB) began renewed studies of the Sänger spaceplane, this time a "piggyback" two-stage-to-orbit horizontal takeoff concept.[*][7]

44.5   See also

- Boost-glide
- Keldysh bomber
- Spacecraft propulsion
- Rocket sled launch

44.6   References

[1] Reuter, Claus (2000). *The V2 and the German, Russian and American Rocket Program*. German - Canadian Museum of Applied History. p. 99. ISBN 9781894643054.

[2] Skoda-Kauba Sk P.14 - Ramjet fighter project

[3] Reuter, C. *The V2 and the German, Russian and American Rocket Program*. CA: German Canadian Museum. pp. 96–97. ISBN 978-1-894643-05-4.

[4] Eugen Sänger; Irene Sänger-Bredt (August 1944). *A Rocket Drive For Long Range Bombers* (PDF). Astronautix.com. Retrieved 2010-04-27.

[5] Duffy, James P (2004). *Target: America—Hitler's Plan to Attack the United States*. Praeger. p. 124. ISBN 0-275-96684-4.

[6] Shayler, David J (2005). *Women in Space—Following Valentina*. Springer Verlag. p. 119. ISBN 1-85233-744-3.

[7] *Sæger II*, Astronautix.

44.7 External links

- Luft '46

- Encyclopedia Astronautica

Chapter 45

Skylon (spacecraft)

For other uses of Skylon, see Skylon (disambiguation).

Skylon is a design for a single-stage-to-orbit spaceplane by the British company Reaction Engines Limited (REL), using SABRE, a combined-cycle, air-breathing rocket propulsion system, potentially reusable for 200 flights. In paper studies, the cost per kilogram of payload carried to low Earth orbit in this way is hoped to be reduced from the current £15,000/kg (as of 2011),[4] including research and development, to around £650/kg, with costs expected to fall much more over time after initial expenditures have amortised.[3] In 2004, the developer estimated the total lifetime cost of the programme to be about $12 billion.[3]

The vehicle design is for a hydrogen-fuelled aircraft that would take off from a conventional runway, and accelerate to Mach 5.4 at 26 kilometres (16 mi) altitude using the atmosphere's oxygen before switching the engines to use the internal liquid oxygen (LOX) supply to take it into orbit.[5] Once in orbit it would release its payload of up to 15 tonnes. The vehicle will be unpiloted, but also be certified to carry passengers. All payloads could be carried in a standardised container compartment. The relatively light vehicle would then re-enter the atmosphere and land on a runway, being protected from the conditions of re-entry by a ceramic composite skin. When on the ground it would undergo inspection and necessary maintenance. If the design goal is achieved, it should be ready to fly again within two days.

As of 2012, only a small portion of the funding required to develop and build Skylon had been secured. The research and development work on the SABRE engine design is proceeding under a small European Space Agency (ESA) grant. In January 2011, REL submitted a proposal to the British government to request additional funding for the project and in April REL announced that they had secured $350 million of further funding contingent on a test of the engine's precooler technology being successful. Testing of the key technologies was successfully completed in November 2012, allowing Skylon's design to advance to its final phase.[6][7] On 16 July 2013 the British government pledged £60M to the project: this investment will provide support at a "crucial stage" to allow a full-scale prototype of the SABRE engine to be built.[8]

If all goes to plan, the first test flights could happen in 2019, and Skylon could be visiting the International Space Station by 2022. It could carry 15 tonnes of cargo to a 300 km equatorial orbit on each trip, and up to 11 tonnes to the International Space Station, almost 45% more than the capacity of the European Space Agency's ATV vehicle.[9]

45.1 Research and development programme

45.1.1 Background and early work

Skylon is based on a previous project of Alan Bond, known as HOTOL.[10] The development of HOTOL began in 1982, at a time when space technology was moving towards reusable launch systems such as the Space Shuttle. In conjunction with British Aerospace and Rolls-Royce, a promising design emerged to which the British government contributed £2 million. However, in 1988, the government withdrew further funding, and development was terminated. Following this setback, Bond decided to set up his own company, Reaction Engines Limited, with the hope of continuing development with private funding.

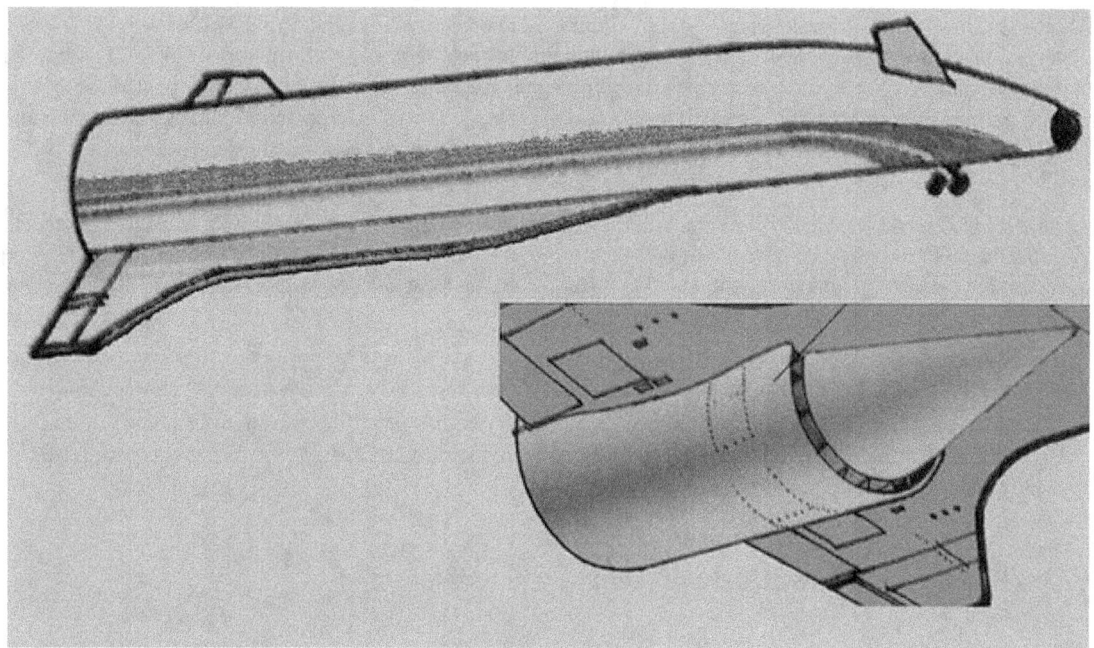

The Skylon was developed from the British HOTOL project.

After securing more funding in the 1990s, the initial design underwent radical revision and, since 2000, Reaction Engines has been working with the University of Bristol to develop an engine design vital to the success of Skylon. The STRICT/STERN designs resulting from this programme were deemed a great success.[*][11] The next stage of development will be to construct a full-sized working prototype of the SABRE Engine.[*][12]

There are several differences compared with HOTOL. Whereas HOTOL would have launched from a rocket sled, to save weight, Skylon uses a conventional retractable undercarriage. Skylon's revised engine design, the SABRE engine, is expected to offer higher performance.[*][13] HOTOL's rear mounted engine gave the vehicle intrinsically poor in-flight stability. Skylon solves this by placing engines at the end of its wings, but further forward and much closer to the vehicle's centre of mass longitudinally. Early attempts to fix this problem had ended up sacrificing much of HOTOL's payload potential, and contributed to the failure of the project.[*][14]

45.1.2 Project brief

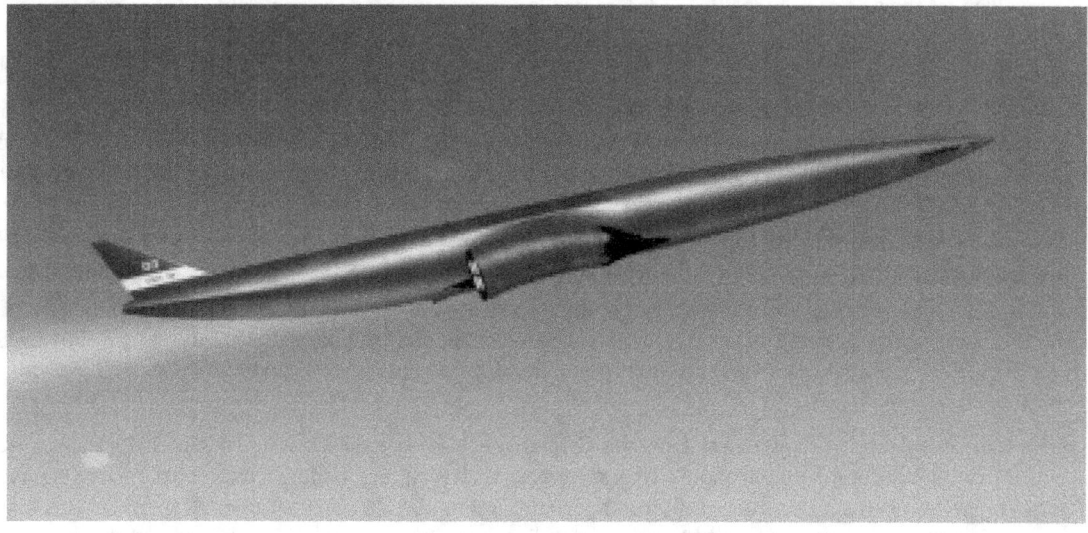

A computer-generated image of the Skylon spaceplane climbing to orbit.

REL intends ultimately to operate as a for-profit commercial enterprise, manufacturing Skylon vehicles for multiple international customers; these customers will operate their fleets directly, with support from REL. While REL intends to manufacture some components, such as the engine precooler, directly, other components have been designed by partner companies and a consortium of various aerospace firms is expected to handle full production of Skylon.[15] According to Management Today, Skylon has been discussed as a possible replacement for NASA's Space Shuttle.[16]

In service, Skylon could potentially lower satellite launch costs from the current £15,000/kg to £650/kg, according to evidence submitted to the UK parliament by Reaction Engines Ltd.[4] Funding for the project from the British government has often been difficult to obtain.[17] Speaking on the topic of Skylon in 2011, David Willetts, the UK Minister of State for Universities and Science, stated:

> The European Space Agency is funding proof of concept work for Skylon from UK contributions. This work is focusing on demonstrating the viability of the advanced British engine technology that would underpin the project. Initial work will be completed in mid 2011 and if the trial is successful, we will work with industry to consider next steps.[4]

45.1.3   Funding and engine development

An unsuccessful request for funding from the British government was issued in 2000. This involved a proposal offering a potentially large return on investment.[18] Subsequent discussions with the British National Space Centre (which later became the UK Space Agency) led to a major funding agreement in February 2009 between the British National Space Centre, European Space Agency (ESA) and REL for €1 million ($1.28 million) to produce a demonstration engine for Skylon by 2011.[19][20][21]

The Technology Demonstration Programme will last approximately 2.5 years and will benefit from another €1 million from ESA.[22] This programme will take Reaction Engines Ltd from a Technology Readiness Level (TRL) of 2/3 up to 4/5.[23] The former UK Minister for Science and Innovation in 2009, Lord Drayson, commented on Skylon in a speech: "This is an example of a British company developing world-beating technology with exciting consequences for the future of space." [4]

As of 2012, the funding required to develop and build the entire craft has not yet been secured, and so current research and development work is focused on the engines, under an ESA grant of €1 million.[24] In January 2011, REL submitted a proposal to the British Government requesting additional funding for the Skylon project.[4] On 13 April 2011, REL announced that the Skylon design had passed several rigorous independent reviews. On 24 May 2011, ESA publicly declared the design to be feasible, having found "no impediments or critical items" in the proposal.[25]

The major milestone of the commencement of static testing of the engine precooler and the SABRE engine was achieved in June 2011, marking the start of Phase 3 in the Skylon development programme.[4] An REL spokesperson announced that they had secured $350 million of further funding, contingent on successful completion of the full-sized precooled jet engine test in June 2011.[26] Engine testing was initiated in June 2011,[27] and was expected to continue to the end of that year.[27] However, testing was delayed until April 2012.[28]

On 9 May 2011, REL stated that a preproduction prototype of the Skylon could be flying by 2016, and the proposed route would be a suborbital flight between the Guiana Space Centre near Kourou in French Guiana and the North European Aerospace Test Range, located in northern Sweden.[29] Pre-orders are expected in the 2011–2013 time frame coinciding with the formation of the manufacturing consortium.[4] On 8 December 2011, Alan Bond, speaking at the 7th Appleton Space Conference, stated that Skylon would enter into service by 2021-2022 instead of 2020 as previously envisaged.[30]

In April 2012, REL announced that the first phase of the precooler test programme had been successfully completed. On 10 July 2012, REL announced that the second of three series of tests has been completed successfully.[31] The test facilities underwent upgrades to allow the third and final phase of testing to proceed.[32] On 13 July 2012, ESA Director-General Jean-Jacques Dordain told Space News that ESA would hold talks with REL to develop a further "technical understanding" .[33]

Following a successful propulsion system test that was audited by ESA's propulsion division in mid-2012, the company announced that it would begin a three-and-a-half-year project to develop and build a test rig of the Sabre engine to prove the engine's performance across its air-breathing and rocket modes.[6] In November 2012, it was announced that a key test of the engine precooler had been successfully completed, and that ESA had verified the precooler's design. The project's development is now allowed to advance to its next phase, which involves the construction and

The precooler rig that tested the heat exchange system of the SABRE engine.

testing of a full-scale prototype engine.[6][34] In June 2013, George Osborne, The Chancellor of the Exchequer stated on his Twitter account that the British government would be giving £60 million towards the further development of the SABRE engine. Osborne's tweet stated: "Just seen SABRE -a rocket engine that cools air from 1000 degrees to −150 in fraction of a second. We're backing the future with £60m funding".[35] This grant was later reduced to £50 million and was approved by the European Commission in August 2015.[36]

In October 2015, BAE Systems entered into an agreement with Reaction Engines where it would invest £20.6 million in Reaction Engines to acquire 20% of its share capital and help develop the SABRE engine.[37][38]

45.2 Technology and design

45.2.1 Overview

See also: Single-stage-to-orbit

Skylon is a fully reusable single stage to orbit (SSTO) vehicle, able to achieve orbit without staging.[39] Proponents of

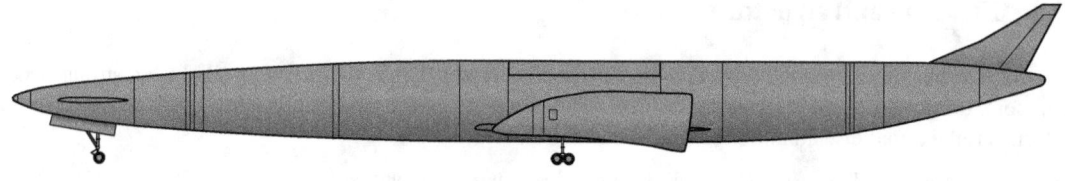

The Skylon spaceplane is designed as a two-engine, "tailless" aircraft, which is fitted with a steerable canard.

SSTO claim that staging causes a number of problems due to its complexity that includes being difficult or impossible to recover and reuse many parts, leading to great expense, and therefore believe that SSTO designs hold the promise of reducing the cost of space-flight.[39] It is intended for Skylon to take off from a specially strengthened runway, fly to low earth orbit, re-enter the atmosphere, and land upon a runway like a conventional aeroplane.[5]

The design of the Skylon C2 features a large cylindrical payload bay, 13 m (42 ft 8 in) long and 4.8 metres (15 ft 9 in) in diameter.[40] It is designed to be comparable with current payload dimensions, and able to support the containerisation of payloads that Reaction Engines hopes for in the future.[40] To an equatorial orbit, Skylon could deliver 15 tonnes (33,069 lb) to a 300 kilometres (186 mi) altitude or 11 tonnes (24,251 lb) to an 800 kilometres (497 mi) altitude.[*][41] Using interchangeable payload containers, Skylon could be fitted to carry satellites or fluid cargo into orbit, or, in a specialised habitation module, up to 30 astronauts in one launch.[*][42][*][43]

45.2.2   SABRE engines

Main article: SABRE (rocket engine)

One of the significant features of the Skylon design is the engine, called SABRE.[*][13][*][44] The engines are designed to operate much like a conventional jet engine to around Mach 5.5 (1,700 m/s),[*][44] 26 kilometres (16 mi) altitude, beyond which the air inlet closes and the engine operates as a highly efficient rocket to orbital speed.[*][44] The proposed SABRE engine is not a scramjet, but a jet engine running combined cycles of a precooled jet engine, rocket engine and ramjet.[*][3] Originally the key technology for this type of precooled jet engine did not exist, as it required a heat exchanger that was ten times lighter than the state of the art.[*][11] Research conducted since then has achieved the necessary performance.[*][13][*][45]

Operating an air-breathing jet engine at velocities of up to Mach 5.5 poses numerous engineering problems.[*][44] Several previous engines proposed by other designers worked well as jet engines but performed poorly as rockets.[*][44] This engine design aims to be a good jet engine within the atmosphere, as well as being an excellent rocket engine outside.[*][44] The problem with operating at Mach 5.5 has been that the air coming into the engine rapidly heats up as it is compressed into the engine; due to certain thermodynamic effects, this greatly reduces the thrust that can be produced by burning fuel.[*][44] Attempts to avoid these issues typically make the engine much heavier (scramjets/ramjets) or greatly reduce the thrust (conventional turbojets/ramjets).[*][44] In either case the end result is an engine that has a poor thrust to weight ratio at high speeds, resulting in an engine that is too heavy to assist much in reaching orbit.[*][44]

The SABRE engine design aims to avoid this by using some of the liquid hydrogen fuel to cool helium in a closed-cycle precooler, which quickly reduces the temperature of the air at the inlet.[*][44] The air is then used for combustion much like in a conventional jet,[*][44] and once the helium has left the pre-cooler it is further heated by the products of the pre-burner, giving it enough energy to drive the turbine and the liquid hydrogen pump.[*][44] Because the air is cooled at all speeds, the jet can be built of light alloys and the weight is roughly halved.[*][44] Additionally, more fuel can be burnt at high speed.[*][44] Beyond Mach 5.5, the air would become unusably hot despite the cooling, so the air inlet closes and the engine relies solely on on-board liquid oxygen and hydrogen fuel as in a normal rocket.[*][44]

Because the engine uses the atmosphere as reaction mass at low altitude, it will have a high specific impulse (around 2,800 seconds), and burn about one fifth of the propellant that would have been required by a conventional rocket.[*][44] Therefore, it would be able to take off with much less total propellant than conventional systems.[*][44] This, in turn, means that it does not need as much lift or thrust, which permits smaller engines, and allows conventional wings to be used.[*][44] While in the atmosphere, using wings to counteract gravity drag is more fuel-efficient than simply expelling propellant (as in a rocket), again reducing the total amount of propellant needed.[*][44]

45.2.3   Fuselage and structure

The fuselage of Skylon is expected to be a carbon-fiber-reinforced polymer space frame; a light and strong structure that supports the weight of the aluminium fuel tanks and to which the ceramic skin is attached.[*][14] Multiple layers of reflective foil thermal insulation fill the spaces of the frame.[*][46]

The currently proposed Skylon model C2 will be a physically large vehicle, with a length of 82 metres (269 ft) and a diameter of 6.3 metres (21 ft).[*][47] Because it will use a low-density fuel, liquid hydrogen, a great volume is needed to contain enough energy to reach orbit. The propellant is intended to be kept at low pressure to minimise stress; a vehicle that is both large and light has an advantage during atmospheric reentry compared to other vehicles due to a low ballistic coefficient.[*][48] Because of the low ballistic coefficient, Skylon would be slowed at higher altitudes where the air is thinner. As a result, the skin of the vehicle would reach only 1,100 K (830 °C).[*][49] In contrast, the smaller Space Shuttle was heated to 2,000 K on its leading edge, and so employed an extremely heat-resistant but fragile silica thermal protection system. The Skylon design does not require such a system, instead opting for using a far thinner yet durable reinforced ceramic skin.[*][3] However, due to turbulent flow around the wings during re-entry, some parts of Skylon would need to be actively cooled.[*][46]

Skylon would employ a highly loaded tightly spaced wheel assembly, to save weight and also interior space when the wheels are retracted into the fuselage.*[40] Because this wheel design distributes the weight of the aircraft and the force of its landing over a smaller area of the runway, it would require a specially designed runway which is strengthened more than usual.*[50] It will possess a retractable undercarriage with high pressure tyres and water-cooled brakes.*[51] If problems were to occur just before a take-off the brakes would be applied to stop the vehicle, the water boiling away to dissipate the heat.*[51] Upon a successful take-off, the water would be jettisoned, thus reducing the weight of the undercarriage by many tons. During landing, the empty vehicle would be far lighter, and hence the water would not be needed.*[51] The payload fraction would be significantly greater than normal rockets and the vehicle should be fully reusable (200 times or more).*[52]

45.3 Specifications (Skylon D1)

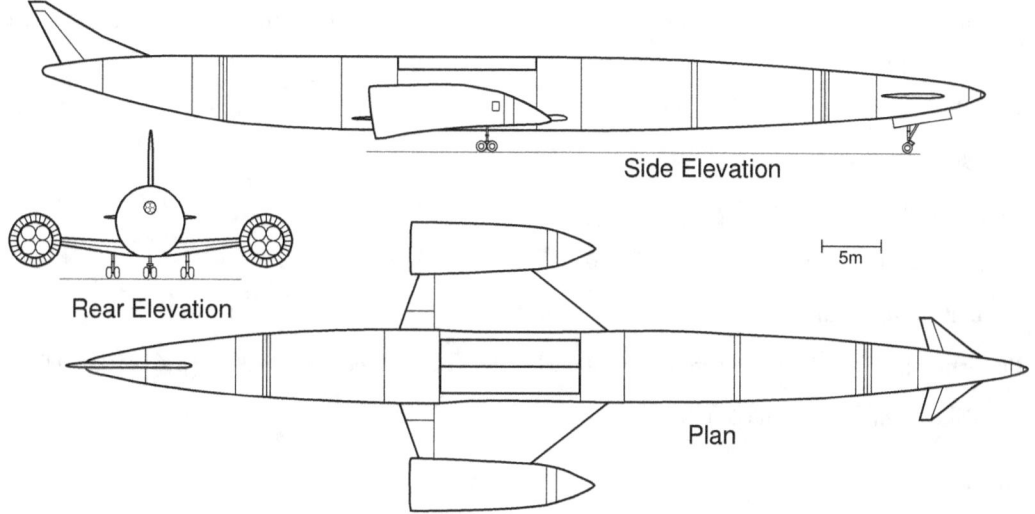

A 3-view drawing of Skylon

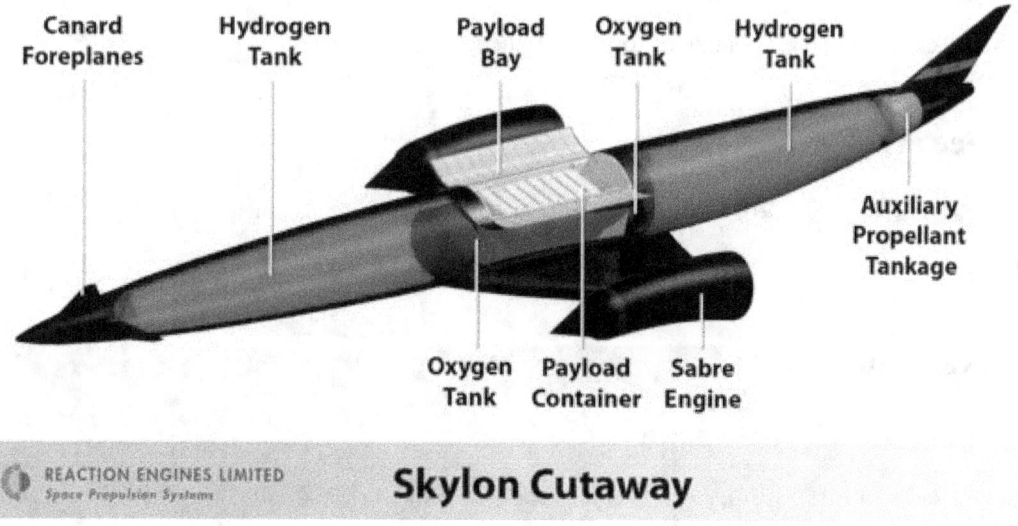

A diagram of Skylon's internal systems.

Data from the Skylon User Manual*[53]

General characteristics

- **Crew:** None, remote controlled from ground

  - The proposed Skylon Personnel/Logistics Module (SPLM) has provision for a Captain.[*][53][*]:43

- **Capacity:** 0

  - up to 24 passengers in the SPLM[*][53][*]:45

  - Potential for up to 30 passengers (in a special passenger module)[*][43]

- **Payload:** 15,000 kg nominal (33,000 lb nominal)

  - 17,000 kg (37,000 lb) to equatorial 160 km (99 mi) orbit from equatorial launch site[*][53][*]:9

  - approx 2,800 kg (6,200 lb) to 98° (sun-synchronous) 600 km (373 mi) orbit from equatorial launch site[*][53][*]:9

- **Length:** 83.133 m[*][53][*]:4 (272.75 ft)

- **Wingspan:** 26.818 m [*][53][*]:4 (87.99 ft)

- **Height:** approx 13.5 m [*][53][*]:4 (44 ft)

- **Empty weight:** 53,400 kg[*][53][*]:6 (117,000 lb)

- **Loaded weight:** 325,000 kg[*][53][*]:6 (717,000 lb)

- **Powerplant:** 2 × SABRE 4 synergistic combined cycle rocket engine, 2,000 kN[*][53][*]:6 (450,000 lbf) each

- **Fuselage diameter:** 6.3 m (20.67 ft)

Performance

- **Maximum speed:** Orbital (air-breathing Mach 5.14, rocket Mach 27.8)[*][53][*]:6

- **Service ceiling:** 28,500 m air-breathing, 90 km SABRE ascent, 600 km exoatmospheric (93,500 ft air breathing, 56 mi rocket ascent, 373 mi exoatmospheric)

- Specific impulse: 4,100 seconds (40,000 N-s/kg)−9,200 seconds (90,000 N-s/kg) air-breathing,[*][53][*]:6 460 seconds (4,500 N-s/kg) rocket,[*][53][*]:6 465.2 seconds (4,562 N-s/kg) orbital[*][53][*]:5

- SABRE engine thrust/weight ratio: up to 14 atmospheric

45.4   See also

- Reaction Engines A2, an REL design for an antipodal airliner using similar engine technology

- Reusable launch system

45.5   References

45.5.1   Citations

[1] Staff (2 November 2015). "BAE invests in space engine firm Reaction Engines". *BBC News*. Retrieved 2015-11-02.

[2] Bond, Alan (2010). "Travelling at the edge of space: Reaction Engines and Skylon in the next 20 years" (video) (lecture). Reaction Engines. Retrieved 9 March 2011. Skip 6 minute irrelevant intro.

[3] "Skylon FAQ". *Frequently Asked Questions*. Reaction Engines. 2010. Retrieved 2011-01-25.

[4] "Skylon Test Date". UK Parliament. 2011. Retrieved 2011-01-27.

[5] Hempsell & Longstaff 2009, p. 5.

[6] "Skylon spaceplane engine concept achieves key milestone". BBC. 28 November 2012. Retrieved 28 November 2012.

[7] "Hypersonic Flight 'Breakthrough' Could Have Us in Tokyo by Lunch". *Wired*. 30 November 2012. Retrieved 1 December 2012.

[8] "UK earmarks £60m for super-fast space rocket engine". *The Guardian* (London). 16 July 2013. Retrieved 19 July 2013.

[9] Clark, Stuart (17 July 2013). "Sabre rocket engine could open up access to space as never before". *The Guardian* (London).

[10] "Reaction Engines Ltd : Company Background". Reaction Engines Limited. Retrieved 2010-09-25.

[11] "Reaction Engines Ltd : Projects STERN and STRICT". Reaction Engines Limited. Retrieved 2010-09-25.

[12] "Reaction Engines Limited :: Technology Demonstration Programme". Reaction Engines Limited. Retrieved 2010-09-25.

[13] Hempsell & Longstaff (2009). *Skylon User Manual*. p. 4.

[14] Hempsell & Longstaff (2009). *Skylon User Manual*. p. 11.

[15] "Broadcast 1203". The Space Show. 6 August 2009. Retrieved 3 December 2012.

[16] Emma Haslett (2011-06-01). "Skylon to Replace Space Shuttle?". *Brits blast off*. Management Today. Retrieved 2011-06-11.

[17] House of Commons: Science and Technology Committee 2007, p. 262.

[18] "Memorandum submitted by the Association of Aerospace Universities". Parliament of the United Kingdom. 2000. Retrieved 2009-07-01.

[19] Rob Coppinger (2009). "Skylon spaceplane engine technology gets European funding". Flight International. Retrieved 2009-07-01.

[20] Jonathan Amos (2009-02-19). "Skylon spaceplane gets cash boost". BBC News. Retrieved 2009-07-01.

[21] Jeremy Hsu (2009). "British Space Plane Concept Gets Boost". space.com. Retrieved 2009-07-01.

[22] "Rockets and Skylon". *20 Years Since HOTOL: Reaction Engines Ltd and SKYLON*. UK Rocketeers. 2009. Retrieved 2010-10-01.

[23] "Reaction Engines Celebrates 20 Years, Looks Forward to Success with Skylon". Parabolic Arc. 2009. Retrieved 2010-09-25.

[24] "The rocket that thinks it's a jet". UK Space Agency. 2009. Retrieved 2010-08-08.

[25] Page, Lewis "ESA: British Skylon spaceplane seems perfectly possible (whizzo robot runway rocketplane cleared to proceed)". *The Register*. 24 May 2011.

[26] "Big Test Looms for British Space Plane Concept". Space.com. 2011. Retrieved 2011-04-18.

[27] Dan Thisdell (2011). "Spaceplane engine tests under way". FlightGlobal. Retrieved 2011-08-06.

[28] "Key tests for Skylon spaceplane project". BBC. 2012-04-27.

[29] "Skylon Phase 3 Development: Q&A". Rocketeers.co.uk. 9 May 2011. Retrieved 3 December 2012.

[30] "Progress on the SKYLON Reusable Spaceplane" (PDF). REL. 8 December 2011. Retrieved 3 December 2012.

[31] "MAJOR ADVANCE TOWARDS THE NEXT JET ENGINE" (PDF). REL. 10 July 2012. Retrieved 3 December 2012.

[32] "Move to open sky for Skylon spaceplane". *BBC News*. 2012-07-11.

[33] "Europe's Next-gen Rocket Design Competition Had Surprise Bidder". Space News. 13 July 2012. Retrieved 3 December 2012.

[34] Thomson, Ian. "European Space Agency clears SABRE orbital engines". *The Register*. 29 November 2012.

[35] *@George_Osborne 27 June 2013*

[36] *"State aid: Commission approves £50 million UK support for the research and development of an innovative space launcher engine"*

[37] Norris, Guy (1 November 2015). "BAE Takes Stake In Reaction Engines Hypersonic Development". *aviationweek.com* (Aviation Week & Space Technology). Retrieved 1 November 2015.

[38] Hollinger, Peggy; Cookson, Clive (2 November 2015). "BAE Systems to pay £20.6m for 20% of space engine group". *CNBC*. Retrieved 2015-11-05.

[39] Varvill & Bond 2003, p. 108.

[40] Hempsell & Longstaff (2009). *Skylon User Manual*. p. 12.

[41] Hempsell & Longstaff (2009). *Skylon User Manual*. p. 7.

[42] "Reaction Engines Ltd : Current Projects : SKYLON – Passenger Capability". Reaction Engines Limited. Retrieved 2010-09-25.

[43] J.L. Scott-Scott; M. Harrison & A.D. Woodrow (2003). "Considerations for Passenger Transport by Advanced Space-planes" (PDF). *Journal of British Interplanetary Society* **56**: 118–126. Retrieved 2009-07-01.

[44] "SABRE engine". *The Sabre Engine*. Reaction Engines Limited. 2010. Retrieved 2011-01-25.

[45] "Revolutionary space engine system for Skylon tested". *BBC News*. 2012-04-27.

[46] Hempsell & Longstaff (2009). *Skylon User Manual*. p. 15.

[47] "Reaction Engines Ltd : Current Projects : SKYLON – The Vehicle". Reaction Engines Limited. Retrieved 2010-09-25.

[48] Hempsell & Longstaff (2009). *Skylon User Manual*. p. 7.

[49] Varvill & Bond 2004, p. 25.

[50] "Skylon Construction". *Skylon-Current Projects*. Reaction Engines Limited. 2010. Retrieved 2011-01-25.

[51] Hempsell & Longstaff (2009). *Skylon User Manual*. p. 21.

[52] Varvill & Bond 2004, p. 22.

[53] Hempsell & Longstaff (2014). "Skylon User Manual" (PDF). Reaction Engines Limited. Retrieved 2015-06-20.

45.5.2 Bibliography

- Hempsell, Mark; Longstaff, Roger (2009). "Skylon User Manual" (PDF). Reaction Engines. pp. 1–21.

- House of Commons Science and Technology Committee (2007). *2007: A Space Policy - Seventh Report of Session 2006-08* **II**. Her Majesty's Stationery Office. ISBN 978-0-215-03509-7.

- Varvill, Richard; Bond, Alan (2003). "A Comparison of Propulsions Concepts for SSTO Reusable launchers" (PDF). *Journal of the British Interplanetary Society* **56**: 108–17. Bibcode:2003JBIS...56..108V.

- Varvill, Richard; Bond, Alan (2004). "The SKYLON Spaceplane" (PDF). *Journal of the British Interplanetary Society* **57**: 22–32. Bibcode:2004JBIS...57...22V.

45.6 External links

- Reaction Engines Limited
- Mark Hempsell from REL on The Space Show talking about Skylon
- World Tiles Skylon model
- Liquid Air Cycle Engine (LACE) rocket equation reasonably well predicts the performance of Skylon
- Video animation TROY Mars Mission Concept
- Skylon Unmanned Reusable Cargo Spacecraft on aerospace-technology.com
- Video - SKYLON - Operations from Reaction Engines Ltd on Vimeo.

Chapter 46

SpaceShipTwo

The **Scaled Composites Model 339 SpaceShipTwo (SS2)** is an air-launched suborbital spaceplane type designed for space tourism. It is manufactured by The Spaceship Company, a California-based company owned by Virgin Galactic.

Together with its mother ship carrier it forms the The Spaceship Company Tier 1b program*[1] where SpaceShipTwo is carried to its launch altitude by a Scaled Composites White Knight Two, before being released to fly on into the upper atmosphere powered by its rocket engine. It then glides back to Earth and performs a conventional runway landing.*[2] The spaceship was officially unveiled to the public on 7 December 2009 at the Mojave Air and Space Port in California.*[3] On 29 April 2013, after nearly three years of unpowered testing, the first one constructed successfully performed its first powered test flight.*[4]

Virgin Galactic plans to operate a fleet of five SpaceShipTwo spaceplanes in a private passenger-carrying service*[5]*[6]*[7]*[8] and has been taking bookings for some time, with a suborbital flight carrying an updated ticket price of US$250,000.*[9] The spaceplane could also be used to carry scientific payloads for NASA and other organizations.*[10]

On 31 October 2014 during a test flight, VSS *Enterprise*, the first SpaceShipTwo craft, broke up in-flight and crashed in the Mojave desert.*[11]*[12]*[13]*[14] A preliminary investigation suggested the feathering system, the ship's descent device, deployed too early.*[15][16] One pilot was killed; the other was treated for a serious shoulder injury*[17] after parachuting from 50,000 feet (15,000 m).*[18] A successor spacecraft, SpaceShipTwo, Serial Number Two, is under construction.

46.1 Design overview

The SpaceShipTwo project is based in part on technology developed for the first-generation SpaceShipOne, which was part of the Scaled Composites Tier One program, funded by Paul Allen. The Spaceship Company licenses this technology from Mojave Aerospace Ventures, a joint venture of Paul Allen and Burt Rutan, the designer of the predecessor technology.

SpaceShipTwo is a low-aspect-ratio passenger spaceplane. Its capacity will be eight people —six passengers and two pilots. The apogee of the new craft will be approximately 110 km (68 mi) in the lower thermosphere, 10 km (6.2 mi) higher than the Kármán line which was SpaceShipOne's target, although the last flight of SpaceShipOne reached a one-time altitude of 112 km (70 mi). SpaceShipTwo will reach 4,200 km/h (2,600 mph), using a single hybrid rocket engine —the RocketMotorTwo.*[19] It launches from its mother ship, White Knight Two, at an altitude of 15,000 metres (50,000 ft), and reaches supersonic speed within 8 seconds. After 70 seconds, the rocket engine cuts out and the spacecraft will coast to its peak altitude. SpaceShipTwo's crew cabin is 3.7 m (12 ft) long and 2.3 m (7.5 ft) in diameter.*[20] The wing span is 8.2 m (27 ft), the length is 18 m (60 ft) and the tail height is 4.6 m (15 ft).*[21]

SpaceShipTwo uses a feathered reentry system, feasible due to the low speed of reentry. In contrast, the Space Shuttle and other orbital spacecraft re-enter at orbital speeds, closer to 25,000 km/h (16,000 mph), using heat shields. SpaceShipTwo is furthermore designed to re-enter the atmosphere at any angle.*[22] It will decelerate through the atmosphere, switching to a gliding position at an altitude of 24 km (15 mi), and will take 25 minutes to glide back to the spaceport.

SpaceShipTwo and White Knight Two are, respectively, roughly twice the size of the first-generation SpaceShipOne

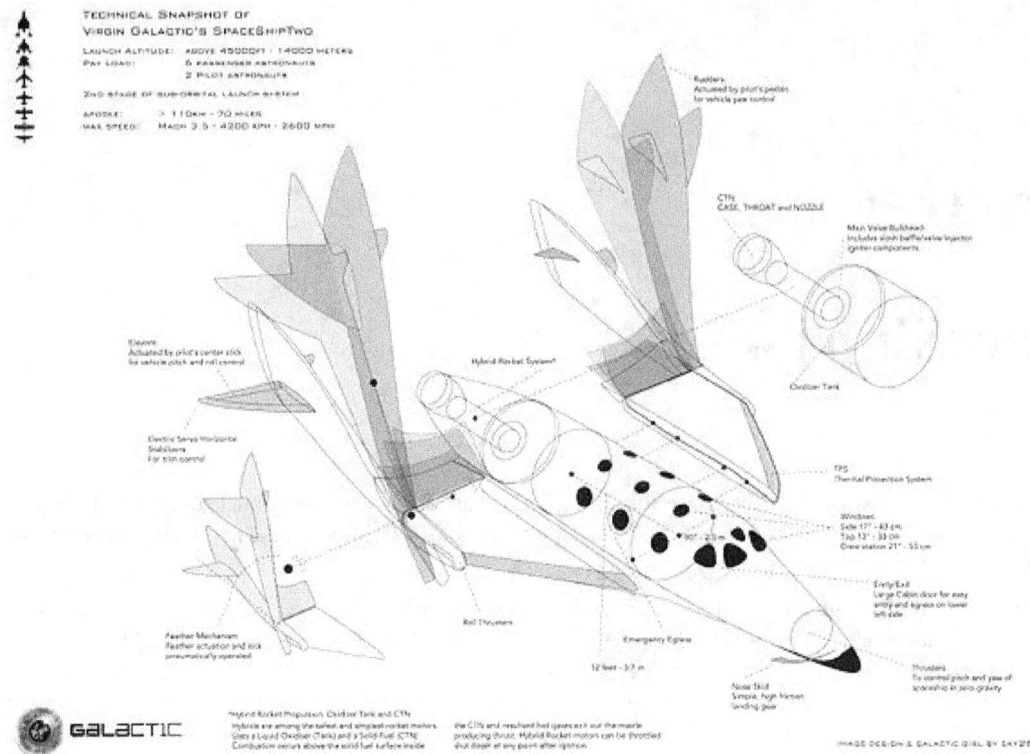

A schematic diagram of SpaceShipTwo.

and mother ship White Knight, which won the Ansari X Prize in 2004. SpaceShipTwo has 43 and 33 cm (17 and 13 in) -diameter windows for the passengers' viewing pleasure,[21] and all seats will recline back during landing to decrease the discomfort of G-forces.[23] Reportedly, the craft can land safely even if a catastrophic failure occurs during flight.[24] In 2008, Burt Rutan remarked on the safety of the vehicle:

> This vehicle is designed to go into the atmosphere in the worst case straight in or upside down and it'll correct. This is designed to be at least as safe as the early airliners in the 1920s ... Don't believe anyone that tells you that the safety will be the same as a modern airliner, which has been around for 70 years.[22]

In September 2011, the safety of SpaceShipTwo's feathered reentry system was tested when the crew briefly lost control of the craft during a gliding test flight. Control was reestablished after the spaceplane entered its feathered configuration, and it landed safely after a 7-minute flight.[25]

46.2   Fleet and launch site

The launch customer of SpaceShipTwo is Virgin Galactic, who have ordered five vehicles.[26][27] The first two were named VSS *Enterprise*[28] and VSS *Voyager*. The "VSS" prefix stands for "Virgin Space Ship". As of November 2014, only VSS *Enterprise* has been flown;[29] it was destroyed in a crash on 31 October 2014. The build of the second aircraft—VSS *Voyager*—is about 65 percent complete as of early November 2014, and Virgin Galactic expects it to be complete in 2015.[30] The third SpaceShipTwo is expected to commence construction by the end of 2015.[31] The WhiteKnightTwo carrying SpaceShipTwo craft take off from the Mojave Air and Space Port in California during testing. Spaceport America (formerly Southwest Regional Spaceport), a US$212 million spaceport in New Mexico, partly funded by the state government,[32] will become the permanent launch site when commercial launches begin.[20]

46.3   Development

On 28 September 2006, Virgin Group founder Sir Richard Branson unveiled a mock-up of the SpaceShipTwo passenger cabin at the NextFest exposition at the Jacob K. Javits Convention Center in New York.[33] The design of the vehicle was revealed to the press in January 2008, with the statement that the vehicle itself was around 60% complete.[20] On 7 December 2009, the official unveiling and rollout of SpaceShipTwo took place. The event involved the first SpaceShipTwo being christened by then-Governor of California Arnold Schwarzenegger as the *VSS Enterprise*.[34]

46.3.1   2007 test explosion

On 26 July 2007, an explosion occurred during an oxidizer flow test at the Mojave Air and Space Port, where early-stage tests were being conducted on SpaceShipTwo's systems. The oxidizer test included filling the oxidizer tank with 4,500 kilograms (10,000 lb) of nitrous oxide, followed by a 15-second cold-flow injector test. Although the tests did not ignite the gas, three employees were killed and three injured, two critically and one seriously, by flying shrapnel.[35]

46.3.2   Rocket engine

The hybrid rocket engine design for SpaceShipTwo has been problematic and caused extensive delays to the flight test program. The original rocket engine design was based on hydroxyl-terminated polybutadiene (HTPB) fuel and nitrous oxide oxidizer, sometimes referred to as an N_2O/HTPB engine.[36][37] It was developed by Scaled Composites subcontractor Sierra Nevada Corporation (SNC) from 2009 to early 2014. In May 2014, Virgin Galactic announced a change to the hybrid engine to be used in SpaceShipTwo, and took the development effort in-house to Virgin Galactic, terminating the contract with Sierra Nevada and halting all development work on the first-generation rocket engine.[38] Virgin then modified the engine design to include a change of the hybrid rocket fuel from a HTPB to a polyamide fuel formulation. In October 2015, Virgin announced that it was considering changing back to the original HTPB fuel.[39]

RocketMotorTwo

Main article: RocketMotorTwo

Between 2005 and 2009, Scaled Composites conducted numerous small-scale rocket tests to evaluate SpaceShipTwo's engine design. After settling on the RocketMotorTwo hybrid rocket design to be developed by Sierra Nevada, the company began performing full-scale hot-fire rocket tests in April 2009.[40] By December 2012, 15 full-scale tests had been successfully conducted,[40][41] and additional ground tests continued into March 2013.[42] In June 2012, the Federal Aviation Administration (FAA) issued a rocket testing permit to Scaled Composites, allowing it to begin SS2 test flights powered by RocketMotorTwo;[43] the first such powered flight took place on 29 April 2013.[44] The Sierra Nevada HTPB-based RocketMotorTwo design generated 60,000 lbf (270 kN) of thrust.[45]

2014 Change of engine manufacturer and hybrid engine fuel

In May 2014, Virgin Galactic took over engine development from Sierra Nevada[38] and announced a change to the fuel to be used in the SpaceShipTwo hybrid rocket engine. Rather than the rubber-based HTPB-fuel engine —engines that had experienced serious engine stability issues on firings longer than approximately 20 seconds— the engine would now be based on a solid fuel composed of a type of plastic called thermoplastic polyamide. The plastic fuel was projected to have better performance (by several unspecified measures) and was projected to allow SpaceShipTwo to make flights to a higher altitude.[46][47][48]

As of May 2014 when the version 2 engine by Virgin Galactic was publicly announced, the engine had already completed full-duration burns of over 60 seconds in ground tests on an engine test stand.[47] The second-generation engine design also required the modification to the SS2 airframe to fit additional tanks in the wings of SpaceShipTwo —one holding methane and the other containing helium—in order to ensure a proper burn and shut-down of the new engine.[49] Additional ground tests were performed of the new engine between May and October 2014.

46.3.3   SpaceShipTwo test flights

Main articles: VSS Enterprise § Flight test program and VSS Voyager § Test flight program
As of October 2014, SpaceShipTwo had conducted 54 test flights.[*][50] The spacecraft has used its "feathered"

SpaceShipTwo in a captive flight configuration underneath White Knight Two, during the runway dedication of Spaceport America in October 2010. VMS Eve is shown carrying VSS Enterprise.

A view of the firing of SpaceShipTwo's rocket engines during its first powered flight in April 2013.

wing configuration during ten of these test flights.[*][50][*][51][*][52]

Testing VSS Enterprise

In September 2012, Virgin Galactic announced that the unpowered subsonic glide flight test program was essentially complete.[*][53] In October 2012, Scaled Composites installed key components of the rocket engine, and Space-ShipTwo performed its first glide flight with the engine installed in December 2012.[*][54][*][55]

The spacecraft's first powered test flight took place on 29 April 2013. SpaceshipTwo reached supersonic speeds in this first powered flight.[56][57] On 5 September 2013, the second powered flight was made by SpaceShipTwo.[58] The first powered test flight of 2014—and third overall—occurred 10 January 2014. The spacecraft reached an altitude of 22,000 metres (71,000 ft) (the highest to date) and a speed of Mach 1.4. The WhiteKnightTwo carrier aircraft released SpaceShipTwo (VSS Enterprise) at an altitude of 14,000 metres (46,000 ft).[59]

October 2014 crash

Main article: VSS Enterprise crash
On October 31, 2014, SpaceShipTwo VSS *Enterprise* suffered an in-flight breakup during a powered flight test,[60][61]

NTSB Go-Team inspects a tail section of VSS Enterprise

resulting in a crash killing one pilot and injuring the other.[11][18] It was coincidentally the first flight to use the new type of fuel, based on nylon plastic grains.[62][63] The crash is believed to have involved a premature deployment of the feathering mechanism, which is normally used to aid in a safe descent. SpaceShipTwo was still in powered ascent when the feathering mechanism deployed. Disintegration was observed two seconds later.[61] The accident is still under investigation.

After the loss of VSS Enterprise

In October 2015 it was reported that the second SpaceShipTwo will make its first flight in 2016.[64]

46.3.4   Costs

SpaceShipTwo's total development costs were estimated at around $400 million in May 2011, a significant increase over the 2007 estimate of $108 million.[65]

46.4   Commercial operation

The duration of the flights will be approximately 2.5 hours, though only a few minutes of that will be in space. The price will initially be $200,000.[66] More than 65,000 would-be space tourists applied for the first batch of 100 tickets. By December 2007, Virgin Galactic had 200 paid-up customers on its books for the early flights, and 95% were passing the 6-8 g centrifuge tests.[67] By the start of 2011, that number had increased to over 400 paid customers,*[68] and to 575 by early 2013.*[69] In April 2013, Virgin Galactic announced that the price for a seat would increase 25 percent to $250,000 before the middle of May 2013,*[69] and would remain at $250,000 "until the first 1,000 people have traveled, so that it matches up with inflation since [Virgin Galactic] started." *[70]

Following 50–100 test flights, the first paying customers were expected to fly aboard the craft in 2014.*[5] Refining the projected schedule in late 2009, Virgin Galactic declined to announce a firm timetable for commercial flights, but did reiterate that initial flights would take place from Spaceport America. Operational roll-out will be based on a "safety-driven schedule" .*[71] In addition to making suborbital passenger launches, Virgin Galactic will market SpaceShipTwo for suborbital space science missions.*[71]

46.4.1 NASA sRLV program

By March 2011, Virgin Galactic had submitted SpaceShipTwo as a reusable launch vehicle for carrying research payloads in response to NASA's suborbital reusable launch vehicle (sRLV) solicitation, which is a part of the agency's Flight Opportunities Program. Virgin projects research flights with a peak altitude of 110 km (68 mi). These flights will provide approximately four minutes of microgravity for research payloads. Payload mass and microgravity levels have not yet been specified.*[2] The NASA research flights could begin during the test flight certification program for SpaceShipTwo.

46.4.2 Future spacecraft

In August 2005, the president of Virgin Galactic stated that if the suborbital service with SpaceShipTwo is successful, the follow-up SpaceShipThree will be an orbital craft. In 2008, Virgin Galactic changed their plans and decided to make it a high-speed passenger vehicle, offering transport through point-to-point suborbital spaceflight.*[72]

46.5 Production

While the first White Knight Two and the first SpaceShipTwo were built by Scaled Composites, The Spaceship Company has responsibility for the manufacture of the second White Knight Two aircraft and the second SS2 spacecraft for Virgin Galactic, as well as additional production craft as other customers for the vehicles emerge.*[73] In October 2010, TSC announced plans to build three White Knight Two aircraft and five SpaceShipTwo spaceplanes.*[74]

46.5.1 List of SS2

- VSS Enterprise (destroyed 31 October 2014)

- VSS Voyager (anticipated in-service date: 2015)

- unnamed tail #3 (anticipated construction start: end 2015)*[31]

46.6 Specifications

Sources: *[75]*[76]

General characteristics

- **Crew:** 2

- **Capacity:** 6 passengers

- **Length:** 18.3 m (60 ft)

- **Wingspan:** 8.3 m (27 ft)

- **Height:** 5.5 m (18 ft – rudders down)

- **Loaded weight:** 9,740 kg (21,428 lb)

- **Powerplant:** 1 × RocketMotorTwo liquid/solid hybrid rocket engine

Performance

- **Maximum speed:** 4,000 km/h (2,500 mph)

- **Service ceiling:** 110,000 m (361,000 ft)

46.7 See also

- Airbus Space and Defence SpacePlane

- Rocketplane XP

- Blue Origin New Shepard

- Dragon (spacecraft)

- Dream Chaser

- List of human spaceflights

- Lynx (spacecraft)

- Orion (spacecraft)

- Private spaceflight

- SpaceShipOne

- Space Shuttle program

- SpaceX

46.8 References

[1] "SpaceShipTwo.net". 2012. Retrieved 16 June 2013.

[2] "sRLV platforms compared". NASA. 7 March 2011. Retrieved 10 March 2011. SpaceShipTwo: Type: HTHL/Piloted

[3] Amos, Jonathan (8 December 2009). "Richard Branson unveils Virgin Galactic spaceplane". BBC News. Retrieved 23 March 2010.

[4] "Sir Richard Branson's Virgin Galactic spaceship ignites engine in flight". BBC. 29 April 2013. Retrieved 29 April 2013.

[5] "Space Ship Completes 24th Test Flight in Mojave". HispanicBusiness.com. 4 April 2013. Retrieved 5 April 2013.

[6] "Virgin Galactic to Launch Passengers on Private Spaceship in 2013". Space.com. 8 June 2012. Retrieved 11 June 2012.

[7] "Virgin Galactic space tourism could begin in 2013". BBC. 26 October 2011.

[8] John Schwartz (23 January 2008). "New Tourist Spacecraft Unveiled". *New York Times*. Retrieved 23 January 2008.

[9] Fly With Us. Virgin Galactic. Retrieved 5 November 2015.

[10] "Virgin spaceship aims to be science lab". BBC. 4 December 2012. Retrieved 4 December 2012.

[11] Chang, Kenneth; Schwartz, John (31 October 2014). "Virgin Galactic's SpaceShipTwo Crashes in New Setback for Commercial Spaceflight". *New York Times*. Retrieved 1 November 2014.

[12] Foust, Jeff (2014-10-31). "SpaceShipTwo Destroyed in Fatal Test Flight Accident". *Space News*. Retrieved 2014-10-31.

[13] "Virgin Galactic's SpaceShipTwo Crashes During Flight Test". October 31, 2014.

[14] Durden, Rick (31 October 2014). "Virgin Galactic's SpaceShipTwo Crashes". *AVweb*. Retrieved 31 October 2014.

[15] Chang, Kenneth (November 3, 2014). "Investigators Focus on Tail Booms in Crash of Space Plane". *New York Times*. Retrieved November 3, 2014.

[16] Melley, Brian (November 3, 2014). "Spaceship's descent device deployed prematurely". *AP News*. Retrieved November 3, 2014.

[17] Klotz, Irene (2014-11-03). "SpaceShipTwo's Rocket Engine Did Not Cause Fatal Crash". *Discovery News*. Retrieved 2014-11-03.

[18] "Virgin Galactic's SpaceShipTwo rocket plane crashes". October 31, 2014.

[19] Scaled Composites LLC. "Project Test Summaries". Scaled.com. Retrieved 5 April 2012.

[20] Rob Coppinger. "Pictures: Virgin Galactic unveils Dyna-Soar style SpaceShipTwo design and twin-fuselage White Knight II configuration". flightglobal.com. Retrieved 23 January 2008.

[21] "Spaceship Unveil Presspack". Virgin Galactic. Retrieved 10 February 2008.

[22] Dignan, Larry (23 January 2008). "Virgin Galactic unveils SpaceShipTwo; Plans open architecture spaceship". *Between the lines*. ZDnet.com. Retrieved 10 February 2008.

[23] Tariq Malik (28 September 2006). "Virgin Galactic Unveils SpaceShipTwo Interior Concept". Space News. Retrieved 6 April 2007.

[24] Peter de Selding. "Virgin Galactic Customers Parting with Their Cash". Space News. Archived from the original on 12 December 2007. Retrieved 6 April 2007.

[25] "Virgin Galactic's private spaceship makes safe landing after tense test flight". Space.com. 17 October 2011. Retrieved 18 October 2011.

[26] "Richard Branson and Burt Rutan Form Spacecraft Building Company". Space.com. 27 July 2005. Retrieved 17 October 2009.

[27] Malik, Tariq (23 January 2008). "Virgin Galactic Unveils Suborbital Spaceliner Design". Space.com. Retrieved 25 January 2008.

[28] "Virgin Galactic to Offer Public Space Flights". Space.com. 27 September 2004. Retrieved 20 December 2007.

[29] "Scale comparison chart of Spaceshipone and Spaceshiptwo". Gizmodo. Retrieved 6 April 2007.

[30] http://www.newsweek.com/virgin-galactic-ceo-sees-new-spacecraft-ready-next-year-281652 (2014-11-02). "Virgin Galactic CEO Sees New Spacecraft Ready Next Year". *Newsweek*.

[31] "Virgin Galactic Gears Up for Building Third SpaceShipTwo". NBC News. 7 February 2015.

[32] New era draws closer: Spaceport dedicates runway on New Mexico ranch. *El Paso Times*. 23 October 2010. Retrieved 25 October 2010. "two-thirds of the $212 million required to build the spaceport came from the state of New Mexico...The rest came from construction bonds backed by a tax approved by voters in Doña Ana and Sierra counties."

[33] Sophie Morrison (30 September 2006). "Buckled up for white knuckle ride". BBC News. Retrieved 6 April 2007.

[34] Richard Branson unveils Virgin Galactic spaceplane. BBC News, 7 December 2009.

[35] Abdollah, Tami and Silverstein, Stuart (27 June 2007). "Test Site Explosion Kills Three". *Los Angeles Times*. Retrieved 27 July 2007.

[36] "Propulsion Systems: multiple-burn, green and low-cost" (PDF). Sierra Nevada. Retrieved 8 March 2013.

[37] "Safe Hybrid Rocket". *Overview – Safety*. Virgin Galactic. 2013. Retrieved 8 March 2013.

[38] "SNC Statement in Response to Inquiries Regarding 10-31-14 Virgin Galactic SpaceShipTwo Incident". *SNC Press Release*. Sierra Nevada Corporation. 2014-10-31. Retrieved 2014-11-01.

[39] Faust, Jeff (14 Oct 2015). "SpaceShipTwo Bounces Back to Rubber Fuel". *spacenews.com*. Retrieved 16 Oct 2015.

[40] RocketMotorTwo Hot-Fire Test Summaries. Scaled.com. Updated 9 August 2012. Retrieved 16 December 2012.

[41] "Virgin Galactic successfully completes SpaceShipTwo glide flight test and rocket motor firing on same day". SpaceRef.com. 28 June 2012.

[42] Richard Branson (5 March 2013). "This isn't sci-fi". Virgin.com. Retrieved 5 March 2013.

[43] "SpaceShipTwo Gets Thumbs Up for Rocket-Powered Flights". *Flying Magazine*. 1 June 2012.

[44] "SpaceShipTwo Test Summaries". Scaled Composites. 29 April 2013. Retrieved 29 April 2013.

[45] "SpaceShipTwo performs first Rocket-Powered Flight". Spaceflight101.com. 29 April 2013. Retrieved 19 May 2013.

[46] Foust, Jeff (2014-05-24). "Virgin Galactic changes fuels as it prepares for its next round of test flights". *NewSpace Journal*. Retrieved 2014-05-25.

[47] Boyle, Alan (2014-05-23). "Virgin Galactic Makes a Switch in SpaceShipTwo's Rocket Motor". *NBC News*. Retrieved 2014-05-24.

[48] "New Fuel to Boost SpaceShip Two". *Aviation Week*. 2014-05-24. Retrieved 2014-05-27.

[49] Messier, Doug (2014-06-30). "WhiteKnightTwo in the Air Over Mojave Today". *Parabolic Arc*. Retrieved 2014-11-01.

[50] Wall, Mike (8 October 2014). "Virgin Galactic's SpaceShipTwo Aces Glide Test Flight". *Space.com*. Retrieved 8 October 2014.

[51] "Feather flight and nitrous vent test success". Virgin Galactic. Retrieved 12 April 2013.

[52] "Virgin Galactic Reaches New Heights in Third Supersonic Test Flight". *virgingalactic.com*. 10 Jan 2014. Retrieved 13 Jan 2014.

[53] Rosenberg, Zach. "Virgin Galactic finishes unpowered flight test". FlightGlobal.com. 13 September 2012. Retrieved 26 September 2012.

[54] "SpaceShipTwo straps on its engine". NBC. 20 December 2012. Retrieved 20 December 2012.

[55] "SpaceShipTwo Fitted With Rocket Propulsion System". *Aviation Week*. 22 October 2012. Retrieved 14 November 2012.

[56] "SpaceShipTwo Test Summaries". Scaled Composites. 8 August 2013. Retrieved 14 August 2013.

[57] "Virgin Galactic Breaks Speed of Sound in First Rocket-Powered Flight of SpaceShipTwo". Virgin Galactic. 29 April 2013. Retrieved 29 April 2013.

[58] "Virgin Galactic's SpaceShipTwo Succeeds In Second Rocket-Powered Flight". Forbes. Retrieved 6 September 2013.

[59] Nancy Atkinson (10 January 2014). "SpaceShipTwo Goes Supersonic in Third Rocket-Powered Test Flight".

[60] "SpaceShipTwo disaster: Industry mourns pilot but vows to keep ... - BakersfieldCalifornian.com". *The Bakersfield Californian*.

[61] "SpaceShipTwo 'Feather' Tail System Deployed Prematurely: NTSB".

[62] Marks, Paul. "Virgin Galactic's Spaceshiptwo in fatal crash". *newscientist.com*. New Scientist. Retrieved 1 November 2014.

[63] "Virgin Galactic's SpaceShipTwo Has Crashed, Possible Casualties". Gizmodo. 31 October 2014. Retrieved 31 October 2014.

[64] http://www.americaspace.com/?p=87222

[65] "A Look at Cost Overuns and Schedule Delays in Major Space Programs". ParabolicArc.com. 4 May 2011. Retrieved 1 April 2012.

[66] "Rich Chinese buying tickets to space". Zee News. Retrieved 6 April 2007.

[67] "Virgin Galactic's timetable for progress". *Spaceflight*. Volume 50. British Interplanetary Society. February 2008. p.48.

[68] "Hold tight: SpaceShipTwo makes near-vertical plunge towards Earth on test flight as space tourism dream edges closer". *Daily Mail* (London). 5 May 2011. Retrieved 11 July 2013.

[69] Messier, Doug (29 April 2013). "Reserve Your SpaceShipTwo Seat Now —Big Price Increase Coming Soon". ParabolicArc.com. Retrieved 30 April 2013.

[70] "Ticket Price for Private Spaceflights on Virgin Galactic's SpaceShipTwo Going Up". Space.com. 30 April 2013. Retrieved 16 June 2013.

[71] Will Whitehorn (27 October 2009). *International Astronautical Congress 2009: Civilian Access to Space* (Video – comments at c. 20:00). Daejeon, South Korea: Flightglobal Hyperbola. Retrieved 19 June 2013.

[72] "SpaceShipThree poised to follow if SS2 succeeds". Flight International. 23 August 2005. Retrieved 6 April 2007.

[73] Norris, Guy (8 July 2011). "An Inside Look At A New Spaceship Factory". *Aviation Week and Space Technology*. Retrieved 8 July 2011.

[74] "Spacecraft factory to break ground in Mojave". *Los Angeles Times*. 8 November 2010. Retrieved 9 November 2010.

[75] Overview – Spaceships. Virgin Galactic. Retrieved 19 June 2013.

[76] "How Virgin Galactic's SpaceShipTwo Passenger Space Plane Works (Infographic)". Space.com. 10 October 2012. Retrieved 14 November 2012.

46.9   External links

- Official Virgin Galactic website

- Official Scaled Composites website

- Virgin Galactic, National Geographic Channel documentary, 2012.

- Formation of The Spaceship Company – Space.com (2005)

- The Birth of SpaceShipTwo – SpaceDaily (2004)

- Space or Bust: Feature article on space tourism – Cosmos Magazine (2005)

- Space Law in Paris – Space Law Probe (2006)

- CNET Images of SS2 mockups – CNET News (2006)

- "VG Powered Flight Updated Drop BRoll". Virgin Galactic via YouTube. 29 April 2013. Shows all 16 seconds of the first-flight rocket firing from three views, and most of the sequence from a fourth view.

Chapter 47

SpaceShipThree

The **Scaled Composites SpaceShipThree** (**SS3**) was a mid-2000s proposed spaceplane to be developed by Virgin Galactic and Scaled Composites, ostensibly to follow SpaceShipTwo.

The mission originally proposed for SpaceShipThree in 2005 was for orbital spaceflight, as part of a program called "Tier 2" by Scaled Composites.*[1]*[2]

By 2008, Scaled Composites had reduced those plans and articulated a conceptual design that would be a point-to-point vehicle traveling outside the atmosphere.*[3] As of 2008, the SpaceShipThree concept spacecraft was conceived to be used for transportation through point-to-point suborbital spaceflight with the spacecraft providing, for example, a two-hour trip on the Kangaroo Route (from London to Sydney or Melbourne).*[3]

Scaled was sold to Northrop Grumman in 2007, and references to further work on a conceptual Scaled SS3 ended after that time.

47.1   References

[1]   "SpaceShipThree poised to follow if SS2 succeeds". *Flight International* (FlightGlobal). 23 August 2005. Retrieved 2008-01-25.

[2] Leonard David (11 August 2006). "Burt Rutan on Civilian Spaceflight, Breakthroughs, and Inside SpaceShipTwo". Space.com.

[3] SpaceShipThree revealed?, FlightGlobal Hyperbola, Rob Coppinger, 29 Feb 2008

47.2   External links

- Space tourism companies aiming for orbit (New Scientist Space, 8/24/2005)

47.3   See also

- StratoLaunch Systems
- Suborbital transport
- Suborbital bomber

Chapter 48

Scaled Composites Tier 1b

Tier 1b's SpaceShipTwo spaceplane (central fuselage) attached to the White Knight Two carrier aircraft.

Tier 1b was the Scaled Composites internal company name of a mid-2000s program to develop a nine-passenger suborbital commercial human spaceflight platform.[*][1]

The Tier 1b program was intended to be an evolutionary development of Scaled Composites Tier One program, which developed the Ansari X-Prize-winning entry in the 2004 spaceflight demonstration contest held by the X-Prize Foundation.[*][1]

The principle hardware envisioned to be developed under the Tier 1b program were the White Knight Two carrier aircraft and the SpaceShipTwo suborbital spaceplane.[*][1]

48.1   See also

- Burt Rutan

- Spaceport America

48.2 References

[1] "SpaceShipThree poised to follow if SS2 succeeds" . *Flight International* (FlightGlobal). 23 August 2005. Retrieved 2008-01-25.

48.3 External links

- Official Virgin Galactic website
- Official Scaled Composites website

Chapter 49

Suborbital spaceplane

A **suborbital spaceplane** is a spaceplane designed specifically for sub-orbital spaceflight. A few projects of civil and military suborbital spaceplanes were in past in Nazi Germany, United States, Soviet Union etc. From the beginning of 21st century, it is expected that this type of spacecraft, as projects of many private companies, will play a key role in early space tourism.

49.1   History

The first ever true suborbital spaceplane was the North American X-15, which first flew above the Kármán line in 1963.

The first privately built and privately funded suborbital spaceplane was Scaled Composites SpaceShipOne, which first flew above the Kármán line in 2004.

49.2   See also

- List of manned spacecraft

- List of private spaceflight companies#Crew and cargo transport vehicles

- Point-to-point sub-orbital spaceflight

Chapter 50

VentureStar

VentureStar was a single-stage-to-orbit reusable launch system proposed by Lockheed Martin and funded by the U.S. government. The goal was to replace the Space Shuttle by developing a re-usable spaceplane that could launch satellites into orbit at a fraction of the cost. While the requirement was for an unmanned launcher, it was expected to carry passengers as cargo. The VentureStar would have had a wingspan of 68 feet, a length of 127 feet, and would have weighed roughly 62,700 pounds.

VentureStar was intended to be a commercial single-stage-to-orbit vehicle that would launch vertically, but return to Earth as an airplane. Flights would have been leased to NASA as needed. After failures with the X-33 subscale technology demonstrator test vehicle, funding was cancelled in 2001.

50.1   Advantages over the Space Shuttle

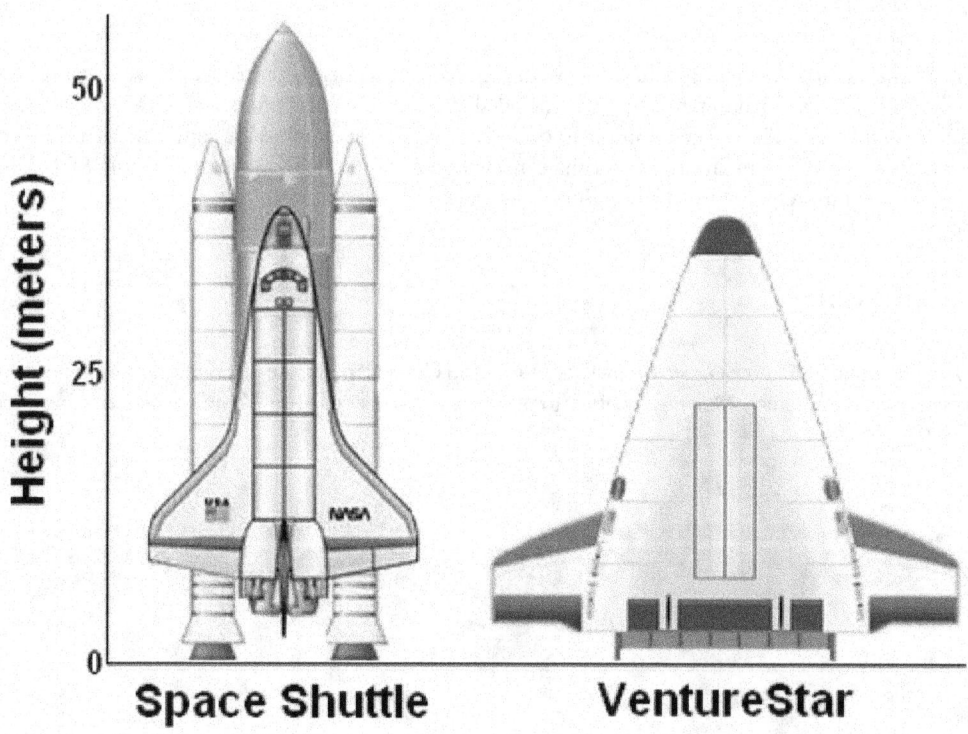

VentureStar would have stood approximately 17 meters shorter than the Space Shuttle.

VentureStar's engineering and design would have offered numerous advantages over the Space Shuttle, representing considerable savings in time and materials, as well as increased safety.[2] VentureStar was expected to launch satellites into orbit at about 1/10 the cost of the Shuttle.

Readying VentureStar for flight would have dramatically differed from that of the Space Shuttle. Unlike the Space Shuttle orbiter, which had to be lifted and assembled together with several other heavy components (a large external tank, plus two solid rocket boosters), VentureStar was to be simply inspected in a hangar like an aeroplane.[2]

Also unlike the Space Shuttle, VentureStar would not have relied upon solid rocket boosters, which had to be hauled out of the ocean and then refurbished after each launch.[2] Furthermore, design specifications called for the use of linear aerospike engines that maintain thrust efficiency at all altitudes. Whereas the Shuttle relied upon conventional nozzle engines which achieve maximum efficiency at only a certain altitude.[2]

VentureStar would have used a new metallic thermal protection system, safer and cheaper to maintain than the ceramic protection system used on the Space Shuttle. VentureStar's metallic heat shield would have eliminated 17,000 between-flight maintenance hours typically required to satisfactorily check (and replace if needed) the thousands of heat-resistant ceramic tiles that compose the Shuttle exterior.[2]

VentureStar was expected to be safer than most modern rockets.[2] Whereas most modern rockets fail catastrophically when an engine fails during flight, VentureStar was intended to have a thrust reserve in each engine in the event of an emergency during flight.[2] For example, if an engine on VentureStar were to have failed during an ascent to orbit, another engine opposite to the failed engine would have shut off to counterbalance the failed thrust, and each of the remaining working engines could then have throttled up so as to safely continue the mission.[2]

Unlike the Space Shuttle, whose solid rocket boosters expended chemical wastes during launch, VentureStar's exhausts would have been composed of only water vapor, since VentureStar's main fuels would have been only liquid hydrogen and liquid oxygen.,[2] meaning that VentureStar would have been environmentally clean.[2] VentureStar's simpler design would have excluded hypergolic propellants and even hydraulics, relying instead upon electrical power for flight controls, doors and landing gear.[2]

Because of its lighter design, VentureStar would have been able to land at almost any major airport in an emergency,[2] whereas the Space Shuttle required much longer runways than available at most public airports.

50.2   Cancellation

The VentureStar program was cancelled due to development cost concerns accompanied by technical problems and failures in the X-33 program, a program which was intended as proof-of-concept for some of the critical technologies needed by the VentureStar. The failure during a test of the X-33's complex, multi-lobe composite-structure cryogenic hydrogen tank was one of the main reasons for the cancellation of both the X-33 and the VentureStar. Ultimately, the VentureStar program required too many technical advances at too high a cost to be viable.

50.3   In fiction

In the 2001 novella and 2015 novel *Lash-Up* by Larry Bond and Chris Carlson, the VentureStar prototype is converted into an armed spacecraft named *Defender* in order to protect US space assets from China, which is using a space gun to destroy GPS satellites.[3][4]

50.4   See also

- Lockheed Martin X-33

- Skylon (spacecraft)

50.5   References

[1]   "AeroSpace Online:X-33 Advanced Technology Demonstrator" . Retrieved 2007-04-23.

[2] "SP-4220 Wingless Flight: The Lifting Body Story (Chapter 9)". *R. Dale Reed (NASA Dryden Flight Research Center, Aerospace and Contract Engineer)*. NASA. August 1997. Retrieved 21 January 2010.

[3] Bond, Larry (2001). "Lash-Up". In Coonts, Stephen. *Combat*. New York: Forge. pp. 149–265. ISBN 0-312-87190-2. OCLC 45066376.

[4] Bond, Larry (2015). *Lash-Up*. With Chris Carlson. New York: Forge. ISBN 978-0-7653-3491-6. OCLC 906798381.

50.6 External links

- Popsci article - October 1996

Chapter 51

SpaceShipTwo, Serial Number Two

SpaceShipTwo *Serial Number Two*[4] (Tail number: **N202VG**[2]) is a SpaceShipTwo-class suborbital rocket-powered manned spaceplane. It is the second *SpaceShipTwo* to be built, and will be used as part of the Virgin Galactic Fleet.

51.1   Overview

The craft has yet to be formally named.[5][6] There was speculation in 2004 that SpaceShipTwo *Serial Number Two* would be named "VSS Voyager".[7] It is commonly referred to as SpaceShipTwo, Serial Number Two.[8][9] Virgin Galactic expects it to be ready for airborne testing no earlier than late 2015.[3][10]

51.2   History

The manufacture of this, the second SpaceShipTwo spacecraft, began in 2012.[2] The spacecraft's registration, N202VG, was filed in September 2014.[11] As of early November 2014, the build of *Serial Number Two* was about 90 percent structurally complete, and 65 percent complete overall. As of April 2015, *Serial Number Two* was approximately 75% complete,[12] and initial ground tests could commence on this craft by the end of 2015.[13] On 21 May 2015, *Serial Number Two* reached the milestone of bearing the weight of the airframe on its own wheels.[14] It was thought that testing could have begun as early as the summer of 2015.[1]

The spacecraft is successor to VSS *Enterprise*, which broke up in-flight in late October 2014, in an accident whose cause is fully understood.

51.2.1   Test flight program

See also: VSS Enterprise § Flight test program

Tail number 2 will undergo a test regimen similar to VSS *Enterprise*, then will embark on testing beyond what Enterprise experienced. The test flights are expected to be fewer, as *Enterprise* has already tested the design's responses under numerous conditions. Testing will start with captive carry, progress to free-flight glide testing, then continue with powered test flights. It is possible that only 2-3 flights under each regime previously tested will be performed, instead of the 5 or 10 that *Enterprise* performed.[15]

51.3   See also

- Rutan Voyager

- Space tourism

51.4 References

[1] Luke Villapaz (5 November 2014). "Virgin Galactic Hopes To Restart SpaceShipTwo Tests In Summer 2015". International Business Times.

[2] Irene Klotz (5 November 2014). "New spaceship restoring hope after Virgin Galactic crash". *Reuters* (SpaceDaily).

[3] "Virgin Galactic CEO Sees New Spacecraft Ready Next Year". *Reuters* (Newsweek). 2 November 2014. Retrieved 2014-11-02.

[4] *One small step for space tourism... Private spaceflight.* 373. *The Economist.* 18 December 2004. Retrieved 2007-12-20.

[5] "Mojave Update: SpaceShipTwo Construction". *virgingalactic.com.* Retrieved 11 September 2015.

[6] Mike Wall (5 May 2015). "Virgin Galactic's Next SpaceShipTwo Spaceliner Is Taking Shape (Photo)". Space.com.

[7] "One small step for space tourism...". The Economist. 16 December 2004.

[8] Rosenberg, Zach. "The Making of SpaceShipTwo". *Air & Space Smithsonian.* Smithsonian Air & Space Magazine. Retrieved 11 September 2015.

[9] Crane, Rachel. "Inside Virgin Galactic's newest passenger spaceship". *Inside Virgin Galactic's Space Tourism Rocket Factory.* CNN. Retrieved 11 September 2015.

[10] Alan Boyle (5 November 2014). "The Next SpaceShipTwo Takes Shape in Virgin Galactic Hangar". NBC News.

[11] "N202VG is Reserved". FAA. 11 September 2014. Retrieved 2014-11-05.

[12] Jonathan O'Callaghan (21 April 2015). "Virgin Galactic will fly again: Company prepares to test new spacecraft after fatal 2014 crash". Daily Main (London).

[13] "Virgin Galactic 'to test new craft by end of year'". The Telegraph (London). 21 April 2015.

[14] Alan Boyle (22 May 2015). "Virgin Galactic's SpaceShipTwo 2.0 Puts 'Weight on Wheels'". NBC News.

[15] Jeff Foust (14 October 2015). "SpaceShipTwo Bounces Back to Rubber Fuel". SpaceNews.

Chapter 52

X-41 Common Aero Vehicle

Initiated in 2003, **X-41** is the designation for a still-classified U.S. military spaceplane. Specifications or photos of the program have not been released to the public yet; as a result not much is known about its goals. It has been described as an experimental manoeuvring re-entry vehicle capable of transporting a 1,000 lb payload on a sub-orbital trajectory at hypersonic speeds and releasing that payload into the atmosphere. The technology required for the X-41 is not yet known and is still undecided by the government. It is believed a new type of hypersonic travel is also being studied for the X-41 that will apparently be able to travel past Mach 7 and perhaps onto Mach 9.

This vehicle is now a part of the FALCON (**F**orce **A**pplication and **L**aunch from **Con**tinental United States) program sponsored by DARPA and NASA.

52.1 External links

- GlobalSecurity.org: X-41

- Spacedaily.com: CAV

- Designation-Systems.net: X-41 CAV

- *Pentagon Has Far-Reaching Defense Spacecraft in Works*, Washington Post, March 16, 2005

Chapter 53

XS-1 (spacecraft)

The **DARPA XS-1** will be a reusable spaceplane/booster to deliver small satellites into orbit for the U.S. Military.[1] The XS-1 is to directly replace the "first stage" of a multistage rocket that will be capable of flying at hypersonic speed at suborbital altitude, enabling one or more expendable upper stages to separate and deploy a payload into low Earth orbit. The XS-1 would then return to Earth, where it could be serviced fast enough to repeat the process at least once every 24 hours.[2][3]

53.1 History

The XS-1 program followed several previous failed attempts to develop a reusable space launch vehicle. The X-30 in the 1980s and X-33 VentureStar in the 1990s never flew due to immature technologies. DARPA's last attempt was the Responsive Access, Small Cargo, Affordable Launch (RASCAL) program in the early 2000s with the goal of placing 300 lb (140 kg) payloads in orbit for less than $750,000.

The XS-1 program was announced in November 2013 at a DARPA industry day. DARPA stated that the XS-1 was more feasible due to better technologies, including light and low-cost composite airframe and tank structures, durable thermal protection, reusable and affordable propulsion, and aircraft-like health management systems.[4] Jess Sponable, the XS-1 program manager, spoke on February 5, 2014 at NASA's Future In-Space Operations group, stating, "The vision here is to break the cycle of escalating space system costs, enable routine space access and hypersonic vehicles." [5]

Three companies were awarded contracts by July 2014 to design a demonstration vehicle. The selected companies were Boeing with Blue Origin, Masten Space Systems with XCOR Aerospace, and Northrop Grumman with Virgin Galactic. Unlike other DARPA programs that were handed off to parts of the United States military once proven successful, this initiative was designed from the start to be a direct partnership between the agency and industry. In August 2015, Boeing, Northrop Grumman, and Masten Space Systems all received additional funding from DARPA to continue their XS-1 design concepts for Phase 1B of the program. The first XS-1 orbital mission could occur as early as 2020.[6]

53.2 Program goals

The goals of the program as of September 2013 were:[2][7] The space plane must carry a 3,000–5,000 lb (1,400–2,300 kg) payload to low Earth orbit for less than a cost of US$5 million per flight,[3] at a rate of 10 or more flights per year; currently, launching that type of payload requires using an Orbital Sciences Corporation Minotaur IV expendable booster, priced at $55 million once per year.

- hypersonic flight to Mach 10 (12,250 km/h) or higher

- fast one-day turnaround time, including flying 10 times in 10 days

- a 1,800 kilograms (4,000 lb) payload on a trajectory to orbit

- launch cost less than 1/10 that of current launch systems, approximately US$5 million per flight*[3]

- unmanned vehicle

- utilize a reusable first stage booster to fly at hypersonic speeds to a suborbital altitude, coupled with one or more expendable upper stages that would separate and deploy a satellite*[8]*[9]

53.3   Entrants

The Boeing Company, Northrop Grumman Aerospace Systems, and Masten Space Systems have Phase 1 conceptual design contracts.

Boeing used its aircraft, spacecraft, and autonomous systems experience to work with its team including Blue Origin. Boeing's design would allow the autonomous booster to carry the second stage and payload to high altitude and deploy them into space. The booster would then return to Earth, where it could be quickly prepared for the next flight by applying operation and maintenance principles similar to modern aircraft, according to Will Hampton, Boeing XS-1 program manager. Drawing on their other innovative technologies, Boeing intends to provide a concept that uses efficient, streamlined ground infrastructure and improves the turnaround time to re-launch this spacecraft for subsequent missions.

Northrop Grumman used its aircraft, spacecraft, and autonomous systems experience to work with its team consisting of Scaled Composites to lead fabrication and assembly, and Virgin Galactic to head commercial spaceplane operations and transition; Virgin Galactic and Scaled Composites both worked on the SpaceShip Two, the world's only commercial spaceline. The team also leveraged technologies developed during related projects for DARPA, NASA, and the U.S. Air Force Research Laboratory to give the government "return on those investments." Their concept included a clean-pad launch using a transporter erector launcher with minimal infrastructure and ground crews, highly autonomous flight operations, and horizontal landing and recovery on standard runways.*[8]

Masten Space Systems had experience in rapid launchings of space vehicles, with their Xombie, Xoie, and Xaero vertical takeoff, vertical landing (VTVL) designs having already met or exceeded the 10 flights in 10 days objective set by the program. Although the company consists of fewer than 20 employees and is headquartered in a small building at the Mojave Air and Space Port, they have spent years flying various small VTVL systems on short hops at the spaceport, serving as test beds for guidance, navigation, and control (GNC) systems designed to safely land spacecraft on the Moon and potentially other planets. Their concept showed a VTVL system taking off vertically from a launch pad with wings and a tail fin. Masten Space Systems partnered with XCOR Aerospace, which would have provided the propulsion system.*[10]

53.4   See also

- Airborne Launch Assist Space Access

- Boeing X-37

- Boeing X-51

- DARPA Falcon Project

- NASA X-43

- Spaceplane

53.5   References

[1] David Axe (2015-08-03). "Pentagon Preps for Orbital War With New Spaceplane". The Daily Beast. Retrieved 2015-08-03.

[2] Foust, Jeff (2013-09-12). "DARPA To Start Reusable Launch Vehicle Program". *Space News*. Retrieved 2013-09-13.

[3] Howell, Elizabeth (1 May 2015). "XS-1: DARPA's Experimental Spaceplane". *Space.com*. Retrieved 2015-05-14.

[4] "Darpa Targets Lower Launch Costs With XS-1 Spaceplane". Aviation Week. 2 December 2013.

[5] "US Military Space Plane aims for 2017 lift off". spacedaily.com. Retrieved 2014-03-21.

[6] DARPA Awards $20 Million for Continued Development of a Military Space Plane - Defense-Update.com, 8 August 2015

[7] "DARPA fires up XS-1 space plane quest | Cutting Edge - CNET News". news.cnet.com. Retrieved 2014-03-21.

[8] Northrop Grumman Developing XS-1 Spaceplane For DARPA - Spacedaily.com, 20 August 2014

[9] DARPA issues first-phase solicitation for XS-1 hypersonic space plane for deploying satellites - Militaryaerospace.com, 15 November 2013

[10] Masten Space Systems Aims High on XS-1 Military Space Plane Project - *Space.com*, 26 August 2014

53.6 External links

- DARPA XS-1 artist concept animation video, 2014.

- Masten Space Systems, Inc. award notice, US$3 million, US government document, June 27, 2014. DARPA XS-1 artist concept animation video

53.7   Text and image sources, contributors, and licenses

53.7.1   Text

- **Spaceplane** *Source:* https://en.wikipedia.org/wiki/Spaceplane?oldid=678495722 *Contributors:* Guppie, Rmhermen, Maury Markowitz, Edward, Patrick, Rlandmann, Astudent, ChrisO~enwiki, Harry Doddema, Xanzzibar, Nadavspi, Wolfkeeper, Vessbot, Jwolfe, N328KF, Rich Farmbrough, Hydrox, Bender235, Mcpusc, Ylee, Huntster, Kross, Dillee1, Radical Mallard, Pauli133, Japanese Searobin, Roylee, AirBa~enwiki, Kosher Fan, Chochopk, Emerson7, Drbogdan, Feydey, FlaBot, SchuminWeb, DVdm, Wavelength, Arado, Hydrargyrum, Dotancohen, Rsrikanth05, Skiffer, Ospalh, Abune, Fram, Chic happens, SmackBot, Nickst, Gilliam, Chris the speller, SeanWillard, WDGraham, Nick Levine, Kelvin Case, Jimbatka, Matt Whyndham, John, Carnby, Euchiasmus, J 1982, RandomCritic, Beetstra, Jiri Svoboda, Peyre, Craigboy, Scarlet Lioness, Raerth, N2e, Cydebot, Fnlayson, Andyjsmith, Al Lemos, Corella, Steelpillow, JAnDbot, OckRaz, Vultur~enwiki, Yomin, Jvhertum, BatteryIncluded, BilCat, Xtifr, Kiore, Jim.henderson, R'n'B, CommonsDelinker, Peterdibble, Fusion7, KTo288, Tdadamemd, Hammersoft, TurboNOMAD, Sdsds, Ultratone85, Riick, Vsst, Ezrado, Zephyrus67, Dolphin51, EoGuy, Iohannes Animosus, DumZiBoT, Addbot, Colt9033, Gwano, Yobot, Edoe, AnomieBOT, Francisco Leandro, ArthurBot, MauritsBot, Coretheapple, Mark Schierbecker, Fotaun, GliderMaven, 光輝十月, Tom.Reding, December21st2012Freak, WPPilot, Defender of torch, RjwilmsiBot, Ajraddatz, ScottyBerg, Racerx11, Kristian Larsen, Mmeijeri, Yattum, Wingman4l7, Quite vivid blur, ChrisCarss Former24.108.99.31, Terra Novus, Planetscared, Rememberway, Incompetence, Senthilvel32, MediaMogulMan, Danim, MerlIwBot, Helpful Pixie Bot, Somatrix, Virtualerian, TCN7JM, HGK745, Haskellelephant, BattyBot, ChrisGualtieri, Mogism, Person Mcpersonjoe, Muchotreeo, Advanceddeepspacepropeller, Frinthruit, Skyhook1 and Anonymous: 61

- **Aerospaceplane** *Source:* https://en.wikipedia.org/wiki/Aerospaceplane?oldid=678708619 *Contributors:* Maury Markowitz, Charles Matthews, Donreed, Woohookitty, Bricktop, Ian Pitchford, Gaius Cornelius, Groovemeister007, Ospalh, Canley, SmackBot, Chris the speller, Cydebot, PhiLiP, Mack2, BilCat, R'n'B, Chiswick Chap, Biscuittin, ClueBot, Addbot, Sayantan m, DexDor and Anonymous: 2

- **Airbus Defence and Space Spaceplane** *Source:* https://en.wikipedia.org/wiki/Airbus_Defence_and_Space_Spaceplane?oldid=675753744 *Contributors:* Chrisjj, Rich Farmbrough, Hektor, Gene Nygaard, MZMcBride, RussBot, Ospalh, ArielGold, SmackBot, Jrockley, JHuwaldt, Craigboy, CmdrObot, N2e, Lokal Profil, Cydebot, Adam Chlipala, BilCat, Don-vip, Ssolbergj, Reedy Bot, Wikigi, DOHC Holiday, Jeroen888, ImageRemovalBot, Tiffe~enwiki, Chaosdruid, XLinkBot, Addbot, Numbo3-bot, Ben Ben, AnomieBOT, Crusoe8181, RjwilmsiBot, H3llBot, SkywalkerPL, BG19bot, Beaucouplusneutre, Carlstak, Andyhowlett, Wiki Thomas, Frinthruit, JGG13, Orduin and Anonymous: 19

- **ASSET (spacecraft)** *Source:* https://en.wikipedia.org/wiki/ASSET_(spacecraft)?oldid=651323415 *Contributors:* Edward, Quadell, Trevor MacInnis, Rich Farmbrough, Dirac1933, Rjwilmsi, SmackBot, Chris the speller, WDGraham, Kelvin Case, Zac67, N2e, Cydebot, Gabeb83, CommonsDelinker, Sirstubby, Petebutt, Editore99, Kitchen Knife, MystBot, Addbot, AnomieBOT, Xosema, Materialscientist, FrescoBot, 777sms, Mddkpp and Anonymous: 12

- **Avatar (spacecraft)** *Source:* https://en.wikipedia.org/wiki/Avatar_(spacecraft)?oldid=688156978 *Contributors:* Wolfkeeper, Oneiros, Gene Nygaard, Jimgeorge, BD2412, Koavf, RussBot, Arado, Muruga86, Ospalh, Fram, Sardanaphalus, SmackBot, SeanWillard, Hibernian, WDGraham, MilborneOne, Nobunaga24, Peyre, WolfgangFaber, CmdrObot, N2e, Cydebot, ANTIcarrot, Aldis90, Chanakyathegreat, BatteryIncluded, S3000, Dudewheresmywallet, Jim.henderson, R'n'B, Ohms law, Hugo999, Vipinhari, AlexNewArtBot, Sniperz11, Antixt, Oldag07, A21sauce, DaddyWarlock, PaulxSA, The Bushranger, Yobot, AnomieBOT, Johnxxx9, Akb amar, Reach Out to the Truth, MegaSloth, M.srihari, John of Reading, Rememberway, Jdcollins13, Jayadevp13, Narayanankrishnan, Frinthruit, Harsh7422, EvilMossman, Theeconomist11, Ohsin and Anonymous: 35

- **Blackstar (spacecraft)** *Source:* https://en.wikipedia.org/wiki/Blackstar_(spacecraft)?oldid=639650154 *Contributors:* Guppie, Ixfd64, Nickshanks, Finlay McWalter, Mustang dvs, Wolfkeeper, Wwoods, Ericg, N328KF, Rich Farmbrough, NeuronExMachina, Sarrica, Eddieuny, PPGMD, TomStar81, Tronno, RoySmith, Reinoutr, GregorB, Rjwilmsi, Ligulem, Wars, Team6and7, StuffOfInterest, RussBot, Hydrargyrum, GeeJo, Emdx, Bill.martin, Malfita, SmackBot, Melchoir, Martylunsford, Grey Shadow, Hibernian, WDGraham, Aerobird, Tomtom9041, Xxxxxxxxxxx, Peyre, ChrisCork, Vinjhed, Benabik, Newsnightmeirion, CmdrObot, N2e, Cydebot, Yomangani, Arcsincostan, Duggy 1138, BilCat, Tarvold, Robbyedge, Ng.j, Maelgwnbot, Hamiltondaniel, Spaz Out Of Hell, DaddyWarlock, MBK004, Mild Bill Hiccup, Niceguyedc, Addbot, The Bushranger, ArthurBot, Jh43785, ChoraPete, Khazar2 and Anonymous: 45

- **Boeing X-20 Dyna-Soar** *Source:* https://en.wikipedia.org/wiki/Boeing_X-20_Dyna-Soar?oldid=678708579 *Contributors:* XJaM, Maury Markowitz, Leandrod, Minesweeper, Egil, Stan Shebs, Rlandmann, Lommer, Mulad, Zoicon5, Aetheling, Reubenbarton, DocWatson42, Greyengine5, Wolfkeeper, Bobblewik, Wmahan, Ericg, N328KF, Brianhe, Rich Farmbrough, Night Gyr, Srbauer, Aqua008, Chairboy, Iridia, Gingko, KBi, Pearle, Amcl, Error 404, Snowolf, Sobolewski, Evil Monkey, Gene Nygaard, Voxadam, Sylvain Mielot, Asav, Mpj17, SDC, Fxer, Marudubshinki, BD2412, Rjwilmsi, Rillian, Jmcc150, Ligulem, SchuminWeb, Mark Sublette, TheDJ, Russavia, Hellbus, Hydrargyrum, Gaius Cornelius, Voidxor, Tony1, Karl Andrews, Haemo, LamontCranston, Little Savage, Petri Krohn, That Guy, From That Show!, SmackBot, Henriok, Setanta747 (locked), Bluebot, Trekphiler, AzaBot, Aces lead, Andy120290, Skaltavista, The PIPE, John, Pete.tian~enwiki, Jaganath, Dumelow, RandomCritic, Xxxxxxxxxxx, Fraantik, Peyre, Pjbflynn, N2e, ShelfSkewed, Cydebot, Fnlayson, Thijs!bot, Epbr123, Ironass, Al Lemos, Nick Number, KidIncredible, Ricnun, Mack2, Mwarren us, Bzuk, Nickpheas, Djroam, A75, BilCat, Archolman, Fusion7, Ndunruh, Ohms law, STBotD, Treisijs, XaHyMaH, Royhouchin, VolkovBot, Amikake3, GimmeBot, Sandtiger1, RobbWiki, 1more, Djmckee1, Ikluft, Hxhbot, OKBot, Binksternet, Sordidatus, Torqtorqtorq, Ktr101, Alexbot, Chaosdruid, PotentialDanger, DumZiBoT, Alihus86, Addbot, DrJos, Artiyom, The Bushranger, Luckas-bot, Yobot, Palamabron, AnomieBOT, Citation bot, GrampaScience, Shadak, Xqbot, Winged Brick, GrouchoBot, NobelBot, Brutaldeluxe, MastiBot, Pastychomper, Enemenemu, Ernie Scribner, EmausBot, WikitanvirBot, Mkinchen, Dominictroc, ZéroBot, Jj98, Ngauger, Helpful Pixie Bot, Trevayne08, BattyBot, Jupiter-4, Dcgibson55, WPGA2345, Ironmungy, KasparBot and Anonymous: 67

- **Boeing X-37** *Source:* https://en.wikipedia.org/wiki/Boeing_X-37?oldid=683959341 *Contributors:* AxelBoldt, Paul Drye, Guppie, Patrick, Rlandmann, PaulinSaudi, Twang, Stewartadcock, Chris-gore, Bogdanb, Wwoods, Fleminra, Bobblewik, Zootalures, Eregli bob, Oneiros, Karl Dickman, Trevor MacInnis, Freakofnurture, N328KF, Pak21, Vsmith, Wk muriithi, Bender235, RJHall, Huntster, Irate~enwiki, Eritain, Pearle, A2Kafir, Alansohn, Walter Görlitz, Thewalrus, Hadlock, ClockworkSoul, Suruena, Vuo, Gene Nygaard, Dan East, Kitch, Dan100, Zntrip, Roylee, Velho, Woohookitty, Poppafuze, Blackeagle, Ekem, Fxer, BD2412, Rjwilmsi, Hiberniantears, Rillian, RobertG, YurikBot, Wormholio, StuffOfInterest, V Brian Zurita, Arado, Kvuo, Bergsten, Reddevil0728, RadioFan, Gaius Cornelius, Los688, WulfTheSaxon, PhilipC, Gadget850, Arthur Rubin, Lynbarn, Heathhunnicutt, Fram, Mikus, Allens, SmackBot, F, Martylunsford, Nickst, Cla68, Bluebot, Kjosmoen, Hibernian, Modest Genius, WDGraham, SheeEttin, Kelvin Case, Mytwocents, NeilFraser, Ohconfucius, SalopianJames, John, Jrvz, MilborneOne, Joshua Scott, Noah Salzman, Meco, Peyre, Xionbox, Osklil, Craigboy, Zephram Stark... Non Stop

Salmanazar, Aremisasling, Abune, SmackBot, Grey Shadow, Jrockley, Oliver.gouldthorpe, Chris the speller, Pcarpent, Aces lead, The PIPE, Gregzsidisin, John, Vampus, MilborneOne, Dl2000, Craigboy, Im.thatoneguy, Raerth, Zeke54, N2e, Cydebot, Arb, Thijs!bot, LionFlyer, Ianare, Pi.1415926535, IanOsgood, Kaleja, Nikbro, BatteryIncluded, BilCat, Taka2007, Techmonk, Abebenjoe, R'n'B, CommonsDelinker, Vlmagee, O.W., Cardinal2, XaHyMaH, Robprain, 28bytes, Voronwae, AJClements, Hardgrovius, AHMartin, Dravecky, Dabomb87, Taliska, Der Golem, NiD.29, U5K0, Kitchen Knife, Arjayay, PistolPete037, Jtle515, XLinkBot, Dilbert2000, Addbot, Lightbot, Anxietycello, Nigelbeameniii, GDK, The Bushranger, Luckas-bot, Yobot, PMLawrence, Tonyrex, AnomieBOT, Xosema, Tucoxn, Arsia Mons, FrescoBot, Dwightfowler, Jonesey95, Ras67, Ripchip Bot, TGCP, EmausBot, John of Reading, Mmeijeri, ZéroBot, Hoeksas, SkywalkerPL, ChiZeroOne, ClueBot NG, Helpful Pixie Bot, Cgruda, BG19bot, Beaucouplusneutre, Jay8g, Frze, Lolle42, D1023319, IluvatarBot, DreamChaserMedia, Blaspie55, Vamc19, BattyBot, America789, Khazar2, JYBot, EagerToddler39, Br'er Rabbit, Mogism, Marywhitney, Xjehl, Hpskiii, Jamesmcmahon0, Shelbystripes, Frinthruit, Mfb, Monkbot, Renee.elia, DRatti, LCPWolf, Cecilia-greentree, James-Ackard, Solardays, Wayfarer1620, LaurenceAryeh and Anonymous: 86

- **Falke (spacecraft)** *Source:* https://en.wikipedia.org/wiki/Falke_(spacecraft)?oldid=671380378 *Contributors:* Hektor, Chris the speller, WDGraham, Paulmcdonald, Addbot, Helpful Pixie Bot and Anonymous: 1

- **Goodyear Meteor Junior** *Source:* https://en.wikipedia.org/wiki/Goodyear_Meteor_Junior?oldid=647289131 *Contributors:* WDGraham, MilborneOne and Tyrol5

- **Hermes (spacecraft)** *Source:* https://en.wikipedia.org/wiki/Hermes_(spacecraft)?oldid=689189909 *Contributors:* Bryan Derksen, Tarquin, Maury Markowitz, (, Ahoerstemeier, Jpatokal, Sunbeam60, Nickshanks, JonathanDP81, AlainV, J.Rohrer, Alan Liefting, Quasarstrider, Lethe, Nick04, Dflock, Geni, Hellisp, Rich Farmbrough, Guanabot, Devil Master, Hektor, Ashley Pomeroy, Dirac1933, Gunter, Dan100, Nuno Tavares, Paldorslate, Josh Parris, Catsmeat, Themanwithoutapast, YurikBot, Wavelength, Gaius Cornelius, Geoffrey.landis, Ariel-Gold, John Broughton, DocendoDiscimus, SmackBot, Nickst, Colonies Chris, Redline, WDGraham, Fuhghettaboutit, John, Peyre, Craigboy, FairuseBot, Cydebot, AntiVandalBot, Ricnun, JAnDbot, IanOsgood, Jatkins, BilCat, Spellmaster, Stephenchou0722, Sinigagl, Rettetast, CommonsDelinker, VolkovBot, TXiKiBoT, Pjoef, SieBot, ImageRemovalBot, MBK004, Patrick Rogel, Trulystand700, DumZiBoT, Addbot, Download, Lightbot, Luckas-bot, Ptbotgourou, Cyan22, GrampaScience, DSisyphBot, Almabot, Fotaun, AmphBot, WikitanvirBot, Mmeijeri, ZéroBot, SkywalkerPL, ChiZeroOne, Penyulap, Rousfo, Drift chambers, 4throck, ZemplinTemplar, Rfassbind, FMASKIEDFJSD, Frinthruit, Appable and Anonymous: 35

- **HOPE-X** *Source:* https://en.wikipedia.org/wiki/HOPE-X?oldid=619436564 *Contributors:* Maury Markowitz, Ropers, Pmsyyz, Huntster, Hektor, BRW, Henry W. Schmitt, Change1211, Nickst, Bluebot, ThreeBlindMice, Cydebot, Honeplus, Thadius856, CommonsDelinker, Hans Dunkelberg, Rei-bot, Ryan shell, Djmckee1, MystBot, Addbot, Yobot, Rubinbot, BenzolBot, Challisrussia, BattyBot, Frinthruit and Anonymous: 15

- **Hopper (spacecraft)** *Source:* https://en.wikipedia.org/wiki/Hopper_(spacecraft)?oldid=674234020 *Contributors:* Bryan Derksen, Maury Markowitz, RickK, Nickshanks, Alan Liefting, Quasarstrider, Ceejayoz, A2Kafir, Hektor, Allen3, Jmcc150, DAJF, Petri Krohn, Sardanaphalus, SmackBot, Skizzik, Radagast83, N2e, Cydebot, Thijs!bot, Escarbot, Jatkins, BatteryIncluded, Sinigagl, Sdsds, Vsst, GirasoleDE, Lightmouse, Enenn, Addbot, Colt9033, Lightbot, DSisyphBot, Fotaun, Lotje, EmausBot, SkywalkerPL, Rememberway, Beaucouplusneutre, Frinthruit and Anonymous: 16

- **HOTOL** *Source:* https://en.wikipedia.org/wiki/HOTOL?oldid=674234085 *Contributors:* Enchanter, Maury Markowitz, Heron, Leandrod, (, [212], Chrisjj, Wolfkeeper, Bobblewik, H1523702, Generica, Hohum, Max rspct, Jheald, Bricktop, GraemeLeggett, Offtherails, SchuminWeb, Mark83, RussBot, Arado, Tungsten, Astral, SmackBot, KVDP, Burbank~enwiki, Chris the speller, JGarry, Aces lead, Siffler~enwiki, Dl2000, Ewulp, N2e, Cydebot, IanOsgood, BilCat, R'n'B, STBotD, Freeman501, Hercule, LG02, RKSimon, Tassedethe, Iancarine, Anxietycello, The Bushranger, Legobot, Luckas-bot, Arsia Mons, GliderMaven, Redrose64, Richhaddon, Sp33dyphil, AvicAWB, SkywalkerPL, Rememberway, El Roih, Elspeth.millar, BG19bot, BendelacBOT, Northamerica1000 and Anonymous: 36

- **Hyflex** *Source:* https://en.wikipedia.org/wiki/Hyflex?oldid=652853904 *Contributors:* Hektor, SmackBot, Nickst, Sam8, WDGraham, Cydebot, Djmckee1, Flyer22 Reborn, Addbot, Gwano, Xosema, サ ヤ, ZéroBot, Frinthruit, My mom is kind!! and Anonymous: 2

- **Intermediate eXperimental Vehicle** *Source:* https://en.wikipedia.org/wiki/Intermediate_eXperimental_Vehicle?oldid=672890607 *Contributors:* Maury Markowitz, Caltrop, Finlay McWalter, Oneiros, (aeropagitica), Hektor, Suruena, Lars T., Vegaswikian, Peter Grey, Gardar Rurak, CambridgeBayWeather, Arthur Rubin, Nickst, Cattus, WDGraham, Stepho-wrs, DinosaursLoveExistence, JorisvS, N2e, Cydebot, Fnlayson, Widefox, Lino Mastrodomenico, TAnthony, Magioladitis, Jatkins, BatteryIncluded, Sinigagl, CommonsDelinker, Katharineamy, PeSHIr, Sdsds, PDFbot, Supercopter, Tomas e, U5K0, Shinkolobwe, Addbot, H92Bot, Yobot, Ptbotgourou, WatcherZero, AnomieBOT, Arkaska, Winged Brick, Nrpf22pr, GrouchoBot, Fotaun, FrescoBot, Tom.Reding, Skyerise, EmausBot, Mmeijeri, ZéroBot, Pippo skaio, Spacegizmo, Tglman, SkywalkerPL, Michaelmas1957, Bibcode Bot, ZemplinTemplar, Rfassbind, Markh89, Mfb, Andyhaslam2310, Monkbot, Strongjam, Rhoark, Anna f64 and Anonymous: 22

- **Keldysh bomber** *Source:* https://en.wikipedia.org/wiki/Keldysh_bomber?oldid=668394538 *Contributors:* Wolfkeeper, BD2412, Salmanazar, Sardanaphalus, Hmains, Derek R Bullamore, Cydebot, Fnlayson, Tec15, Nemissimo, Pleasantville, Lightmouse, Foofbun, Addbot, Bazook, The Bushranger, Yobot, Le Bao, Monkbot and Anonymous: 5

- **Kliper** *Source:* https://en.wikipedia.org/wiki/Kliper?oldid=689236125 *Contributors:* Bryan Derksen, XJaM, Leandrod, Edward, Nikai, Nickshanks, Maver1ck, Dflock, Archie, Bobblewik, Wmahan, Now3d, Imroy, Rich Farmbrough, Slipstream, Sperling, Sarrica, Aranel, Chairboy, Bobbis, Fjl, Hektor, Andrew Gray, Mmmready, Sobolewski, Hadlock, Yogi de, Dan100, Adrian.benko, Japanese Searobin, Hegen~enwiki, Siafu, Bricktop, Chochopk, Lawman~enwiki, Palica, Ilya, BD2412, Miq, Josh Parris, Rjwilmsi, Vegaswikian, Chekaz, Isthatyou, FlaBot, Themanwithoutapast, GagHalfrunt, YurikBot, TexasAndroid, RussBot, Hydrargyrum, Shaddack, Kantokano, Exir Kamalabadi, Howcheng, VinnyCee, Voidxor, Ilmaisin, Ageekgal, Mikus, Nixer, Philip Stevens, SmackBot, Sam8, Chris the speller, Lubos, DHN-bot~enwiki, WDGraham, Stepho-wrs, Nuaetius, Zen611, TheManWithNoName, SashatoBot, Dr. Sunglasses, John, RandomCritic, Luke Maurits, Peyre, Iridescent, Craigboy, Henrickson, Mattbr, N2e, Cydebot, Duccio, Thijs!bot, Akradecki, Ricnun, Storkk, Dariuz~enwiki, SEREGA784, MetsBot, NERV~enwiki, R'n'B, Ash sul, VolkovBot, Ilyaroz, Moonriddengirl, KGyST, Hamiltondaniel, Micov, Explicit, ImageRemovalBot, Bwfrank, DFRussia, U5K0, Dj manton, Chaosdruid, DumZiBoT, Addbot, Jim10701, LemmeyBOT, Lightbot, Yobot, LilHelpa, Xqbot, Johnxxx9, Fortdj33, Mgabiz, SporkBot, Wingman4l7, ChiZeroOne, Danim, Gotech8, Appable and Anonymous: 86

- **Lockheed L-301** *Source:* https://en.wikipedia.org/wiki/Lockheed_L-301?oldid=678708818 *Contributors:* Grim~enwiki, BD2412, Ground Zero, SmackBot, Bluebot, MilborneOne, Robofish, Sketch051, Citizenposse, Cydebot, Fnlayson, Pascal.Tesson, Gioto, Cm1701, Jay-Duck, BilCat, R'n'B, Petebutt, Maelgwnbot, DaddyWarlock, The Bushranger, Forcep caliper, 777sms, Black Yoshi, Will Beback Auto and Anonymous: 7

(Can), Chris the speller, Bluebot, MalafayaBot, TheFeds, Colonies Chris, Trekphiler, Aerobird, Kelvin Case, MJBurrage, Snowmanradio, Andy120290, Check-Six, Mytwocents, Ken keisel, BiggKwell, John, Vgy7ujm, Paul Raveling, AllStarZ, Bollinger, Fraantik, Dragos muresan, Quaeler, Osklil, Pjbflynn, Will Pittenger, JForget, CmdrObot, Van helsing, N2e, WeggeBot, Cydebot, Fnlayson, Peripitus, Crowish, Mike1942f, Tunheim, Thijs!bot, DulcetTone, Al Lemos, Dawkeye, Danarmstrong, DMcAlinden, Bzuk, Magioladitis, VoABot II, Father Goose, Swpb, Elsp, Jatkins, Catgut, Elentirmo, BilCat, LorenzoB, Archolman, Zg, SU Linguist, Tdadamemd, LordAnubisBOT, Plasticup, 350z33, Ndunruh, Ohms law, Potatoswatter, KylieTastic, DH85868993, DorganBot, Treisijs, Nat682, Dorftrottel, Tttecumseh, VolkovBot, Thomas.W, Jeff G., Amikake3, EH101, Sdsds, GimmeBot, Jashgerard, JhsBot, Razvan NEAGOE, RobbWiki, Buffs, Crvarvaro, 1-555-confide, Thunderbird2, SieBot, BotMultichill, Zephyrus67, Caltas, Allansplace, Moonraker12, Lightmouse, AMCKen, Moletrouser, Martin Velek, Afernand74, Kumioko (renamed), Mojoworker, Hamiltondaniel, Anyeverybody, Farcross, Alpha Centaury, Randy Kryn, YSSYguy, Ackshatt, MBK004, ClueBot, SalineBrain, Binksternet, Foxj, Dantheman 30, EoGuy, FieldMarine, Wwheaton, HDP, Mild Bill Hiccup, Niceguyedc, Brett Buck, Nimbus227, Alexbot, Apddraig, Human.v2.0, Posix memalign, Sun Creator, Mheimbecker, Wikiuser100, WikHead, Mimarx, Alexius08, Addbot, Xtreme001, Landon1980, Canejas, DrJos, Pelzig, Ericg33, Jaydec, Lightbot, Apteva, The Bushranger, Legobot, Drpickem, Luckas-bot, Jhswalwell, Donfbreed, Jason Recliner, Esq., AnomieBOT, Tryptofish, Jim1138, RadioBroadcast, Ckruschke, Xqbot, .45Colt, Ad Meskens, SpaceHistory101, Fotaun, FrescoBot, Jc3s5h, Silvyrian, Bioboy28, MastiBot, RazielZero, Coronium, FoxBot, Telerace, Dinamik-bot, WPPilot, 777sms, Hinko13, Threewms, DexDor, EmausBot, JustinTime55, Sp33dyphil, Bullmoosebell, L Kensington, Mikhail Ryazanov, ClueBot NG, Dwc89, Nordwestlicht, Helpful Pixie Bot, John Cummings, Rousfo, Soerfm, Zedshort, Caypartisbot, BattyBot, ChrisGualtieri, Dexbot, Makecat-bot, Fycafterpro, Shelbystripes, Marigold100, HamiltonFromAbove, Frinthruit, Ronrosano, Olenyash, TerryAlex, Stephen Mraz, Maruthish, Lelaleopard11, Hyperclassic, DalotekAffair, Buddyboy521, Ranaumar92 and Anonymous: 228

- **Orbital Sciences X-34** *Source:* https://en.wikipedia.org/wiki/Orbital_Sciences_X-34?oldid=675812758 *Contributors:* Angela, Rlandmann, Jeandré du Toit, Tempshill, Quasarstrider, Bobblewik, Keith Edkins, Karl Dickman, N328KF, Pak21, Huntster, Pearle, Nick Moss, Gene Nygaard, Fxer, BD2412, Mike Peel, Arado, Los688, Pil56, Sardanaphalus, SmackBot, WikiuserNI, Nickst, Bluebot, WDGraham, John, Craigboy, N2e, Cydebot, Fnlayson, Heinz-bert, Clh288, Rod57, Trashbag, Gibson Flying V, Ikluft, Centaur327, Pxma, Addbot, Awatral, Zorrobot, חובבשירה, The Bushranger, Yobot, AnomieBOT, Xosema, MattTheGreat13, Citation bot, Trueravenfan, GrouchoBot, Sarcastic ShockwaveLover, FrescoBot, D'ohBot, Puppier, Spacejulien, Emericpro, WikitanvirBot, Brt0000, Chesipiero, Wdchk, Andyhowlett, Marigold100, Adirlanz, Xsjlepling and Anonymous: 25

- **Orbital Space Plane Program** *Source:* https://en.wikipedia.org/wiki/Orbital_Space_Plane_Program?oldid=628463245 *Contributors:* Alex.tan, Maury Markowitz, Frecklefoot, Rlandmann, Alba, Wolfkeeper, 朝彦, Dcfleck, Andrew Gray, Evil Monkey, Pauli133, Dan100, Nightscream, Srleffler, YurikBot, Hairy Dude, Gaius Cornelius, Exir Kamalabadi, Rrrob, SmackBot, WDGraham, Aces lead, Craigboy, Joseph Solis in Australia, N2e, Fnlayson, Rod57, Jeepday, Ohms law, Addbot, Lightbot, Yobot, AnomieBOT, LittleWink, Mmeijeri, H3llBot, Monkbot and Anonymous: 10

- **Programme for Reusable In-orbit Demonstrator in Europe** *Source:* https://en.wikipedia.org/wiki/Programme_for_Reusable_In-orbit_ Demonstrator_in_Europe?oldid=674330103 *Contributors:* Gogo Dodo, War wizard90, Yobot, Josve05a, SkywalkerPL, Vilhelm.s, Sulfurboy, Matiia, Truthveyor, Oba Musiliu, Olojoru 1, Pavantechs, God1234567891011, Sandrabbit, Jamessmith62, Rauljayesh, Deansoran, Cyan0, Juneerowe, Aliman77, TMozeS, Lokenathji153, Anna f64, Sbunderlein, Beker64, Skysugar and Cow Supplier Tamilnadu

- **Project 921-3** *Source:* https://en.wikipedia.org/wiki/Project_921-3?oldid=654872415 *Contributors:* Huntster, Sukiari, Anthony Appleyard, Gene Nygaard, Woohookitty, Rjwilmsi, SmackBot, Neo-Jay, Nautilator, N2e, Cydebot, LionFlyer, Ricnun, LorenzoB, Rod57, Ohms law, Discovery103, Eckrantz, Gwano, Yobot, Guy1890, AnomieBOT, Chen Guangming, Nasa-verve, Trust Is All You Need, PigFlu Oink, Full-date unlinking bot, Updatehelper, Polylepsis, Midas02, H3llBot, Makecat, Gold Hat, ClueBot NG, Vosplay, 4throck, Quant18, Tony Mach, Frinthruit and Anonymous: 10

- **Prometheus (spacecraft)** *Source:* https://en.wikipedia.org/wiki/Prometheus_(spacecraft)?oldid=639608210 *Contributors:* Hektor, Racklever, Craigboy, N2e, Niceguyedc, Addbot, Jafeluv, AnomieBOT, XZise, Trappist the monk, Mmeijeri, ZéroBot, 321Kepler and Anonymous: 3

- **Rocketplane XP** *Source:* https://en.wikipedia.org/wiki/Rocketplane_XP?oldid=675754088 *Contributors:* Rmhermen, Patrick, Rlandmann, Rich Farmbrough, DS1953, Krellis, Christopher Thomas, RedBLACKandBURN, SmackBot, MichaelSH, Chris the speller, JHuwaldt, Morio, Craigboy, ShimaKatase, Raerth, N2e, Cydebot, After Midnight, Bzuk, Violentbob, Jatkins, BilCat, Redguard101, Tiyoringo, Sdsds, Petebutt, Fibo1123581321, FerdinandFrog, Wikiluck06, Addbot, Dfaulk, Lightbot, The Bushranger, Winged Brick, NorthnBound, Full-date unlinking bot, 777sms, Beaucouplusneutre, JGG13, MedainSalch and Anonymous: 14

- **Rockwell X-30** *Source:* https://en.wikipedia.org/wiki/Rockwell_X-30?oldid=676817959 *Contributors:* Carlj7, Axeman, Wolfkeeper, Oneiros, Dabarkey, Karl Dickman, Trevor MacInnis, N328KF, Rich Farmbrough, Pak21, Qutezuce, Night Gyr, Amcl, GeorgeStepanek, Gene Nygaard, GraemeLeggett, Emerson7, BD2412, Drbogdan, Rapier Shade, Rillian, Ahunt, RussBot, Arado, RadioFan, Logawi, SmackBot, Armeria, Bluebot, Joema, Kleuske, John, Ewulp, N2e, Cydebot, Fnlayson, Heinz-bert, Grant76, Honeplus, Felix Portier~enwiki, AntiVandalBot, Bzuk, Benstown, Schoowru, Jatkins, BilCat, Tdadamemd, Wind of Night, Ndunruh, VolkovBot, TXiKiBoT, GimmeBot, Petebutt, Farcross, Badger Brock, Nimbus227, Mickman1234, MystBot, Addbot, OlEnglish, Zorrobot, The Bushranger, Luckasbot, Xqbot, Fotaun, FrescoBot, 777sms, EmausBot, WikitanvirBot, Wingman4l7, Cgruda, Beaucouplusneutre, Glenncal, ChrisGualtieri, Andyhowlett, John Simpson54, Frinthruit and Anonymous: 46

- **Rotary Rocket** *Source:* https://en.wikipedia.org/wiki/Rotary_Rocket?oldid=688211398 *Contributors:* Bryan Derksen, Maury Markowitz, Patrick, Maximus Rex, HarryHenryGebel, Finlay McWalter, Ke4roh, Steve Leach, Giftlite, Wolfkeeper, FleaPlus, Bobblewik, Oneiros, Karl Dickman, Savuporo, JTN, RJHall, Cmdrjameson, Alphax, A.T.M.Schipperijn, Zippanova, BRW, Greg Kuperberg, Bricktop, Someone42, BD2412, Rjwilmsi, Vegaswikian, Ahunt, StuffOfInterest, RussBot, Arado, Logawi, Bozoid, Jhinman, Ilmaisin, SmackBot, ProveIt, Septegram, Bluebot, Modest Genius, Aces lead, A5b, Jafafa Hots, John, Jaganath, Kavanagh~enwiki, George100, Benabik, Mellery, N2e, Cydebot, Akradecki, Born2flie, Harryzilber, IanOsgood, Magioladitis, Melkor23, JaGa, Wolfy9005, Bricology, Speaker to wolves, UnitedStatesian, Riick, Ikluft, Lightmouse, SallyForth123, Kitchen Knife, DumZiBoT, Addbot, Lightbot, The Bushranger, Yobot, Amirobot, AnomieBOT, Ulric1313, FrescoBot, Vehement, Zalnas, Mysticyx, Terra Novus, Chesipiero, Hupaleju, Snotbot, Helpful Pixie Bot, BattyBot, ChrisGualtieri, Khazar2, Dobie80, Frinthruit, Fgphd, Jonpedwards and Anonymous: 27

- **Saenger (spacecraft)** *Source:* https://en.wikipedia.org/wiki/Saenger_(spacecraft)?oldid=650251510 *Contributors:* Carlossuarez46, PamD, Ezrado, Niceguyedc, Tom.Reding, Wgolf and FatalGravity

- **Sharp Edge Flight Experiment** *Source:* https://en.wikipedia.org/wiki/Sharp_Edge_Flight_Experiment?oldid=671375603 *Contributors:* Jrcrin001, N2e, Marcric, WikHead, Addbot, Hahc21, LaaknorBot, Yobot, Rubinbot, Nordwin, LilHelpa, Thehelpfulbot, Dwassel, 4throck, Mauerquadrant and Anonymous: 6

- **Silbervogel** *Source:* https://en.wikipedia.org/wiki/Silbervogel?oldid=668349766 *Contributors:* Rmhermen, Maury Markowitz, Caltrop, Leandrod, Rlandmann, Smallweed, Huxley75, Greyengine5, Sonance, N328KF, Guanabot, Tronno, Johnteslade, G026r, Pauli133, Dennis Bratland, Ae7flux, BD2412, Rjwilmsi, FlaBot, Ground Zero, RussBot, Arado, Witan, Hydrargyrum, Salmanazar, Petri Krohn, Geoffrey.landis, Fourohfour, Sardanaphalus, SmackBot, SeanWillard, JanCeuleers, The PIPE, Mion, LanternLight, Vgy7ujm, Mitternacht90, CmdrObot, Cydebot, Thijs!bot, SomeStranger, NByz, HolyT, Mark Grant, Nemissimo, Achero, Magioladitis, R'n'B, Amikake3, Bachcell, VVVBot, AMCKen, Riisipuuro, Foofbun, Jtle515, DumZiBoT, Addbot, AnomieBOT, Xufanc, Xqbot, GrouchoBot, Fotaun, Akasanof, Swawdsd, Ninja Auditor, Ardeshirmehta, RjwilmsiBot, Le Bao, EmausBot, John of Reading, ZéroBot, Huston84, Helpful Pixie Bot, Frinthruit, Monkbot and Anonymous: 28

- **Skylon (spacecraft)** *Source:* https://en.wikipedia.org/wiki/Skylon_(spacecraft)?oldid=689481516 *Contributors:* Bryan Derksen, Topbanana, Chrisjj, Witbrock, Cyrius, Fennec, Wolfkeeper, Bobblewik, Bcameron54, Oneiros, N328KF, Rich Farmbrough, Qutezuce, Huntster, Utopia104, Hohum, Tony Sidaway, Gene Nygaard, Siafu, Woohookitty, Jeff3000, Someone42, GregorB, Tmassey, CharlesC, Frankie1969, GraemeLeggett, Rjwilmsi, Astronaut, The Rambling Man, Hairy Dude, RussBot, Arado, Gaius Cornelius, Ospalh, EEMIV, CLW, Arthur Rubin, Paul White, RG2, Groyolo, Sardanaphalus, SmackBot, Mirokado, Chris the speller, Bluebot, WDGraham, GDM, Will Beback, Starlionblue, Korval, Peyre, Akademy, Dl2000, Michaelbusch, Craigboy, Jafet, N2e, Rmallins, Cydebot, ANTIcarrot, Brad101, Matthewakisan, Malleus Fatuorum, S Marshall, Ingolfson, Magioladitis, Jatkins, BatteryIncluded, NERV~enwiki, Loberttp, KTo288, Hans Dunkelberg, Qwidjib0, TXiKiBoT, Dojarca, Hqb, Peter Erwin, McM.bot, Andy Dingley, Benjwgarner, Thunderbird2, Ohiostandard, Ezrado, Caltas, Rob83uk, Lord British, MarkMLl, Denisarona, Finetooth, U5K0, Auntof6, Kitchen Knife, Gbrims, SoHome, Rickremember, Tarheel95, WikHead, MystBot, Addbot, Ashanda, Sparrer, Eshmo~enwiki, 84user, Lightbot, Anxietycello, WikiDreamer Bot, The Bushranger, Luckas-bot, Yobot, AnomieBOT, Archon 2488, Xqbot, DataWraith, Off2riorob, Nathanielvirgo, Jezhotwells, Frodz, Eugene-elgato, Rosetta2004, GliderMaven, FrescoBot, Kyteto, Richhaddon, Tom.Reding, Dinamik-bot, 777sms, Jfmantis, Updatehelper, RjwilmsiBot, John of Reading, GA bot, JRJJ, Quantanew, Shining.Star, Dewritech, Challisrussia, Realgem, Grondilu, ZéroBot, A2soup, Allforrous, Tchad49, Alpha Quadrant, Gniniv, SkywalkerPL, AndyTheGrump, ChiZeroOne, Terra Novus, Rememberway, ClueBot NG, Michaelmas1957, Helpful Pixie Bot, Bibcode Bot, Virtualerian, Beaucouplusneutre, BendelacBOT, Wordlet, Blaspie55, Zedshort, Cyberbot II, ChrisGualtieri, Singe onion, MSUGRA, Leptus Froggi, Maddit, Ruby Murray, Jamesmcmahon0, Lasith011, 17A Africa, Anythingcouldhappen, WPGA2345, Unician, Space Craze and Anonymous: 124

- **SpaceShipTwo** *Source:* https://en.wikipedia.org/wiki/SpaceShipTwo?oldid=689182628 *Contributors:* Raduga~enwiki, Edward, Patrick, Tango, Rlandmann, Shizhao, Chrisjj, Drxenocide, Owain, Sverdrup, Fennec, Wolfkeeper, Everyking, Joconnor, Ericg, SYSS Mouse, N328KF, NeuronExMachina, Rama, Adam850, Sarrica, Violetriga, Flyskippy1, Ascorbic, Cacophony, Anarchofascist, Richi, Pearle, Ommnomnomgulp, Equinoxe, Ahruman, Fivetrees, BastiaanNaber, Cromwellt, M3tainfo, Alai, Adrian.benko, Jcooper95, Jamsta, LoopZilla, Zrenneh, Benbest, Tabletop, GregorB, SCEhardt, Rchamberlain, Cornince, GraemeLeggett, Emerson7, Dmountain, Yuriybrisk, Drbogdan, Rjwilmsi, Hiberniantears, Hack-Man, Mike Peel, Vegaswikian, SchuminWeb, Arnero, A.K.Karthikeyan, Fosnez, Ahunt, Bgwhite, Xela, Hairy Dude, Freiberg, Hydrargyrum, Gaius Cornelius, Bovineone, Peterlean, Anomie, Jakash, Adelphos~enwiki, Megapixie, Jamesmcguigan, Ravedave, Change1211, KennethUrban, Eduardo89, Mjroots, Deuar, Sardanaphalus, SmackBot, Reedy, Anarchist42, Jrockley, SarcasticDwarf, Amux, Chris the speller, Bluebot, Hibernian, Emurphy42, WDGraham, Trekphiler, Frap, MJBurrage, Init~enwiki, Alexmcfire, Britmax, Aces lead, Ssnseawolf, SalopianJames, John, Freewol, Pat Payne, MilborneOne, Tlesher, Trebuchetguy, Xxxxxxxxxxx, Peyre, Dl2000, Dragos muresan, Craigboy, Joseph Solis in Australia, Kavanagh~enwiki, ShimaKatase, Eluchil404, Raerth, N2e, Chmee2, Cydebot, Fnlayson, Future Perfect at Sunrise, Gogo Dodo, MikeLacey, Thijs!bot, Sagaciousuk, Dafydd Williams, Curtisarends, Dawnseeker2000, Hmrox, Yonatan, Akradecki, Ran4, Yellowdesk, Steelpillow, Andysimo123, Harryzilber, IanOsgood, Z22, AJRG, JamesBWatson, Violentbob, Ahecht, Bombofin, BilCat, Saganaki-, Wikianon, RP88, PrestonH, Fpbear, Darin-0, Colincbn, Jb2005, JulesVerne, Resplendent, Nikthestunned, Larryisgood, ColdCase, Flyingidiot, Kyle the bot, TXiKiBoT, Dojarca, MatthewHaywood, Pah246, Carmen56, Garrett.Curley, Andy Dingley, Dirkbb, Wagaf-d, Avinesh, Coolstuff03, SieBot, Bodyn, Jack Merridew, KGyST, Jarh, GregCampbellUSA, Rocketchemist, Bonesy96, Chrisdicknson, Danio, Bibliophylax, Prmaddox, Martarius, Kennvido, Xertoz, Niceguyedc, BMello1618, Sethton, Ktr101, CohesionBot, Alexbot, Flightsoffancy, Iac74205, Alejandrocaro35, Chaosdruid, C628, XLinkBot, D1ma5ad, Addbot, CanadianLinuxUser, Tide rolls, Lightbot, Anxietycello, زرشک, Yobot, JoTan, Evans1982, AnomieBOT, Myself488, Materialscientist, Lkt1126, GB fan, LilHelpa, Xqbot, Ita140188, Patcharlton1992, Erik9bot, Aaronb121, FrescoBot, D'ohBot, Zolarketh, 777sms, Yeng-Wang-Yeh, RjwilmsiBot, Ripchip Bot, Johnjstimsoniii, Becritical, Galactic Penguin SST, Rbartelds, JDPurches, Challisrussia, Mmeijeri, Nzdoc, ZéroBot, Recorder111, Green Lane, H3llBot, Jayrose12, Mentalmerlin, Batox~enwiki, Accotink2, Gold Hat, Fanyavizuri, Hoeksas, ChrisCarss Former24.108.99.31, ChiZeroOne, ClueBot NG, Michaelmas1957, Giggett, IPadFanboy, Wllmevans, Oddbodz, Helpful Pixie Bot, MarkMysoe, BG19bot, Beaucouplusneutre, Op47, Donriop, Cyberbot II, Pink.up, SD5bot, Qxukhgiels, RandomLittleHelper, Joeinwiki, Mikerodneyfox, Howicus, Daydreamers, Shelbystripes, Machinegoesbing, NorthBySouthBaranof, Harry bainbridge, MWagonseller, Stephanie Bowman, Moblecl, Ronrosano, JGG13, Inphynite, Lonew8, Space 123, TheHeroWolf, ECCole 00 and Anonymous: 214

- **SpaceShipThree** *Source:* https://en.wikipedia.org/wiki/SpaceShipThree?oldid=676580347 *Contributors:* GTBacchus, Finlay McWalter, Wolfkeeper, Iceberg3k, SYSS Mouse, NeuronExMachina, Yamla, Alai, Benbest, Emerson7, Mendaliv, Vegaswikian, YurikBot, Ergzay, Change1211, Aremisasling, Deuar, Sardanaphalus, SmackBot, Grey Shadow, Jrockley, WDGraham, Raerth, N2e, Anidnmeno, Cydebot, Thijs!bot, Sagaciousuk, Dawnseeker2000, IanOsgood, Magioladitis, Jatkins, BilCat, Saganaki-, Threedots dead, Dhaluza, JulesVerne, Hydrokevin, Keraunoscopia, Jefflayman, Trulystand700, Svgalbertian, Addbot, Jgreep, The Bushranger, GrouchoBot, DrilBot, Recorder123, ChiZeroOne, Qxukhgiels and Anonymous: 18

- **Scaled Composites Tier 1b** *Source:* https://en.wikipedia.org/wiki/Scaled_Composites_Tier_1b?oldid=663416359 *Contributors:* Chrisjj, Wolfkeeper, Harej, Vegaswikian, Bgwhite, SmackBot, Jrockley, WDGraham, Doodle77, Rory096, Xxxxxxxxxxx, N2e, Cydebot, Yellowdesk, GrahamHardy, Sdsds, Petebutt, Sintaku, Niceguyedc, Lightbot, AnomieBOT, DrilBot, Chesipiero, Michaelmas1957 and Anonymous: 13

- **Suborbital spaceplane** *Source:* https://en.wikipedia.org/wiki/Suborbital_spaceplane?oldid=474305207 *Contributors:* Chrisjj, Alinor, SmackBot, Jrockley, Hmains, IanOsgood, T.Neo, Fotaun, ChiZeroOne and Anonymous: 2

- **VentureStar** *Source:* https://en.wikipedia.org/wiki/VentureStar?oldid=687344145 *Contributors:* Bryan Derksen, Gsl, Rlandmann, Mulad, Sverdrup, Greyengine5, Wwoods, Everyking, Jason Quinn, Kuralyov, Sarrica, E Pluribus Anthony redux, Chairboy, Cwolfsheep, Giraffedata, Pearle, AndromedaRoach, Alai, Jason Palpatine, Bricktop, Bluemoose, Jivecat, The wub, SchuminWeb, Billso, Knife Knut, YurikBot, PileOnades, RussBot, JihemD, CambridgeBayWeather, Kb1koi, Neum, Change1211, Wolbo, Bozoid, Wangi, Aceslead, Aremisasling, Ninly, Closedmouth, Sardanaphalus, SmackBot, Carbonix, Bluebot, Hibernian, Rpspeck, WDGraham, Kelvin Case, Aces lead, Joema, Gildir, Salamurai, Craigboy, Cast2007, Starbuzz3d, N2e, Cydebot, Fnlayson, Gogo Dodo, Clh288, Thijs!bot, Alphachimpbot, Ingolfson, Teraflop122, Ohms law, Teyrana, Malinaccier, Lucamauri, Oldag07, SidewinderX, MystBot, Addbot, Dawynn, PFSLAKES1,

Lightbot, Torin, Yobot, DemocraticLuntz, Full-date unlinking bot, Airbuilder7, Wingman4l7, Donner60, SkywalkerPL, Terra Novus, Chesipiero, ClueBot NG, Intermittentgardener, BG19bot, Andyhowlett, Monkbot, Crystallizedcarbon, Mike J. Henry, Tyman03, Bbar18, Clairelynncrosby, Solarflare4774, Aja286, ChicagoRapPolitic, Stinkysocks11, Hfslt, Http.lindseey, Mott1234567, Colonnam14, The 6 Official, ColCS, Bsmith5003, Hartch01 and Anonymous: 49

- **SpaceShipTwo, Serial Number Two** *Source:* https://en.wikipedia.org/wiki/SpaceShipTwo%2C_Serial_Number_Two?oldid=689579757 *Contributors:* Atlan, Andrewa, Benbest, GraemeLeggett, Vegaswikian, Canley, Sardanaphalus, Nickst, Skizzik, WDGraham, MJBurrage, Jpagel, Roguegeek, MilborneOne, FairuseBot, N2e, Cydebot, KylieTastic, Jb2005, Petebutt, MarshallKe, Qsaw, Rreagan007, Addbot, Yobot, Obersachsebot, Allanlw, DrilBot, Full-date unlinking bot, 777sms, AmericaIsNumberOne, ChuispastonBot, Chesipiero, BG19bot, Cyberbot II, Shelbystripes, Comto67, Ronrosano and Anonymous: 23

- **X-41 Common Aero Vehicle** *Source:* https://en.wikipedia.org/wiki/X-41_Common_Aero_Vehicle?oldid=668333635 *Contributors:* Rlandmann, DmitryKo, Pak21, Remuel, Alai, Sylvain Mielot, Fxer, Marudubshinki, BD2412, Arado, Bergsten, Los688, Daveswagon, Ravedave, Change1211, Alexthegreater, Hibernian, WDGraham, TheGerm, Will Beback, Fraantik, Heqs, Cydebot, Heinz-bert, Escarbot, Akradecki, Ohms law, GimmeBot, Petebutt, Addbot, Zorrobot, Xqbot, WikitanvirBot, Dougmcdonell and Anonymous: 5

- **XS-1 (spacecraft)** *Source:* https://en.wikipedia.org/wiki/XS-1_(spacecraft)?oldid=677984161 *Contributors:* Bearcat, Necrothesp, Arado, WDGraham, Derek R Bullamore, Kavanagh~enwiki, N2e, Fnlayson, Hebrides, BatteryIncluded, BilCat, The Bushranger, Yobot, AnomieBOT, Cnwilliams, Jmostly, Doyna Yar, Bockser, BattyBot, America789, Snow in July, JAGofc and Anonymous: 4

53.7.2   Images

- **File:ASSET_THOR.jpg** *Source:* https://upload.wikimedia.org/wikipedia/commons/5/54/ASSET_THOR.jpg *License:* Public domain *Contributors:* Transferred from en.wikipedia to Commons. *Original artist:* ?

- **File:ASSET_THOR2.jpg** *Source:* https://upload.wikimedia.org/wikipedia/commons/f/f3/ASSET_THOR2.jpg *License:* Public domain *Contributors:* Transferred from en.wikipedia to Commons. *Original artist:* ?

- **File:ASSET_USAF2.JPG** *Source:* https://upload.wikimedia.org/wikipedia/en/f/f3/ASSET_USAF2.JPG *License:* PD *Contributors:* Image taken from the NASA-published work, "Testing Lifting Bodies at Edwards" by Robert G. Hoey *Original artist:* ?

- **File:ASV-3_ASSET_Lifting_Body.jpg** *Source:* https://upload.wikimedia.org/wikipedia/commons/a/a8/ASV-3_ASSET_Lifting_Body.jpg *License:* Public domain *Contributors:* http://www.nationalmuseum.af.mil/shared/media/photodb/photos/050328-F-1234P-005.jpg *Original artist:* US air Force

- **File:AVATAR-1.JPG** *Source:* https://upload.wikimedia.org/wikipedia/en/e/e7/AVATAR-1.JPG *License:* Fair use *Contributors:* http://www.nal.res.in/nal50/incast/incast/01-Invited%20Talk%20Full%20papers/INCAST%202008-%20IT12.pdf *Original artist:* ?

- **File:Acap.svg** *Source:* https://upload.wikimedia.org/wikipedia/commons/5/52/Acap.svg *License:* Public domain *Contributors:* Own work *Original artist:* F l a n k e r

- **File:Aero-stub_img.svg** *Source:* https://upload.wikimedia.org/wikipedia/commons/4/44/Aero-stub_img.svg *License:* Public domain *Contributors:* This is a remake of Aero-stub img.png: <a href='//commons.wikimedia.org/wiki/File:Aero-stub_img.png' class='image'><img alt='Aero-stub img.png' src='https://upload.wikimedia.org/wikipedia/commons/b/b6/Aero-stub_img.png' width='32' height='32' data-file-width='20' data-file-height='20' /></a> *Original artist:* Bobarino

- **File:Aerospaceplane.jpg** *Source:* https://upload.wikimedia.org/wikipedia/commons/5/53/Aerospaceplane.jpg *License:* Public domain *Contributors:* http://www.pr.afrl.af.mil/history_wpafb.html History of Propulsion Research, United States Air Force *Original artist:* Unknown

- **File:Ambox_current_red.svg** *Source:* https://upload.wikimedia.org/wikipedia/commons/9/98/Ambox_current_red.svg *License:* CC0 *Contributors:* self-made, inspired by Gnome globe current event.svg, using Information icon3.svg and Earth clip art.svg *Original artist:* Vipersnake151, penubag, Tkgd2007 (clock)

- **File:Ambox_important.svg** *Source:* https://upload.wikimedia.org/wikipedia/commons/b/b4/Ambox_important.svg *License:* Public domain *Contributors:* Own work, based off of Image:Ambox scales.svg *Original artist:* Dsmurat (talk · contribs)

- **File:Antonov_An-225_with_Buran_at_Le_Bourget_1989_Manteufel.jpg** *Source:* https://upload.wikimedia.org/wikipedia/commons/a/aa/Antonov_An-225_with_Buran_at_Le_Bourget_1989_Manteufel.jpg *License:* GFDL 1.2 *Contributors:* http://www.airliners.net/photo/Untitled-(Antonov-Design/Antonov-An-225-Mriya/1240864/L/ *Original artist:* Ralf Manteufel

- **File:Atlantis_is_landing_after_STS-30_mission.jpg** *Source:* https://upload.wikimedia.org/wikipedia/commons/2/26/Atlantis_is_landing_after_STS-30_mission.jpg *License:* Public domain *Contributors:* http://grin.hq.nasa.gov/ABSTRACTS/GPN-2000-000667.html *Original artist:* NASA

- **File:AvatarTD.JPG** *Source:* https://upload.wikimedia.org/wikipedia/commons/e/e8/AvatarTD.JPG *License:* CC BY-SA 3.0 *Contributors:* Transferred from en.wikipedia to Commons by Roland zh using CommonsHelper. *Original artist:* Johnxxx9 at English Wikipedia

- **File:Aviacionavion.png** *Source:* https://upload.wikimedia.org/wikipedia/commons/6/68/Aviacionavion.png *License:* Public domain *Contributors:*

- Turkmenistan.airlines.frontview.arp.jpg *Original artist:* Turkmenistan.airlines.frontview.arp.jpg: elfuser

- **File:Aviation_Week_03-06-2006_cover.jpg** *Source:* https://upload.wikimedia.org/wikipedia/en/d/d2/Aviation_Week_03-06-2006_cover.jpg *License:* ? *Contributors:* ? *Original artist:* ?

- **File:BOR-2.jpg** *Source:* https://upload.wikimedia.org/wikipedia/commons/5/54/BOR-2.jpg *License:* CC BY 2.5 *Contributors:* No machine-readable source provided. Own work assumed (based on copyright claims). *Original artist:* No machine-readable author provided. Jno~commonswiki assumed (based on copyright claims).

- **File:BOR-4S.jpg** *Source:* https://upload.wikimedia.org/wikipedia/commons/e/e2/BOR-4S.jpg *License:* CC BY 2.5 *Contributors:* No machine-readable source provided. Own work assumed (based on copyright claims). *Original artist:* No machine-readable author provided. Jno~commonswiki assumed (based on copyright claims).

- Original by Hektor

- **File:ESA_logo.svg** *Source:* https://upload.wikimedia.org/wikipedia/commons/8/80/ESA_logo.svg *License:* Public domain *Contributors:* http://esamultimedia.esa.int/multimedia/ESA_Logo/logotype.html (EPS file) *Original artist:* ESA

- **File:Edit-clear.svg** *Source:* https://upload.wikimedia.org/wikipedia/en/f/f2/Edit-clear.svg *License:* Public domain *Contributors:* The *Tango! Desktop Project. Original artist:*
  The people from the Tango! project. And according to the meta-data in the file, specifically: "Andreas Nilsson, and Jakub Steiner (although minimally)."

- **File:Emoji_u1f52e.svg** *Source:* https://upload.wikimedia.org/wikipedia/commons/6/6c/Emoji_u1f52e.svg *License:* Apache License 2.0 *Contributors:* https://code.google.com/p/noto/ *Original artist:* Google

- **File:File_006.jpeg** *Source:* https://upload.wikimedia.org/wikipedia/commons/3/32/File_006.jpeg *License:* CC BY-SA 3.0 *Contributors:* Own work *Original artist:* Kb5urq

- **File:Flag_of_Canada.svg** *Source:* https://upload.wikimedia.org/wikipedia/en/c/cf/Flag_of_Canada.svg *License:* PD *Contributors:* ? *Original artist:* ?

- **File:Flag_of_France.svg** *Source:* https://upload.wikimedia.org/wikipedia/en/c/c3/Flag_of_France.svg *License:* PD *Contributors:* ? *Original artist:* ?

- **File:Flag_of_German_Reich_(1935–1945).svg** *Source:* https://upload.wikimedia.org/wikipedia/commons/9/99/Flag_of_German_Reich_ %281935%E2%80%931945%29.svg *License:* Public domain *Contributors:* Own work *Original artist:* Fornax

- **File:Flag_of_Germany.svg** *Source:* https://upload.wikimedia.org/wikipedia/en/b/ba/Flag_of_Germany.svg *License:* PD *Contributors:* ? *Original artist:* ?

- **File:Flag_of_India.svg** *Source:* https://upload.wikimedia.org/wikipedia/en/4/41/Flag_of_India.svg *License:* Public domain *Contributors:* ? *Original artist:* ?

- **File:Flag_of_Japan.svg** *Source:* https://upload.wikimedia.org/wikipedia/en/9/9e/Flag_of_Japan.svg *License:* PD *Contributors:* ? *Original artist:* ?

- **File:Flag_of_Romania.svg** *Source:* https://upload.wikimedia.org/wikipedia/commons/7/73/Flag_of_Romania.svg *License:* Public domain *Contributors:* Own work *Original artist:* AdiJapan

- **File:Flag_of_Russia.svg** *Source:* https://upload.wikimedia.org/wikipedia/en/f/f3/Flag_of_Russia.svg *License:* PD *Contributors:* ? *Original artist:* ?

- **File:Flag_of_Switzerland.svg** *Source:* https://upload.wikimedia.org/wikipedia/commons/f/f3/Flag_of_Switzerland.svg *License:* Public domain *Contributors:* PDF Colors Construction sheet *Original artist:* User:Marc Mongenet

Credits:

- **File:Flag_of_Ukraine.svg** *Source:* https://upload.wikimedia.org/wikipedia/commons/4/49/Flag_of_Ukraine.svg *License:* Public domain *Contributors:* ДСТУ 4512:2006 - Державний прапор України. Загальні технічні умови

  SVG: 2010

  Original artist: України

- **File:Flag_of_the_People'{}s_Republic_of_China.svg** *Source:* https://upload.wikimedia.org/wikipedia/commons/f/fa/Flag_of_the_People% 27s_Republic_of_China.svg *License:* Public domain *Contributors:* Own work, http://www.protocol.gov.hk/flags/eng/n_flag/design.html *Original artist:* Drawn by User:SKopp, redrawn by User:Denelson83 and User:Zscout370

- **File:Flag_of_the_Soviet_Union.svg** *Source:* https://upload.wikimedia.org/wikipedia/commons/a/a9/Flag_of_the_Soviet_Union.svg *License:* Public domain *Contributors:* http://pravo.levonevsky.org/ *Original artist:* СССР

- **File:Flag_of_the_United_Kingdom.svg** *Source:* https://upload.wikimedia.org/wikipedia/en/a/ae/Flag_of_the_United_Kingdom.svg *License:* PD *Contributors:* ? *Original artist:* ?

- **File:Flag_of_the_United_States.svg** *Source:* https://upload.wikimedia.org/wikipedia/en/a/a4/Flag_of_the_United_States.svg *License:* PD *Contributors:* ? *Original artist:* ?

- **File:Flight_16P_taxi_pre_launch_photo_D_Ramey_Logan.jpg** *Source:* https://upload.wikimedia.org/wikipedia/commons/b/b0/Flight_ 16P_taxi_pre_launch_photo_D_Ramey_Logan.jpg *License:* CC BY-SA 4.0 *Contributors:* Own work *Original artist:* D. Ramey Logan (WPPilot)

- **File:Folder_Hexagonal_Icon.svg** *Source:* https://upload.wikimedia.org/wikipedia/en/4/48/Folder_Hexagonal_Icon.svg *License:* Cc-by-sa-3.0 *Contributors:* ? *Original artist:* ?

- **File:Gemini_paraglider.JPG** *Source:* https://upload.wikimedia.org/wikipedia/commons/3/36/Gemini_paraglider.JPG *License:* Public domain *Contributors:* NASA via Flight Global *Original artist:* NASA

- **File:Goodyear_Meteor_Junior,_Smithsonian.jpg** *Source:* https://upload.wikimedia.org/wikipedia/commons/3/36/Goodyear_Meteor_ Junior%2C_Smithsonian.jpg *License:* Public domain *Contributors:* Smithsonian Institution *Original artist:* Smithsonian Institution

- **File:HOTOL.JPG** *Source:* https://upload.wikimedia.org/wikipedia/commons/d/d1/HOTOL.JPG *License:* Public domain *Contributors:* Own work *Original artist:* KVDP

- **File:Hermes_Spaceplane_ESA.jpg** *Source:* https://upload.wikimedia.org/wikipedia/en/7/73/Hermes_Spaceplane_ESA.jpg *License:* Fair use *Contributors:*
  http://www.esa.int/esa-mmg/mmg.pl?b=b&keyword=hermes&single=y&start=17 *Original artist:* ?

- **File:ISS_Crew_Return_Vehicle.jpg** *Source:* https://upload.wikimedia.org/wikipedia/commons/1/1b/ISS_Crew_Return_Vehicle.jpg *License:* Public domain *Contributors:* Armstrong Photo Gallery: Home - info - pic *Original artist:* NASA / Carla Thomas

- **File:IXV_drop-test_model_behind.jpg** *Source:* https://upload.wikimedia.org/wikipedia/commons/8/8a/IXV_drop-test_model_behind. jpg *License:* CC BY-SA 4.0 *Contributors:* Own work *Original artist:* Suruena

- **File:Orbital_Space_Plane_Concepts.jpg** *Source:* https://upload.wikimedia.org/wikipedia/commons/6/61/Orbital_Space_Plane_Concepts. jpg *License:* Public domain *Contributors:* http://web.archive.org/web/20040214180027/http://www.slinews.com/nasaconcepts.html *Original artist:* NASA/MSFC

- **File:PRIDE_concept_servicing_satellite.jpg** *Source:* https://upload.wikimedia.org/wikipedia/en/f/f2/PRIDE_concept_servicing_satellite. jpg *License:* Fair use *Contributors:* **Original publication**: ESA Space in Images

  Immediate source: http://www.esa.int/spaceinimages/Images/2012/11/PRIDE_mission2 *Original artist:* ESA - J.Huart

- **File:People_icon.svg** *Source:* https://upload.wikimedia.org/wikipedia/commons/3/37/People_icon.svg *License:* CC0 *Contributors:* Open-Clipart *Original artist:* OpenClipart

- **File:Pilot_Neil_Armstrong_with_X-15_-1_(9458061153).jpg** *Source:* https://upload.wikimedia.org/wikipedia/commons/9/9c/Pilot_ Neil_Armstrong_with_X-15_-1_%289458061153%29.jpg *License:* Public domain *Contributors:* Pilot Neil Armstrong with X-15 #1 *Original artist:* US Air Force

- **File:Portal-puzzle.svg** *Source:* https://upload.wikimedia.org/wikipedia/en/f/fd/Portal-puzzle.svg *License:* Public domain *Contributors:* ? *Original artist:* ?

- **File:Precooler_Rig.jpg** *Source:* https://upload.wikimedia.org/wikipedia/en/d/db/Precooler_Rig.jpg *License:* Fair use *Contributors:* http://www.bbc.co.uk/news/science-environment-13506289 *Original artist:* ?

- **File:Question_book-new.svg** *Source:* https://upload.wikimedia.org/wikipedia/en/9/99/Question_book-new.svg *License:* Cc-by-sa-3.0 *Contributors:*
  Created from scratch in Adobe Illustrator. Based on Image:Question book.png created by User:Equazcion *Original artist:*
  Tkgd2007

- **File:RU072_10.jpg** *Source:* https://upload.wikimedia.org/wikipedia/commons/6/60/RU072_10.jpg *License:* Public domain *Contributors:* [1] *Original artist:* Stamp issuing authority - MARKA Publishing & Trading Centre. Printer - Association GOZNAK of the Ministry of Finance of the Russian Federation

- **File:RocketSunIcon.svg** *Source:* https://upload.wikimedia.org/wikipedia/commons/d/d6/RocketSunIcon.svg *License:* Copyrighted free use *Contributors:* Self made, based on File:Spaceship and the Sun.jpg *Original artist:* Me

- **File:Rocketplane_xp_concept.jpg** *Source:* https://upload.wikimedia.org/wikipedia/en/b/bf/Rocketplane_xp_concept.jpg *License:* Fair use *Contributors:* http://www.rocketplane.com/press/20071026a.html *Original artist:* Rocketplane XP official webpage

- **File:Roscosmos_logo_ru.svg** *Source:* https://upload.wikimedia.org/wikipedia/commons/d/da/Roscosmos_logo_ru.svg *License:* Public domain *Contributors:* Official site of the Russian Federal Space Agency *Original artist:* Russian Federal Space Agency

- **File:Rotary-Rocket-logo.png** *Source:* https://upload.wikimedia.org/wikipedia/en/1/1c/Rotary-Rocket-logo.png *License:* Fair use *Contributors:*
  The logo may be obtained from Rotary Rocket.
  Original artist: ?

- **File:Rotary-rocket-hangars.jpg** *Source:* https://upload.wikimedia.org/wikipedia/commons/5/5c/Rotary-rocket-hangars.jpg *License:* CC-BY-SA-3.0 *Contributors:* No machine-readable source provided. Own work assumed (based on copyright claims). *Original artist:* No machine-readable author provided. Bricktop assumed (based on copyright claims).

- **File:Rotaryrocket-061114-01-8.jpg** *Source:* https://upload.wikimedia.org/wikipedia/commons/0/0f/Rotaryrocket-061114-01-8.jpg *License:* CC BY 2.5 *Contributors:* Own work *Original artist:* Alan Radecki

- **File:Roton-bat-cave.jpg** *Source:* https://upload.wikimedia.org/wikipedia/commons/4/4b/Roton-bat-cave.jpg *License:* CC BY 2.5 *Contributors:* ? *Original artist:* ?

- **File:SHEFEX_II_-_assembled.jpg** *Source:* https://upload.wikimedia.org/wikipedia/commons/e/ea/SHEFEX_II_-_assembled.jpg *License:* CC BY 3.0 de *Contributors:* http://www.dlr.de/dlr/desktopdefault.aspx/tabid-10186/274_read-1344/ *Original artist:* Deutsches Zentrum für Luft- und Raumfahrt

- **File:SRBsepfromDiscovery07042006.png** *Source:* https://upload.wikimedia.org/wikipedia/commons/a/ad/SRBsepfromDiscovery07042006. png *License:* Public domain *Contributors:* photo from video released by NASA into public domain *Original artist:* NASA

- **File:SS2_First_Launch.jpg** *Source:* https://upload.wikimedia.org/wikipedia/en/f/f0/SS2_First_Launch.jpg *License:* Fair use *Contributors:* **Original publication**: facebook.com

  Immediate source: https://www.facebook.com/photo.php?fbid=10151599408249297&set=a.10150653086679297.403080.171270329296& type=1&theater *Original artist:* Virgin Galactic

- **File:SS2_and_VMS_Eve.jpg** *Source:* https://upload.wikimedia.org/wikipedia/commons/a/ad/SS2_and_VMS_Eve.jpg *License:* CC BY-SA 3.0 *Contributors:* http://www.virgingalactic.com/extranet/file/spaceship-unveil/monday-7th-dec/ss2-and-vms-eve-preview4/ *Original artist:* Virgin Galactic/Mark Greenberg

- **File:STS-41-B_MMU.jpg** *Source:* https://upload.wikimedia.org/wikipedia/commons/4/4c/STS-41-B_MMU.jpg *License:* Public domain *Contributors:* http://grin.hq.nasa.gov/IMAGES/LARGE/GPN-2000-001156.jpg *Original artist:* NASA

- **File:STS-61-A_crew_in_Spacelab_D-1.jpg** *Source:* https://upload.wikimedia.org/wikipedia/commons/8/8b/STS-61-A_crew_in_Spacelab_ D-1.jpg *License:* Public domain *Contributors:*

- http://nix.larc.nasa.gov/info?id=STS61A-01-030 *Original artist:* NASA

- **File:STS-73_landing.jpg** *Source:* https://upload.wikimedia.org/wikipedia/commons/2/21/STS-73_landing.jpg *License:* Public domain *Contributors:* http://spaceflight.nasa.gov/gallery/images/shuttle/sts-73/html/sts073-s-047.html *Original artist:* NASA

- **File:Samsung_Galaxy_S5_Vector.svg** *Source:* https://upload.wikimedia.org/wikipedia/commons/1/15/Samsung_Galaxy_S5_Vector. svg *License:* CC BY-SA 3.0 *Contributors:* Own work *Original artist:* Rafael Fernandez

- **File:Scaled-wk-070711-08-16.jpg** *Source:* https://upload.wikimedia.org/wikipedia/commons/7/71/Scaled-wk-070711-08-16.jpg *License:* GFDL *Contributors:* Own work *Original artist:* Alan Radecki Akradecki

- **File:Schematic_diagram_of_Virgin_Galactic'{}s_SpaceShipTwo.jpg** *Source:* https://upload.wikimedia.org/wikipedia/en/e/e6/Schematic_ diagram_of_Virgin_Galactic%27s_SpaceShipTwo.jpg *License:* Public domain *Contributors:*
  http://www.image.net/virgingalactic *Original artist:*
  Virgin Galactic

- **File:Seal_of_the_US_Air_Force.svg** *Source:* https://upload.wikimedia.org/wikipedia/commons/2/23/Seal_of_the_US_Air_Force.svg *License:* Public domain *Contributors:* SVG created from this image *Original artist:* Arthur E. DuBois, according to [1]

- **File:Sevilla_Expo_92-Projecto_ESA-1992_05_05.jpg** *Source:* https://upload.wikimedia.org/wikipedia/commons/9/90/Sevilla_Expo_ 92-Projecto_ESA-1992_05_05.jpg *License:* CC BY-SA 3.0 *Contributors:* Self-photographed. *Original artist:* Daniel Villafruela.

- **File:Silbervogel.jpg** *Source:* https://upload.wikimedia.org/wikipedia/commons/8/8f/Silbervogel.jpg *License:* Public domain *Contributors:* ? *Original artist:* ?

- **File:Skylon.svg** *Source:* https://upload.wikimedia.org/wikipedia/commons/1/13/Skylon.svg *License:* CC BY 3.0 *Contributors:* Own work *Original artist:* GW Simulations

- **File:Skylon_climbing.jpg** *Source:* https://upload.wikimedia.org/wikipedia/en/c/cc/Skylon_climbing.jpg *License:* Fair use *Contributors:*
  http://www.reactionengines.co.uk/images/skylon/library/skylon_climb_1l.jpg *Original artist:*
  Adrian Mann [1]

- **File:Skylon_colour.svg** *Source:* https://upload.wikimedia.org/wikipedia/commons/3/33/Skylon_colour.svg *License:* Attribution *Contributors:* Own work *Original artist:* me

- **File:Skylon_diagram.jpg** *Source:* https://upload.wikimedia.org/wikipedia/en/0/03/Skylon_diagram.jpg *License:* Fair use *Contributors:*
  http://www.theregister.co.uk/2011/05/24/skylon_esa_report/page2.html *Original artist:* ?

- **File:Skylon_front_view.jpg** *Source:* https://upload.wikimedia.org/wikipedia/en/e/ea/Skylon_front_view.jpg *License:* Fair use *Contributors:*
  http://dvice.com/archives/2011/05/skylon-spacepla-1.php *Original artist:* ?

- **File:Soyuz,_Space_Shuttle,_Buran_comparison.svg** *Source:* https://upload.wikimedia.org/wikipedia/commons/2/2b/Soyuz%2C_Space_ Shuttle%2C_Buran_comparison.svg *License:* Public domain *Contributors:* http://ston.jsc.nasa.gov/collections/TRS/_techrep/RP1357. pdf *Original artist:* NASA

- **File:Soyuz_TMA-6_spacecraft.jpg** *Source:* https://upload.wikimedia.org/wikipedia/commons/7/70/Soyuz_TMA-6_spacecraft.jpg *License:* Public domain *Contributors:* http://spaceflight.nasa.gov/gallery/images/station/crew-10/html/iss010e24875.html *Original artist:* NASA

- **File:SpaceShipOne_Nose.jpg** *Source:* https://upload.wikimedia.org/wikipedia/commons/c/c8/SpaceShipOne_Nose.jpg *License:* CC BY-SA 2.5 *Contributors:* ? *Original artist:* ?

- **File:Space_Shuttle_concepts.jpg** *Source:* https://upload.wikimedia.org/wikipedia/commons/3/39/Space_Shuttle_concepts.jpg *License:* Public domain *Contributors:* NASA-Website *Original artist:* NASA

- **File:Space_Shuttle_diagram.jpg** *Source:* https://upload.wikimedia.org/wikipedia/commons/8/8a/Space_Shuttle_diagram.jpg *License:* Public domain *Contributors:* http://mix.msfc.nasa.gov/abstracts.php?p=1861 *Original artist:* NASA/MSFC

- **File:Symbol_list_class.svg** *Source:* https://upload.wikimedia.org/wikipedia/en/d/db/Symbol_list_class.svg *License:* Public domain *Contributors:* ? *Original artist:* ?

- **File:Sänger_Raumtransporter.JPG** *Source:* https://upload.wikimedia.org/wikipedia/commons/b/b0/S%C3%A4nger_Raumtransporter. JPG *License:* CC BY 3.0 *Contributors:* Own work *Original artist:* Palatinatian

- **File:Telecom-icon.svg** *Source:* https://upload.wikimedia.org/wikipedia/commons/4/4e/Telecom-icon.svg *License:* Public domain *Contributors:* ? *Original artist:* ?

- **File:Text_document_with_red_question_mark.svg** *Source:* https://upload.wikimedia.org/wikipedia/commons/a/a4/Text_document_ with_red_question_mark.svg *License:* Public domain *Contributors:* Created by bdesham with Inkscape; based upon Text-x-generic.svg from the Tango project. *Original artist:* Benjamin D. Esham (bdesham)

- **File:Twin_Linear_Aerospike_XRS-2200_Engine_PLW_edit.jpg** *Source:* https://upload.wikimedia.org/wikipedia/commons/8/8c/Twin_ Linear_Aerospike_XRS-2200_Engine_PLW_edit.jpg *License:* CC BY-SA 3.0 *Contributors:* http://nix.ksc.nasa.gov/info;jsessionid= 8a9ofkryn0iy?id=MSFC-0103149&orgid=11 *Original artist:* NASA Marshall Space Flight Center (NASA-MSFC) [1], Papa Lima Whiskey (restoration credit)

- **File:US-Satellite.svg** *Source:* https://upload.wikimedia.org/wikipedia/commons/2/2f/US-Satellite.svg *License:* Public domain *Contributors:* Created myself using Inkscape, incorporates PD File:Earth clip art.svg and File:Flag of the United States.svg *Original artist:*
  <a href='//en.wikipedia.org/wiki/User:GW_Simulations' class='extiw' title='en:User:GW Simulations'>G</a><a href='//commons. wikimedia.org/wiki/User_talk:GW_Simulations' title='User talk:GW Simulations' class='mw-redirect'>W</a> ⋯ (User • Talk • EN)

- **File:USSR-Satellite.svg** *Source:* https://upload.wikimedia.org/wikipedia/commons/4/4f/USSR-Satellite.svg *License:* Public domain *Contributors:* Created myself using Inkscape, incorporates PD File:Earth clip art.svg and File:Flag of the Soviet Union.svg *Original artist:*
  <a href='//en.wikipedia.org/wiki/User:GW_Simulations' class='extiw' title='en:User:GW Simulations'>G</a><a href='//commons. wikimedia.org/wiki/User_talk:GW_Simulations' title='User talk:GW Simulations' class='mw-redirect'>W</a> ⋯ (User • Talk • EN)

- **File:VentureStar_Shuttle_Comparison.PNG** *Source:* https://upload.wikimedia.org/wikipedia/commons/8/87/VentureStar_Shuttle_ Comparison.PNG *License:* Public domain *Contributors:* NASA *Original artist:* Kelvin Case

- **File:White_Knight_Two_and_SpaceShipTwo_from_directly_below.jpg** *Source:* https://upload.wikimedia.org/wikipedia/commons/ 4/41/White_Knight_Two_and_SpaceShipTwo_from_directly_below.jpg *License:* CC BY 2.0 *Contributors:* Flickr: WK2/SS2 from directly below *Original artist:* Jeff Foust

- **File:Wiki_letter_w.svg** *Source:* https://upload.wikimedia.org/wikipedia/en/6/6c/Wiki_letter_w.svg *License:* Cc-by-sa-3.0 *Contributors:* ? *Original artist:* ?

- **File:Wikinews-logo.svg** *Source:* https://upload.wikimedia.org/wikipedia/commons/2/24/Wikinews-logo.svg *License:* CC BY-SA 3.0 *Contributors:* This is a cropped version of Image:Wikinews-logo-en.png. *Original artist:* Vectorized by Simon 01:05, 2 August 2006 (UTC) Updated by Time3000 17 April 2007 to use official Wikinews colours and appear correctly on dark backgrounds. Originally uploaded by Simon.

- **File:World'{}s_First_Five_Spaceplanes.PNG** *Source:* https://upload.wikimedia.org/wikipedia/commons/e/ef/World%27s_First_Five_Spaceplanes.PNG *License:* CC BY-SA 2.5 *Contributors:* Transferred from en.wikipedia to Commons by The Bushranger using CommonsHelper.

  (Original text : *Primarily a personal artistic creation, created by myself, Kelvin Case. Additional derivative work was from these following U.S. government and Wikipedia images. (Furthermore, for a summary of additional visual arts reference images reviewed, see* Licensing *below.)*

  Original artist: Kelvin Case at English Wikipedia

- **File:X-15A2_NB-52B_3.jpg** *Source:* https://upload.wikimedia.org/wikipedia/commons/4/4d/X-15A2_NB-52B_3.jpg *License:* Public domain *Contributors:* http://www.dfrc.nasa.gov/Gallery/Photo/X-15/HTML/EC68-1889.html *Original artist:* NASA

- **File:X-15_Pilots_-_GPN-2000-000143.jpg** *Source:* https://upload.wikimedia.org/wikipedia/commons/6/60/X-15_Pilots_-_GPN-2000-000143.jpg *License:* Public domain *Contributors:*

- Armstrong Photo Gallery: Home - info - pic *Original artist:* NASA

- **File:X-15_and_B-52_Mother_ship.jpg** *Source:* https://upload.wikimedia.org/wikipedia/commons/f/fb/X-15_and_B-52_Mother_ship.jpg *License:* Public domain *Contributors:* https://www.flickr.com/photos/tom-margie/1572816152/ *Original artist:* United States Air Force

- **File:X-15_flying.jpg** *Source:* https://upload.wikimedia.org/wikipedia/commons/a/a3/X-15_flying.jpg *License:* Public domain *Contributors:* ? *Original artist:* ?

- **File:X-15_in_flight.jpg** *Source:* https://upload.wikimedia.org/wikipedia/commons/d/d6/X-15_in_flight.jpg *License:* Public domain *Contributors:* Current Upload: cropped from http://www.dfrc.nasa.gov/Gallery/Photo/X-15/HTML/EC88-0180-1.html (direct link) (date reference) *Original artist:* NASA

- **File:X-24C_Configuration_January_1977.jpg** *Source:* https://upload.wikimedia.org/wikipedia/commons/3/34/X-24C_Configuration_January_1977.jpg *License:* Public domain *Contributors:* http://ntrs.nasa.gov/archive/nasa/casi.ntrs.nasa.gov/19790008668_1979008668.pdf *Original artist:* Harry G. Combs, et al

- **File:X-30_NASP_1.jpg** *Source:* https://upload.wikimedia.org/wikipedia/commons/1/18/X-30_NASP_1.jpg *License:* Public domain *Contributors:* [1] *Original artist:* NASA

- **File:X-30_NASP_2.jpg** *Source:* https://upload.wikimedia.org/wikipedia/commons/3/32/X-30_NASP_2.jpg *License:* Public domain *Contributors:* [1] *Original artist:* NASA

- **File:X-30_NASP_3.jpg** *Source:* https://upload.wikimedia.org/wikipedia/commons/5/5a/X-30_NASP_3.jpg *License:* Public domain *Contributors:* http://nix.nasa.gov/info?id=EL-2001-00432 (direct link) *Original artist:* James Schultz

- **File:X-30_NASP_4.jpg** *Source:* https://upload.wikimedia.org/wikipedia/commons/f/fa/X-30_NASP_4.jpg *License:* Public domain *Contributors:* http://nix.nasa.gov/info?id=EL-1996-00226 (direct link) *Original artist:* NASA

- **File:X-30_futuristic_nasa.jpg** *Source:* https://upload.wikimedia.org/wikipedia/commons/1/17/X-30_futuristic_nasa.jpg *License:* Public domain *Contributors:* http://web.archive.org/web/2/http://ails.arc.nasa.gov/Images/Space/AC86-0699-2.html (direct link) *Original artist:* NASA/Ames Research Center

- **File:X-33_Liquid_Hydrogen_Multi-Lobed_Tank_Failure.png** *Source:* https://upload.wikimedia.org/wikipedia/commons/c/cf/X-33_Liquid_Hydrogen_Multi-Lobed_Tank_Failure.png *License:* Public domain *Contributors:* http://alpha.tamu.edu/public/jae/misc/tankreport.pdf *Original artist:* NASA

- **File:X-37_spacecraft,_artist'{}s_rendition.jpeg** *Source:* https://upload.wikimedia.org/wikipedia/commons/7/74/X-37_spacecraft%2C_artist%27s_rendition.jpeg *License:* Public domain *Contributors:* http://www.nasa.gov/centers/marshall/multimedia/photogallery/photos/photogallery/x37/x37.html (direct link; DFRC file) *Original artist:* NASA/Marshall Space Flight Center

- **File:X-38.webm** *Source:* https://upload.wikimedia.org/wikipedia/commons/5/56/X-38.webm *License:* CC BY-SA 3.0 *Contributors:* Own work *Original artist:* Kb5urq

- **File:X-38TestModel.JPG** *Source:* https://upload.wikimedia.org/wikipedia/commons/8/81/X-38TestModel.JPG *License:* CC-BY-SA-3.0 *Contributors:* Originally uploaded on en.wikipedia *Original artist:* Originally uploaded by Slammer111 (Transferred by Grondemar)

- **File:X-38_Crew.jpg** *Source:* https://upload.wikimedia.org/wikipedia/commons/2/27/X-38_Crew.jpg *License:* Public domain *Contributors:* NASA photo JSC2000E15221 *Original artist:* NASA

- **File:X-38_Landing.jpg** *Source:* https://upload.wikimedia.org/wikipedia/commons/4/43/X-38_Landing.jpg *License:* Public domain *Contributors:* NASA *Original artist:* NASA

- **File:X-38_Project_Team.jpg** *Source:* https://upload.wikimedia.org/wikipedia/commons/9/93/X-38_Project_Team.jpg *License:* Public domain *Contributors:* National Aeronautics and Space Administration *Original artist:* NASA

- **File:X-38_Ship_-2_Release_from_B-52_-_GPN-2000-000196.jpg** *Source:* https://upload.wikimedia.org/wikipedia/commons/c/ce/X-38_Ship_-2_Release_from_B-52_-_GPN-2000-000196.jpg *License:* Public domain *Contributors:*

- Armstrong Photo Gallery: Home - info - pic *Original artist:* NASA / DFRC / Carla Thomas

- **File:X-38_research_aircraft_fifth_test_drop_flight.ogg** *Source:* https://upload.wikimedia.org/wikipedia/commons/6/69/X-38_research_aircraft_fifth_test_drop_flight.ogg *License:* Public domain *Contributors:* http://www.dfrc.nasa.gov/gallery/Movie/X-38/Medium/EM-0038-10.mov (gallery from movie collection) *Original artist:* NASA

53.7.3 Content license